THE
ROBOT BUILDER'S BONANZA:
99 INEXPENSIVE ROBOTICS PROJECTS

GORDON MCCOMB

TAB Books
Division of McGraw-Hill

New York San Francisco Washington, D.C. Auckland Bogotá
Caracas Lisbon London Madrid Mexico City Milan
Montreal New Delhi San Juan Singapore
Sydney Tokyo Toronto

To my daughter, Mercedes, who still believes that a
robot is a ''metal man who serves drinks at parties.''

© 1987 by **Gordon McComb**.
Published by TAB Books.
TAB Books is a division of McGraw-Hill, Inc.

pbk		18	19	20	21	22	23	24	25	DOC/DOC	9	9	8	7	6
hc		3	4	5	6	7	8	9	10	11	DOH/DOH	8	8		

Library of Congress Cataloging-in-Publication Data

McComb, Gordon.
   The robot builder's bonanza.

   Includes index.
   1. Robotics. I. Title.
TJ211.M418   1987   629.8'92   87-5040
ISBN 0-8306-0800-1
ISBN 0-8306-2800-2 (pbk.)

Contents

Acknowledgments

Only until after you've climbed the mountain can you look behind you and see the vast distance that you've covered, and remember those you've met along the way that made your trek a little easier. Now that this book is finally finished, after the many miles of weary travel, I look back to those that helped me turn it into a reality, and offer my heartfelt thanks: To my good friends Ian Simpson, Eli Hollander, and Steve York, Brint Rutherford and the editors at TAB BOOKS, my agent Bill Gladstone, and my wife Jennifer Meredith.

Introduction

The word *robot* is defined as a mechanical device that is capable of performing human tasks or behaving in a human-like manner. No argument here. The description certainly fits.

To the robotics experimenter, however, the word robot has a completely different meaning. A robot is a special brew of motors, solenoids, wires, and assorted electronic odds-and-ends, a marriage of mechanical and electronic gizmos.

Taken together, the parts make a personable creature that can vacuum the floor, serve drinks, protect the family against intruders and fire, entertain, educate, and lots more. In fact, there's almost no limit to what a well-designed robot can do.

Robotics, like rocketry, television, and countless other technology-based industries, started small. But growth and progress have been slow. Robotics is still a cottage industry, even considering the special-purpose automatons now in wide use in the car industry. The science of personal robotics—the R2-D2 and C-3PO kind of "Star Wars" fame—is even smaller—an infant in a brand new family on the block. All that this means is, for the robotics experimenter, there is plenty of room for growth. There are a lot of discoveries yet to be made.

INSIDE *THE ROBOT BUILDER'S BONANZA*

The Robot Builder's Bonanza takes an educational but fun approach to designing working robots. Its modular projects take you from building basic motorized platforms to giving the machine a brain—and teaching it to walk and talk and obey commands.

If you are interested in mechanics, electronics, or robotics, you'll find this book a treasure chest of information and ideas on making thinking machines. The projects include all the necessary information on how to construct the essential building blocks of a personal robot. Suggested alternative approaches, parts lists, and possible sources of electronic and mechanical components are also provided where appropriate.

Several good books have been written on how to design and build your own robot. These have been aimed at making just one or two fairly sophisticated automatons, and at a great price. Because of the complexity of the robots detailed in these books, they require a fairly high level of expertise and pocket money on your part.

This book is different. Its modular *cookbook* approach offers a mountain of practical, easy to follow, and inexpensive robot experiments. Taken together, the modular projects in *The Robot Builder's Bonanza* can be combined to create several different types of highly intelligent and workable robots, of all shapes and sizes—rolling robots, walking robots, talking robots, you name it. You can mix and match projects as desired.

How to Use This Book

This text could be divided into seven sections. Each section would cover a major component of the common personal or hobby (as opposed to commercial or industrial) robot. The sections would be:

- Robot Basics—What you need to get started; setting up shop; how and where to buy robot parts.
- Body and Frame—Robots made of plastic, wood, and metal; working with common metal stock; converting toys into robots.
- Power and Locomotion—Using batteries; powering the robot; working with dc and stepper motors; gear trains; walking robot systems; special robot locomotion systems.
- Appendages—Building robot arms and hands; adding the sense of touch.
- Eyes, Ears, and Mouth—Speech synthesis; music and sound effect generation; sound detection; robot eyes; smoke, flame, and heat detection.
- Navigation—Collision detection; collision avoidance; ultrasonic ranging; infrared beacon systems; track guidance, navigation.
- Electronic Control—Infrared remote control; ultrasonic remote control; radio links; robot control with computer parallel port; computer system bus interface; on-board computer systems.

The chapters tell you how to build the many parts that go into the typical hobby robot. Most chapters present one or more projects that you can duplicate for your own robot creations. Whenever practical, I designed the components as discrete building blocks, so that you can combine the blocks in just about any configuration you desire. The robot you create will be uniquely yours, and yours alone.

If you have some experience in electronics, mechanics, or robot building in general, you can skip around and read only those chapters that provide the information you're looking for. The chapters are very much stand-alone modules, like the robot designs presented. This allows you to pick and choose, using your time to its best advantage.

If you're new to robot building, and the varied disciplines that go into it, you should take a more pedestrian approach and read as much of the book as possible. In this way, you'll get a thorough understanding of how robots tick. When you finish the book, you'll know the kind of robot(s) you'll want to make, and how you'll make them.

Conventions Used in This Book

You need little advance information before you can jump head-first into this book, but you should take note of a few conventions I've used in the description of electronic parts, and the schematic diagrams for the electronic circuits.

TTL integrated circuits are referenced by their standard 74XX number. The "LS" identifier is assumed. I built most of the circuits using LS TTL chips, but the projects should work with the other TTL family chips: the standard (non-LS) chips, as well as those with the S, ALS, and C identifiers. If you use a type of TTL chip other than LS, you should consider current consumption, fanout, and other design criteria. These may affect the operation or performance of the circuit.

The chart in Fig I-1 details the conventions used in the schematic diagrams. Note that non-connected wires are shown by a direct cross or lines, or a broken line. Connected wires are only shown by the connecting dot.

Details on the specific parts used in the circuits are provided in the Parts List tables that accompany the schematic. Refer to the Parts List for information on resistor and capacitor type, tolerance, and watt or voltage rating.

In all full circuit schematics, the parts are referenced by component type and followed by a number.

- U means an integrated circuit (IC).
- R means a resistor or potentiometer (variable resistor).
- C means a capacitor.
- D means a diode, a zener diode, and sometimes a light sensitive phototransistor.
- Q means a transistor and sometimes a light sensitive photodiode.
- LED means a light emitting diode (most any visible LED will do, unless the Parts List specifically calls for an infrared LED).
- XTL or XTAL means a crystal or ceramic resonator.
- Finally, S means a switch; RL means a relay, SPKR means a speaker, TR means a transducer (usually ultrasonic), MIC means a microphone.

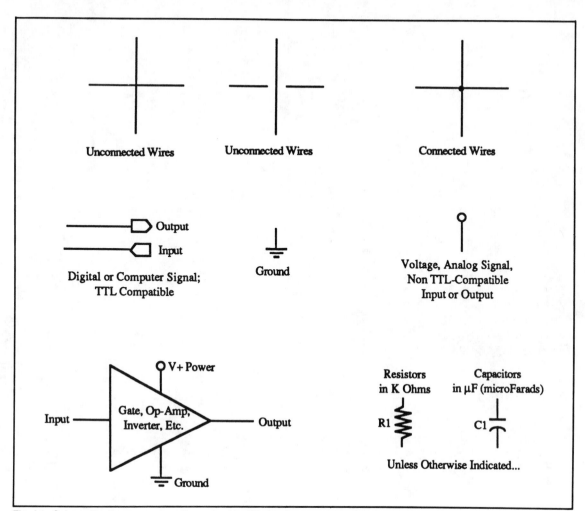

Fig. I-1. Schematic diagram conventions used in this book.

Chapter 1

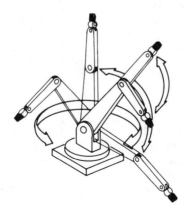

The Robot Experimenter

There he sits, as he's done countless long nights before, alone and deserted in a dank and musty basement. With each strike of his ball-peen hammer comes an ear-shattering bong, and an echo that seems to ring forever. Slowly, his creation takes shape and form—it first started as an unrecognizable blob of metal and plastic, then it transformed into an eerie silhouette, then. . .

Brilliant and talented, but perhaps a bit crazed, he is before his time: a social outcast, a misfit that belongs neither to science nor fiction. He is the robot experimenter, and all he wants to do is make a mechanical creature that serves drinks at parties and wakes him up in the morning.

Maybe this is a rather dark view of the present-day hobby robotics experimenter. But though you may find a dash of the melodramatic in it, the picture is not entirely unrealistic. It's a view many outsiders to the robot building craft hold, one that's over 100 years old; when the prospects of building a human-like machine first came within technological grasp. It's a view that will continue for another 100 years, perhaps beyond.

Like it or not, if you're a robot experimenter, you are an odd-ball, an egg-head, and—yes, let's get it all out—a little on the weird side!

In a way, you're not unlike Victor Frankenstein, the old-world doctor from Mary Wollstonecraft Shelley's immortal 1818 horror thriller. Instead of robbing graves at the still of night, you rob electronic stores, flea markets, surplus outlets, and other specialty shops in your unrelenting quest—your thirst—for all kinds and sizes of motors, batteries, gears, wires, switches, and other odds-and-ends. Like Dr. Frankenstein, you galvanize life from these "dead" parts.

If you have yet to build your first robot, you're in for a wonderful experience. Watching your creation scoot around the floor or table can be exhilarating, breathtaking. Those around you may not immediately share your excitement, but you know that you've built something—however humble—with your own hands and ingenuity.

If you're one of the lucky few that has already assembled a working robot, then you know of the excitement I speak of. You know how thrilling it is to see your robot obey your commands, as if it were a trusted dog. You know the time and effort that went into constructing your mechanical marvel, and although others may not always appreciate it, especially when it marks up the kitchen floor with its rubber tires, you are satisfied with the accomplishment, and look forward to the next challenge.

If you have built a robot, you also know of the heartache and frustration that's inherent in the process. You know that not every design works, and that even a simple design flaw can cost weeks of work, not to mention ruined parts. This book will help you—beginner and experienced robot maker alike—avoid these mistakes.

THE BUILDING BLOCK APPROACH

One of the best ways to experiment with—and learn about hobby robotics is to construct individual robot components, then combine the completed modules to make a finished, fully functional machine. For maximum flexibility, these modules should be interchangeable. You should be able to choose locomotion system "A" to work with appendage system "B," and operate the mixture with control system "C"—or any variation thereof.

The robots you create are made from building blocks, so making changes and updates are relatively simple and straightforward tasks. When designed and constructed properly, the building blocks, as shown in diagram form in Fig. 1-1, may be shared among a variety of robots. It's not unusual to reuse parts as you experiment with new robot designs.

Most of the building block designs presented in the following chapters are complete, working subsystems. The majority operate without ever attaching them to a mechanical mainframe or control computer. The way you interface the modules is up to you, and will require some forethought and attention on your part (I'm not doing *all* the work, you know!). Feel free to experiment with each subsystem, altering it and improving upon it as you see fit. When it works the way you want, incorporate it into your robot, or save it for a future project.

BASIC SKILLS

What skills do you need as a robot experimenter? Certainly, if you are already well versed in electronics and mechanical design, you are on your way to becoming a robot experimenter extraordinnaire. But an intimate knowledge of neither electronics nor mechanical design is absolutely necessary.

All you really need to start yourself in the right direction as a robot experimenter is a basic familiarity with electronic theory and mechanics. The rest you can learn as you go. If you feel that you're lacking in either beginning electronics or mechanics, pick up a book or two on these subjects at the bookstore or library. See Appendix

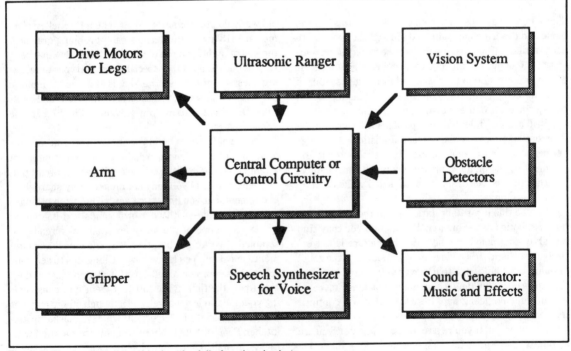

Fig. 1-1. The basic building blocks of a fully functional robot.

B for a selected list of suggested further reading.

Electronics Background

Study analog and digital electronic theory, and learn the function of resistors, capacitors, transistors, and other common electronic components. Your mastery of the subject does not need to be extensive, just enough so that you can build and troubleshoot electronic circuits for your robot. You'll start out with simple circuits with a minimum of parts, and go from there. As your skills increase, you'll be able to design your own circuits from scratch, or at the very least, customize existing circuits to match your needs.

Schematic diagrams are a kind of recipe for electronic circuits. The designs in this book, as well as most any book that deals with electronics, are in schematic form. If you don't know already, you owe it to yourself to learn how to read a schematic diagram. There are really only a dozen or so common schematic symbols, and memorizing them takes just one evening of concentrated study. A number of books have been written on how to read schematic diagrams. Once again you are referred to Appendix B for a list of suggested reading.

Sophisticated robots use a computer or microprocessor to control their actions. This book is not as computer-oriented as some, but a number of projects described throughout this book require at least some knowledge of computers and the way computers manipulate data. The popularity of personal computers means that there are plenty of instructional guides on all facets of this growing science. If you're new or relatively new to the field, start with a beginning computer book, then move up to more advanced texts. Don't start with a book on Assembly Language programming or microprocessor interfacing techniques; an introductory guide will get you going in the right direction, and the rest will follow naturally.

Mechanical Background

The majority of us are far more comfortable with the mechanical side of hobby robot building than the electronic side. After all, you can see gears meshing and pulleys moving as a robot glides across the floor; you can't see electrons flowing within a circuit. It's far easier to see how a mechanical drive train works than to see how an ultrasonic ranging system works. Whether or not you are comfortable with mechanical design, you do not need to possess a worldly knowledge of mechanical theory. You should be comfortable with mechanical and electro-mechanical components such as motors, solenoids, and chain drives. This book provides some mechanical theory as it pertains to robot building, but you may want to supplement your learning with books or study aides.

The Workshop Aptitude

To be a successful robot builder, you must be comfortable working with your hands, and thinking problems through from start to finish. You should know how to use common shop tools and have some basic familiarity in working with wood, lightweight metals, and plastic. Once more, if you feel your skills aren't up to par, read up on the subject and try your hand at a simple project or two.

With experience comes confidence, and with both comes more professional results. Work on it long enough, and the robots you build may be indistinguishable from store-bought models (in appearance, not capability; yours will undoubtedly be far more sophisticated!).

The Two Most Important Skills

Two important skills that you can't develop from reading books are patience and the willingness to learn. Both are absolutely essential if you want to build your own working robots. Without these skills, you can't do anything, at least anything well, and you will most assuredly give up in desperation and frustration.

Give yourself time to experiment with your projects. Don't rush into things, because you are bound to make mistakes. If a problem continues to nag at you, put the project aside, letting it sit for a few days, even a few weeks. You'll be surprised how the solution to many of your most challenging problems will come when you least expect them: while dozing off for the night, while showering, while driving to work. Keep a small notebook handy and jot down your ideas so you won't forget them.

If trouble persists, perhaps you need to bone up on the subject before you can adequately tackle the problem. Take the time to study, to learn more about the various sciences and disciplines involved. While you are looking for ways to combat the current dilemma, you are increasing your general robot-building knowledge. Research is never in vain.

THE MIND OF THE ROBOT EXPERIMENTER

Robot experimenters have a unique way of looking at things. Nothing is taken for granted.

- At a restaurant, it's the robot experimenter who

collects the carcasses of lobster and crabs, to learn how nature has designed articulated joints where the muscles and tendons are inside the bone. Perhaps the articulation and structure of a lobster leg can be duplicated in the design of a robotic arm.

- At a county fair, it's the robot experimenter who studies the way the egg-beater ride works, watching the various gears mesh and spin in perfect unison. Perhaps the gear train can be duplicated in an unusual robot locomotion system.
- At a phone booth, it's the robot experimenter who listens to the dial tones that are emitted when buttons are pressed. These tones, the experimenter knows, trigger circuitry at the phone company office to call a specific telephone among the millions in the world. Perhaps these or similar tones can be used to remotely control a robot.
- At work on the computer, it's the robot experimenter who rightly assumes that if a computer can control a printer or plotter through an interface port, the same computer and interface can be used to control a robot.
- When taking a snapshot at a family gathering, it's the robot experimenter who studies the inner workings of the automatic focus system of the camera. The camera uses ultrasonic sound waves to measure distance, and automatically adjusts its lens to keep things in focus. The same system should be adaptable to a robot, enabling it to judge distances and see with sound.

The list could go on and on. The point? All around us, from Mother Nature's designs to the latest electronic gadgets, are an infinite number of ways to make better and more sophisticated robots. Uncovering these solutions requires extrapolation—figuring out how to apply one design and make it work in another application, then experimenting with the contraption until everything works.

FOLLOW THE YELLOW BRICK ROAD

Now you know what is expected of you, and how—as a robot experimenter—you must begin to look at the world around you. Knowing your own strengths and weaknesses goes a long way in helping you develop skills so that even the most complex robot designs are within your grasp.

The following chapter discusses what makes up a robot, how the parts function and interrelate to one another, the differences between so-called smart and dumb machines, and the dream called the self-contained, autonomous robot.

Chapter 2

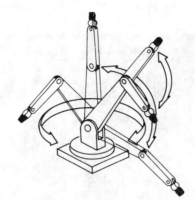

Anatomy of a Robot

We are fortunate. The human body is, all things considered, a nearly perfect machine: it is intelligent (usually!), it can lift heavy loads, it can move itself around, and it has built-in protective mechanisms to feed itself when hungry, avoid danger, and sense pain so it can avoid further damage. Other living creatures on this earth possess the same or similar functions, though not always in such advanced form.

Robots are often modeled after humans, if not in form, then at least in function. For decades, scientists and experimenters have tried to duplicate the human body, to create machines with intelligence, strength, mobility, and auto sensory mechanisms. That goal has not yet been realized, and in fact, probably never will be.

Nature provides a striking model for robot experimenters to mimic, and it is up to us to take up the challenge. Some, but by no means all, of nature's mechanisms—human or otherwise—can be duplicated to some extent in the robot shop. Robots can be built with eyes to see, ears to hear, a mouth to speak, and appendages and locomotion systems of one kind or another to manipulate the environment and explore surroundings.

This is fine theory; what about real life? Exactly what makes up a real hobby robot? What basic parts are necessary for a machine to have before it can be given the ti-

tle robot? Let's take a close look at the anatomy of a robot in this chapter, and the kinds of materials used to construct one. For the sake of simplicity, not every robot subsystem in existence will be covered, just the components that are most often found in hobby and personal robots.

SELF-CONTAINED VERSUS TETHERED

People like to argue the definition of a real robot. One side says that a robot is a completely *self-contained, autonomous* (self-governed) machine that needs only occasional instructions from its master to set it about its various tasks. A self-contained robot includes its own power system, brain, wheels (or legs) and manipulating devices such as claws or hands. The robot does not depend on any other mechanism or system to perform its tasks. It's complete, in and of itself.

The other side says that a robot is anything that moves under its own motor power, *for the purpose of performing near-human tasks* (this is, in fact, the definition of the word "robot" in most dictionaries). The mechanism that does the actual task is the robot itself; the support electronics or components may be separate. The link between robot and control components might be a wire, a beam of infrared light, even a radio signal.

5

In the 1969 experimental robot shown in Fig. 2-1, for example, a man sat inside the mechanism and operated it, almost as if driving a car. The purpose of the four-legged lorry was not to create a self-contained robot, but to further the development of *cybernetic anthropomorphous machines*, otherwise known as *cyborgs*. Hardly a useless robotic endeavor.

The semantics of robot design won't be argued here

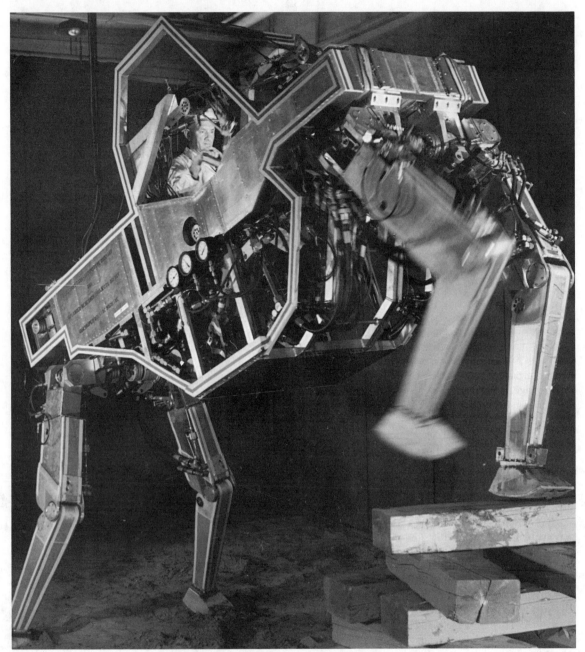

Fig. 2-1. This GE qudraped was controlled by a human operator who sat inside it. The robot was developed in the late 1960's under a contract with the U.S. government. Photo courtesy General Electric.

6

(this is a workbook after all, not a critique on theory), but it's still necessary to establish some basic levels of robot characteristics: what makes a robot a robot, and just not another machine? For the purposes of this book, let's consider a robot as any device that—in one way or another—mimics human or animal function. The way that the robot does this is of no concern; the fact that it does it at all is enough.

The functions that are of interest to the robot builder run a wide gamut: from listening to sounds and acting on them, to talking, walking or moving across the floor, to picking up objects, and sensing special conditions such as heat, flames, or light.

When we talk about a robot, it could very well be a self-contained automaton that takes care of itself, perhaps even programming its own brain, learning from its surroundings and environment. Or, it could be a small motorized cart, operated by a strict set of pre-determined instructions, repeating the same task over and over again until its batteries wear out, or it could be a radio controlled arm that you operate manually from a control panel. All are no less robots than the others, though some are more useful and flexible. As you'll discover in this chapter and the others that follow, the level of complexity of your robot creations is completely up to you.

THE BODY

Like the human body, the body of a robot contains all its vital parts. It's the superstructure that prevents its electronic and electro-mechanical guts from spilling out. Robot bodies go by many names, including *frame* and *chassis*, but the idea is the same.

Skeletal Structures

In nature, and in robotics, there are two general types of support frames: endoskeleton and exoskeleton.

- *Endoskeleton* support frames are the kind found in the majority of nature's critters—including humans, mammals, reptiles, and most fish (note that the majority is species, not actual numbers of living organisms). The skeletal structure is on the inside, the organs, muscles, body tissues, and skin are on the outside. The endoskeleton is a characteristic of vertebrates.
- *Exoskeleton* support frames are the bones on the outside of the organs and muscles. Common exoskeletal creatures are spiders, all shell fish such as lobsters and crabs, and an endless variety of insects.

Which is better, endoskeleton or exoskeleton? Both. It all depends on the living conditions of the animal and its eating and survival tactics. The same is true of robots.

The main structure of the robot is generally a wood or metal frame, and is constructed a little like the frame

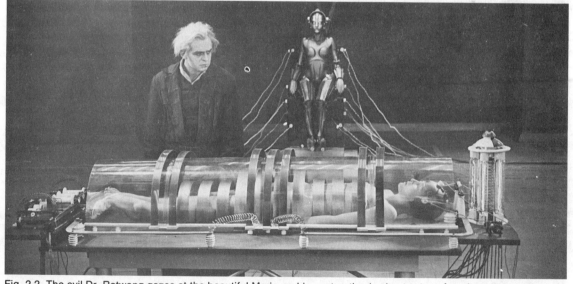

Fig. 2-2. The evil Dr. Rotwang gazes at the beautiful Maria, as his contraption is about to transform her shape to the cold steel robot who sits in the background. Today, the robot could effectively be made from materials other than steel, like plastic and aluminum. From the movie classic ''Metropolis.'' Photo courtesy Museum of Modern Art/Film Archives.

of a house, with a bottom, top, and sides. This gives the automaton a boxy or cylindrical shape, though any shape is possible. It could even be human form, like the robot in Fig. 2-2. For a machine, however, the body of man and woman is terribly inefficient.

On the frame of the robot are attached motors, batteries, electronic circuit boards, and other necessary components. In this way, the main support structure of the robot can be considered an exoskeleton, because it is outside the major organs. The design further lacks a central spine, a characteristic of endoskeletal systems, and something most of us think about when trying to model robots after humans. In many cases, a shell is sometimes placed over the robot, but the skin is for looks only (sometimes protection of the internal components), not support.

This is not to say that some robots are not designed with endoskeletal structures, but most such creatures are reserved for high-tech research and development projects and science fiction films. For the most part, the main body of your robots will have an exoskeleton support structure, because they are cheaper to build, stronger, and less prone to problems.

Not all of your hobby robot must be exoskeletal in nature. Robot arms and hands are generally designed as endoskeletal structures. Wooden dowels or metal rods are often used as the bone of the arm. Attached to the bone are motors, cams, cables, and other devices to affect movement of joints. If the robot has legs, the legs usually follow a similar design approach. The combination of exoskeletal and endoskeletal systems is something unique to robotics, and doesn't occur in nature.

Size and Shape

The size and shape of the robot can vary greatly, and size alone does not determine the intelligence of the machine, nor its capabilities. Homebrew robots are generally the size of a small dog, although some are as small as an aquarium turtle and a few as large as Arnold Schwarzenegger.

The overall shape of the robot is generally dictated by the internal components that make up the machine, but most designs fall into one of the following "categories":

- *Turtle.* Turtle robots—or turtlebots—are simple and compact, designed primarily for tabletop robotics. Turtlebots get their name because their body somewhat resembles the shell of a turtle.
- *Toy car,* or scooter-type robots are small automations with wheels, often built on the chassis of a

radio controlled car. The radio control aspect is usually employed to operate the converted car/robot.
- *Garbage can.* Looking a lot like the famous R2-D2 of "Star Wars" fame, garbage can robots are short and stout, and most are circular in shape. They closely resemble their namesake; in fact, some hobby robots are actually built from metal and plastic trash cans! Despite the euphemistic title, garbage can robots represent an extremely workable design approach. From now on in this book, garbage can robots will be called rovers.
- *Appendage.* Appendage designs are used specifically with robotic arms, whether the arm is attached to a robot, or is a stand-alone mechanism.
- *Android.* Android robots are specifically modeled after the human form, and is the type most people picture when talk turns to robots. Realistically, android designs are the most restrictive and least workable.

Flesh and Bone

In the 1926 movie classic "Metropolis," an evil scientist, Dr. Rotwang, transforms a cold and calculating robot into the body of a beautiful woman. This film, generally considered to be the first science fiction cinema epic, also set the psychological stage for later movies, particularly those of the 1950s and 1960s. The shallow and stereotyped character of Dr. Rotwang, shown in the movie still in Fig. 2-3, was to pop up many more times in later films. The shapely robotrix changed form for these other films, but not character. Robots have often been depicted as metal creatures with hearts as cold as the steel in their bodies.

Which brings us to an interesting point: Are all real robots made of heavy-gauge steel, stuff so thick that bullets, disinto-ray guns, even atomic bombs can't penetrate? Indeed, while metal of one kind or another is a major component in robot bodies, the list of usable materials is much larger and diverse. Hobby robots can be easily constructed from aluminum, steel, tin, wood, plastic, or a combination of them all.

- *Aluminum.* Aluminum is the best all-around robot building material because it is exceptionally strong for its weight. Aluminum is easy to cut and bend using ordinary shop tools. It is commonly available in long lengths of various shapes, but it's somewhat expensive.
- *Steel.* Although sometimes used in the structural

Fig. 2-3. The "android" design of robots is the most difficult to achieve, not only because of its bipedal (two leg) structure, but because of its distribution of weight towards the mid- and top-sections of the body. Photo courtesy of The Museum of Modern Art/Film Stills Archive.

frame of a robot because of its strength, steel is difficult to cut and shape without special tools. Stainless steel is sometimes used for precision components, like arms and hands, and also for parts that require more strength than a lightweight metal (such as aluminum) can provide. Expensive.

- *Tin* and *iron*. Tin and iron are a common hardware metal, often used to make angle brackets, sheet metal (various thicknesses from 1/32″ on up), and (when galvanized) nail plates for house framing. Tin and iron are stronger than aluminum, but about twice as heavy for a piece the same size. Cost: cheap.

- *Wood*. Surprise! Wood is an excellent material for robot bodies, although you may not want to use it in all your designs. Wood is easy to work with, can be sanded and sawed to any shape, doesn't conduct electricity (avoids short circuits) and is available everywhere. Disadvantage: Wood is rather weak for its weight, so you need fairly large pieces to provide stability. You must make your robot designs larger to accommodate the bulk of the wood.

- *Plastic*. Everything is going plastic these days, including robots. Pound for pound, plastic has more strength than most metals, yet is easier to work with. You can cut and shape it, drill it, even glue it. Effective use of plastic requires some special tools, and availability of extruded pieces might be somewhat scarce unless you live near a well-stocked plastic specialty store. Mail order is an alternative.

POWER SYSTEMS

We eat food which is processed by the stomach and intestines to make fuel for our muscles, bones, skin, and the rest of our body. While you could probably design a digestive system for a robot, and feed it hamburgers, french fries, and other semi-radioactive foods, an easier way to generate power for the purpose of making your robot go is to take a trip to the store and buy a set of dry-cell batteries. Connect the batteries to the robot's motors, circuits, and other parts, and you're all set.

Types of Batteries

There are a number of different types of batteries, and Chapter 9, "All About Batteries," goes into more detail about them. But here are a few quick details:

Batteries generate dc and come in two distinct categories: rechargable and non-rechargable (let's forget the non-descriptive terms like storage, primary, and secondary). *Non-rechargable* batteries include the standard zinc and alkaline cells you buy at the drug store, as well as special purpose lithium and mercury cells for calculators, smoke detectors, watches, and hearing aids. These have limited use in power-hungry robotics, because replacing the batteries after a few hours or days use can be an expensive proposition.

Rechargable batteries include nickle-cadmium (Ni-Cad), gel-cell, and sealed lead-acid cells. Ni-Cad batteries are a popular choice because they are relatively easy to find, come in popular household sizes ("D," "C," etc.) and can be recharged many hundreds of times using an inexpensive recharger. Gel-cell and lead-acid batteries provide longer-lasting power, but they are heavy and bulky. Both gel-cell and lead-acid cells are most often used in medium and large robot systems.

Alternative Power Sources

Batteries are required in a fully self-contained robot, because there can be no power cord connecting the automaton to an electrical socket. That doesn't mean other power sources, including ac, can't be used in some of your robot designs. On the contrary, stationary robot arms don't have to move around the room; they are designed to be placed about the perimeter of the work place, and perform within this predefined area. The motors and control circuits may very well run off ac power, thus freeing you from replacing batteries, and worrying about operating times and recharging periods.

This doesn't mean that ac power is the preferred method. While it's true that ac motors pack a bigger wollop than dc motors of the same size, working with ac poses greater shock hazards. Should you ever decide to make your robot independent, you must exchange all the ac motors for dc ones. Electronic circuits ultimately run off dc power, even when the equipment is plugged into an ac outlet. The ac is converted to dc, and usually reduced in voltage (5 and 12 volts is common) before being applied to the circuits. The motors share this dc voltage.

One alternative to batteries in an all-dc robot system is to construct an ac operated power station that provides your robot with regulated dc juice. The power station converts the ac to dc, and provides a number of different voltage levels for the various components in your robot, including the motors.

Because the entire robot is designed to operate under dc power, converting it later to batteries is much easier. You can also use the power station instead of batteries when experimenting with your robot designs on the workbench. Read more about the power station in Chapter 10, "Build an Experimenter's Power Supply Station."

Pressure Systems

Two other forms of robotic power, not discussed in depth in this book, are hydraulic and pneumatic. *Hydraulic* power uses oil or fluid pressure to move linkages. You've seen hydraulic power at work if you've ever watched a bulldozer go about moving dirt from pile to pile. And, while you drive, you use it every day when you press down on the brake pedal. Similarly, *pneumatic* power uses air pressure to move linkages. Pneumatic systems are cleaner than hydraulic systems, but all things considered, they aren't as powerful.

Both hydraulic and pneumatic systems must be pressurized to work, and this pressurization is most often done by a pump. The pump is driven by an electric motor, so in a way, robots that use hydraulics or pneumatics are fundamentally electrical.

Hydraulic and pneumatic systems are rather difficult to effectively implement, but they provide an extra measure of power over dc and ac motors. With a few hundred dollars in surplus pneumatic cylinders, hoses, fittings, solenoid valves, and a pressure supply (battery-powered pump, air tank, regulator), you could conceivably build a hobby robot that picks up chairs, bicycles, even people!

LOCOMOTION SYSTEM

Some robots aren't designed to move around. These include robotic arms, which manipulate objects placed within a work area. But these are exceptions rather than the rule. Most hobby robots are designed to get around in this world. They do so in a variety of ways, from wheels to legs to tank tracks. In each case, the locomotion system is driven by a motor, which turns a shaft, cam, or lever. This motive force affects forward or backward movement.

Wheels

Wheels are the most popular method of providing robot mobility. There may be no animals on this earth that use wheels to get around, but for us robot builders, it's the simple and foolproof choice.

Robot wheels can be just about any size, dictated only by the dimensions of the robot and your outlandish imagination. Turtle robots usually have small wheels, less than two or three inches in diameter. Medium sized rover-type robots use wheels with diameters up to seven or eight inches. A few unusual designs call for bicycle wheels, which despite their size, are lightweight but very sturdy.

Robots can have just about any number of wheels, although two is the most common. The robot is balanced on the two wheels by one or two free-rolling casters, or perhaps even a third swivel wheel. Four and six-wheel robots are also around.

Legs

Only a small percentage of robots—particularly the hobby kind—are designed with legs. First, there is the question of the number of legs, and how the legs provide stability when the robot is in motion or when it's standing still. Then there is the question of how the legs propel the robot forward or backward—and more difficult still!—the question of how to turn the robot so it can navigate a corner.

Tough questions, yes, but insurmountable? No way! Legged robots are a challenge to design and build, but they provide an extra level of mobility that wheeled robots can't. Wheel-based robots may have a difficult time navigating through rough terrain, but leg-based robots can easily walk right over small ditches and obstacles.

A few daring robot experimenters have come out with two-legged robots, but the difficulties in assuring balance and control make these designs largely impractical. Four-legged robots are easier to balance, but locomotion and steering are hard to achieve.

I've found robots with six legs (called a hexapod) are able to walk at brisk speeds without falling, and are more than capable of turning corners, bounding over uneven terrain, and making the neighborhood dogs and cats hide for cover.

Tracks

The basic design of track-driven robots is pretty simple: two tracks, one on each side of the robot, act as giant wheels. The tracks turn, like wheels, and the robot lurches forward or backward. For maximum traction, each track is about as long as the robot itself.

Track drive is practical for many reasons, the least of which is the ability to mow through all sorts of obstacles, like rocks, ditches, and potholes. Given the right track material, traction is excellent, even on slippery surfaces like snow, wet concrete, or a clean kitchen floor.

Alas, all is not rosy when it comes to track-based robots. Unless you plan on using your robot exclusively outdoors, you should probably stay away from track drive. Making the drive work can be harder than implementing wheels or even legs.

The biggest problem is what the tracks do to floors and carpets. If the track is cleated (with plastic or metal), it tends to dig into carpeting and other floorings. A robot running around a carpeted house quickly becomes an unwelcome guest—along with, perhaps, the poor soul who built it!

Rubber marks appear on smooth surfaces when the robot turns a corner. The reason: in a track drive system, steering is accomplished by powering just one of the tracks. This causes the tracks to slip in the direction of the unpowered side, and that slip leaves unsightly rubber marks. Outside, on grass or dirt, the damage is minimal, so outside is where a tracked robot belongs.

ARMS AND HANDS

The ability to manipulate objects is a trait that has enabled humans, as well as a few other creatures in the animal kingdom, to manipulate the environment.

Without our arms and hands, we wouldn't be able to use tools, and without tools we wouldn't be able to build houses, cars, and . . . hummm, robots. It makes sense, then, to provide arms and hands to our robot creations,

Fig. 2-4. A robotic arm from General Electric spot welds with great precision. Photo courtesy General Electric.

to enable them to manipulate objects and use tools. A commercial industrial robot welding arm is shown in Fig. 2-4.

Human arms can be duplicated in a robot with just a couple of motors, some metal rods, and a few ball bearings. Add a gripper to the end of the robot arm and you've created a complete arm/hand module.

Not all robot arms are modeled after the human appendage. Some look more like forklifts than arms, and a few use retractable push-rods to move a hand or gripper toward or away from the robot.

Stand-alone or Built-in

Some arms are themselves a complete robot. Car manufacturing robots are really arms that can reach in just about every possible direction with incredible speed and accuracy. You can build a stand-alone robotic arm

11

trainer, which can be used to manipulate objects within a defined workspace, or you can build an arm and attach it to your robot. Some arm/robot designs concentrate on the arm part much more than the robot part. They are, in fact, little more than arms on wheels.

Grippers

Robot hands are commonly referred to as *grippers*, or sometimes end effectors. We'll stick with the simpler-sounding hands and grippers in this book. Robot grippers come in a variety of styles; few are designed after the human counterpart. A functional robot claw can be built with just two fingers. The fingers close like a vise and can exert, if desired, a surprising amount of pressure.

SENSORY DEVICES

Imagine a world without light, sound, touch, smell, or taste. Without these senses, we'd be nothing more than an inanimate machine, like the family car or the living-room television. Our senses are an integral part of our lives—if not life itself.

It makes good sense to build at least one of these senses into our robot designs, because the more senses a robot has, the more it can interact with its environment. That makes the robot better able to go about its business on its own, which allows for more sophisticated tasks. Sensitivity to sound is the most common sensory system given to robots. The reason: sound is easy to detect, and unless you're trying to listen for a specific kind of sound, circuits for sound detection are simple and straightforward.

Sensitivity to light is also common, but the kind of light is usually restricted to a slender band of infrared, for the purpose of sensing the heat of a fire or for navigating through a room by way of an invisible light beam.

Robot eyesight is a completely different matter. The visual scene must be electronically rendered into a form the circuits on the robot can accept, and the machine must be programmed to understand and act on the shapes it sees. A great deal of experimental work is going on right now to allow robots to distinguish objects, but true robot vision is far off, even for well-financed scientists.

In robotics, the sense of touch is most often confined to pressure sensors attached to the tips of fingers in the robot's hand. The more the fingers of the hand close in around the object, the greater the pressure. The pressure information is relayed to the robot's brain, which then decides if the correct amount of pressure is being exerted.

There are a number of commercial products avail-

able that register pressure of one kind or another, but most are expensive. Simple pressure sensors can be constructed cheaply and quickly, and though they aren't as accurate as commercially manufactured pressure sensors, they are more than adequate for hobby robotics.

The senses of smell and taste aren't generally implemented in robot systems, though some security robots designed for industrial use are outfitted with a gas sensor that, in effect, smells the presence of dangerous toxic gas.

OUTPUT DEVICES

Output devices are components that relay information from the robot to the outside world. A common output device in a computer-controlled robot (discussed below) is the video screen. As with a personal computer, the robot communicates with its master by flashing messages on the video screen.

Another popular robotic output device is the speech synthesizer. In the 1968 movie "2001: A Space Odyssey," Hal the computer talks to its shipmates in a soothing but electronic voice. The idea of a talking computer was a rather novel concept at the time of the movie, but today voice synthesis is commonplace. Computerized speech can be added to any robot design at little cost.

There are two main types of voice synthesizers: fixed and open vocabulary. Fixed vocabulary synthesizers have built-in word lists, and say only those words in the list. The speech is most often digitized from a real voice and permanently implanted in the voice chip, much in the same way that music is recorded and stored on an audio compact disc. The number of words varies from one speech chip to another, but it's generally in the range of 150 to 200 words. The vocabulary includes words like "and," "the," "ready," numerals, and all the letters of the alphabet.

Conversely, open vocabulary synthesizers can be programmed to speak any word. The speech output is made up of many distinct sounds, called phonemes, and by stringing these sounds together, it's possible to say complete words and sentences. The speech quality of open vocabulary synthesizers isn't as good as that obtained from fixed vocabulary types, but it's easy to understand once you get used to it.

Many hobby robots are built with sound and music generators. These generators are sometimes used as warning signals, but by far, the application of speech, music, and sound is for entertainment purposes. Somehow, a robot that wakes you up to an electronic rendition of Bach seems a little more human.

SMART VERSUS DUMB ROBOTS

There are smart robots and there are dumb robots, but the difference really has nothing to do with intelligence. Even considering the science of artificial intelligence, all self-contained autonomous robots are unintelligent, no matter how sophisticated the electronic brain that controls it. Intelligence is not a measurement of computing capacity, but the ability to reason, to figure out how to do something by examining all the variables and choosing the best course of action, perhaps even coming up with something entirely new. Only humans and a few rare mammals (such as dolphins) seem able to conceptualize ideas based on brand new information.

In this book, the difference between dumb and smart depends on the ability to take two or more pieces of data and decide on a pre-programmed course of action. Usually, a *smart* robot is one that is controlled by a computer, although some amazingly sophisticated actions can be built into an automaton that contains no computer. A *dumb* robot is one that blindly goes about its task, never taking the time to analyze its actions and what impact they may have.

Using a computer as the brains of a robot provides a great deal of operating flexibility. Unlike a control circuit, which is wired according to a schematic plan and performs a specified task, a computer can be electronically rewired using *software* instructions—programs.

To be effective, the computer must be connected to all the control and feedback components of the robot. This includes the drive motors, the motors that control the arm, the speech synthesizer, the pressure sensors, and so forth. Connecting a computer to a robot is a demanding task, one requiring many hours of careful work. This book presents some computer-based control projects, as well as some alternate methods for giving non-computer based robot creations some extra smarts.

Note that this book does not tell you how to construct a computer. Rather than build a specially designed computer for your robot, the projects in this book call on using ready-built personal computers. Some computers, particularly the portable variety, can be permanently integrated with one of your larger robot projects.

The Knee-Bone is Connected To The . . .

On the surface, there doesn't seem to be much to robots. In fact, many appear terribly simple, almost embarrassingly so. Only by constructing one yourself do you realize that a robot is a sophisticated machine, where everything must work in perfect synchronism. When something goes wrong, the hulking mass of metal, wood, plastic, and electronic parts grinds to a halt. The art of robot building goes beyond constructing an arm here and a leg there, but making sure that everything fits together properly, that everything works together as it should.

Chapter 3

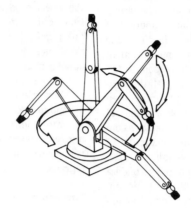

Tools and Supplies

Take a long look at your the tools in your garage or workshop. You probably have all the implements necessary to build your own robots. Unless your robot designs require a great deal of precision (and most hobby robots don't), a common assortment of hand tools are all that's really required to construct robot bodies, arms, drive systems, and more.

Most of the hardware, parts, and supplies are things you probably already have, left over from old projects from around the house. The pieces you don't have can be readily purchased at a hardware store and a few specialty stores around town or through the mail.

This chapter discusses the basic tools and supplies needed for hobby robot building, and how you might use them. You should consider this chapter a guide only; suggestions for tools and supplies are just that—suggestions. By no means should you feel that you must own each tool mentioned in this chapter, or have on hand all the parts and supplies.

The concept of this book is to provide you with ways you can build robots from discrete modules, in keeping with the open-end design, and you are free to exchange parts in the modules as you see fit. Some supplies and parts may not be readily available to you, and it's up to you to consider alternatives, and how to work these al-

ternatives into the design. Ultimately, it will be your task to take a trip to the hardware store, collect various miscellaneous items, and go home to hammer out a unique creation that's all your own.

CONSTRUCTION TOOLS

Construction tools are the things you use to fashion the frame and other mechanical parts of the robot. These include hammer, screwdriver, saw, and so forth. Tools for the assembly of the electronic subsystems are discussed later on.

Basic Tools

No robot workshop is complete without the following:

- *Claw hammer*, used for just about anything you can think of.
- *Rubber mallet*, for gently bashing pieces together that resist going together; also for forming sheet metal.
- *Screwdriver assortment*, including various sizes of flat-head and Philips-head screwdrivers. A few long-blade screwdrivers are handy to have, as well

as a ratchet driver. Get a screwdriver magnetizer/demagnetizer; it lets you magnetize the blade so it attracts and holds screws for easier assembly.

- *Hacksaw*, to cut anything. The hacksaw is the staple of the robot builder. Get an assortment of blades. Coarse-tooth blades are good for wood and PVC pipe plastic; fine-tooth blades are good for copper, aluminum, and light gauge steel.
- *Miter box*, to cut straight lines. Buy a good miter box and attach it to your work table (avoid wood miter boxes; they don't last). You'll also use the box to cut stock at near-perfect 45 degree angles, helpful when building robot frames.
- *Wrenches*, all types. Adjustable wrenches are helpful additions to the shop but careless use can strip nuts. The same goes for long-nosed pliers, useful for getting at hard to reach places. A pair or two of Vice-Grips helps you hold pieces for cutting and sanding. A set of nutdrivers make it easy to attach nuts to bolts.
- *Measuring tape*. A six- or eight-foot steel measuring tape is a good choice. Also get a cloth tape at a fabric store for measuring things like chain and cable lengths.
- *Square*, for making sure that pieces you cut and assemble from wood, plastic, and metal are square.
- *File assortment*, to smooth the rough edges of cut wood, metal, and plastic (particularly important when working with metal because the sharp unfinished edges can cut you).
- *Drill motor*. Get one that has a variable speed control (reversing is nice, but not absolutely necessary). If the drill you have isn't variable speed, buy a variable speed control for it. You need to slow the drill when working with metal and plastic. A fast drill motor is good for wood only. The size of the chuck is not important, since most of the drill bits you'll be using will fit a standard 1/4-inch chuck.
- *Drill bit assortment*. Good, sharp ones only. If yours are dull, have them sharpened (or do it yourself with a drill bit sharpening device), or buy a new set.
- *Vise*, for holding parts while you drill them, nail them, and otherwise torment them. An extra large vice isn't required, but you should get one that's big enough to handle the size of pieces you'll be working with. A rule of thumb: A vice that can't close around a three inch block of metal or wood is too small.
- *Safety goggles*. Wear them when hammering, cut-

ting, drilling, and any other time when flying debris could get in your eyes. Be sure you use the goggles. A shred of aluminum sprayed from a drill bit while drilling a hole can rip through your eye, permanently blinding you. No robot project is worth that.

If you plan on building your robots from wood, you may want to consider adding rasps, wood files, coping saws, and other woodworking tools to your toolbox. Working with plastic requires a few extras, as well, including a burnishing wheel to smooth the edges of the cut plastic (a flame from a cigarette lighter also works, but is harder to control), a strip-heater for bending, and special plastic drill bits. These bits have a modified tip that aren't as likely to rip through the plastic material. Bits for glass can be used as well. Small plastic parts can be cut and scored using a sharp razor knife or razor saw, available at hobby stores.

Optional Tools

There are a number of tools you can use to make your time in the robot shop more productive and less time consuming. A *drill press* helps you drill better holes, because you have more control over the angle and depth of each hole. Be sure to use a drill press vice to hold the pieces. Never use your hands.

A *table saw* or *circular* saw makes cutting through large pieces of wood and plastic easier. Use a guidefence, or fashion one out of wood and clamps, to ensure a straight cut. Be sure to use a fine-tooth saw blade if cutting through plastic. Using a saw designed for general wood cutting will cause the plastic to shatter.

A motorized *hobby tool*, such as the model in Fig. 3-1, is much like a hand held router. The bit spins very fast (25,000 rpm and up), and you can attach a variety of wood, plastic, and metal working bits to it. The better hobby tools, such as those made by Dremel and Weller, have adjustable speed controls. Use the right bit for the job. For example, don't use a wood rasp bit with metal or plastic, because the flutes of the rasp will too easily fill with metal and plastic debris.

A *nibbling tool* is a fairly inexpensive accessory (under $20) that lets you nibble small chunks from metal and plastic pieces. The maximum thickness depends on the bite of the tool, but it's generally about 1/16-inch. Use the tool to cut channels, enlarge holes, and so forth.

A *tap and die set* lets you thread holes and shafts to accept standard size nuts and bolts. Buy a good set. A cheap assortment of taps and dies is more trouble than

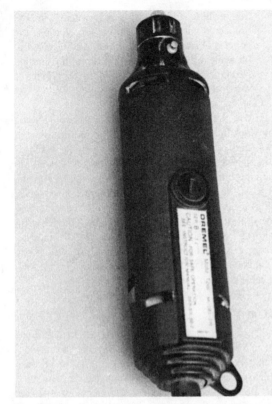

Fig. 3-1. A motorized hobby tool is ideal for drilling, sanding, and shaping small parts.

it's worth. The bargain brands use soft, unforged metal, causing you to work harder than you should. They also yield less-than-professional results.

A *thread size gauge*, made of stainless steel, may be expensive, but it helps you determine the size of any standard SAE or metric bolt. It's a great accessory for tapping and dieing. Most gauges can be used when chopping threads off bolts with a hacksaw, providing a cleaner cut.

A *brazing tool* or *small welder* lets you spot weld two metal pieces together. These tools are designed for small pieces only. They don't provide enough heat to adequately weld pieces larger than a few inches in size. Be sure that extra fuel and oxygen cylinders or pellets are readily available for the brazer or welder you buy. There's nothing worse than spending $30 to $40 for a home welding set, only to discover that supplies are not available for it. Be sure to read the instructions that accompany the welder and observe all precautions.

ELECTRONIC TOOLS

Constructing electronic circuit boards, or wiring the power system of your robot, requires only a few standard tools. A *soldering iron* leads the list. For maximum flexibility, invest in a modular soldering pencil, the kind that lets you change the heating element. For routine electronic work, you should get a 25 to 30 watt heating element. Anything higher may damage electronic components. A 40 or 50 watt element can be used for wiring switches, relays, and power transistors. Stay away from "instant-on" soldering irons. They put out far too much heat for most any application other than soldering large gauge wires.

Supplement your soldering iron with these accessories:

- *Soldering stand*, for keeping the soldering pencil in a safe, upright position.
- *Soldering tip assortment*. Get one or two small tips for intricate printed circuit board work, and a few larger sizes for routine soldering chores.
- *Solder*. And not just any kind of solder, but the resin or flux core type. Acid core and silver solder should never be used on electronic components.
- *Sponge*, for cleaning the soldering tip during use. Keep the sponge damp and wipe the tip clean every few joints.
- *Heat sink*, for attaching to sensitive electronic components during soldering. The heat sink draws the excess heat away from the component, and helps prevent damage to it.
- *Desoldering vacuum tool*, to soak up molten solder. Used to get rid of excess solder, to remove components, or redo a wiring job.
- *Dental picks*, for scraping, cutting, forming, and gouging into the work.
- *Resin cleaner*. Apply the cleaner after soldering is complete to remove excess resin.
- *Solder vise* or third hand. The vise holds together pieces to be soldered, leaving you free to work the iron and feed the solder.

VOLT-OHM METER

A *volt-ohm meter* is used to test voltage levels and the impedance of circuits. This moderately priced electronic tool is the basic requirement for working with electronic circuits of any kind. If you don't already own a volt-ohm meter you should seriously consider buying one. The cost is rather minimal considering the usefulness of the device.

There are many volt-ohm meters (or VOMs) on the market today. For work on robotics, you don't want a cheap model and you don't need an expensive one. A meter of intermediate quality is sufficient and does the job admirably. The price for such a meter is between $30 and $75 (it tends to be on the low side of this range). Meters are available at Radio Shack and most electronics outlets. Shop around and compare features and prices.

Digital or Analog

There are two general types of VOMs available today: digital and analog. The difference is not that one meter is used on digital circuits and the other on analog circuits. Rather, digital meters employ a numeric display not unlike a digital clock or watch. Analog VOMs use the older fashioned—but still useful—mechanical movement with a needle that points to a set of graduated scales.

Digital VOMs used to cost a great deal more than the analog variety, but the price difference has evened out recently. Digital VOMs, such as the one shown in Fig.

3-2, are fast becoming the standard; in fact, it's hard to find a decent analog meter anymore.

Analog VOMs are traditionally harder to use, because you must select the type and range of voltage you are testing, find the proper scale on the meter face, then estimate the voltage as the needle swings into action. Digital VOMs, on the other hand, display the voltage in clear numerals, and with a greater precision than most analog meters.

Automatic Ranging

As with analog meters, some digital meters require you to select the range before it can make an accurate measurement. For example, if you are measuring the voltage of a 9-volt transistor battery, you set the range to the setting closest to, but above, 9 volts (with most meters it is the 20 or 50 volt range). Auto-ranging meters don't require you to do this, so they are inherently easier to use. When you want to measure voltage, you set the meter to volts (either ac or dc) and take the measurement. The meter displays the results in the readout panel.

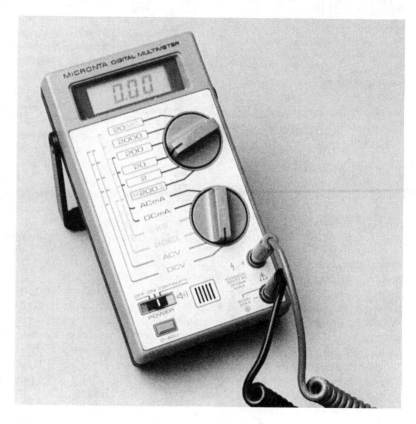

Fig. 3-2. A volt-ohm meter checks resistance, voltage, and current. This model is digital, and has a 3 1/2 digit liquid crystal readout.

Accuracy

Little of the work you'll do with robot circuits require a meter that's super-accurate. A VOM with average accuracy is more than enough. The accuracy of a meter is the minimum amount of error that can occur when taking a specific measurement. For example, the meter may be accurate to 2,000 volts, ±0.8 percent. A 0.8 percent error at the kinds of voltages used in robots—typically 5 to 12 volts dc—is only 0.096 volts!

Digital meters have another kind of accuracy. The number of digits in the display determines the maximum resolution of the measurements. Most digital meters have 3 1/2 digits, so it can display a value as small as .001 (the half digit is a "1" on the left side of the display). Anything less than that is not accurately represented; then again, there's little cause for accuracy higher than this when working with a robot.

Functions

Digital VOMs vary greatly in the number and type of functions they provide. At the very least, all standard VOMs let you measure ac volts, dc volts, milliamps, and ohms. Some also test capacitance and opens or shorts in discrete components like diodes and transistors. These additional functions are not absolutely necessary for building general-purpose robot circuits, but they are handy to have when troubleshooting a circuit that refuses to work.

The maximum ratings of the meter when measuring volts, milliamps, and resistance also varies. For most applications, the following maximum ratings are more than adequate:

Dc Volts	1,000	volts
Ac Volts	500	volts
Dc Current	200	milliamps
Resistance	2	megohms

One exception to this is when testing current draw for the entire robot or for motors. All but the smallest dc motors draw in excess of 200 milliamps, and the entire robot is likely to draw two or more amps. Obviously, this is far out of range of most digital meters. You need to get a good assessment of current draw, to anticipate the type and capacity of batteries, but to do so, you'll need either a meter with a higher dc current rating (digital or analog) or a special purpose ac/dc current meter. You can also use a resistor in series with the motor and apply Ohm's Law to calculate the current draw.

Meter Supplies

Meters come with a pair of test leads—one black and one red—each equipped with a needle-like metal probe. The quality of the test leads is usually minimal, so you may want to purchase a better set. The coiled kind are handy. They stretch out to several feet yet recoil to a manageable length when not in use.

Standard leads are fine for most routine testing, but some measurements may require the use of a clip lead. These attach to the end of the regular test leads and have a spring loaded clip on the end. You can clip the lead in place so your hands are free to do other things. The clips are insulated to prevent short circuits.

Meter Safety and Use

Most applications of the meter involve testing low voltages and resistance, both of which are relatively harmless to humans. Sometimes, however, you may need to test high voltages—like the input to a power supply—and careless use of the meter can cause serious bodily harm. Even when you're not actively testing a high voltage circuit, dangerous currents might still be exposed.

Proper procedure for meter use involves setting the meter beside the unit under test, making sure it is close enough so that the leads reach the circuit. Plug in the leads and test the meter operation by first selecting the resistance function setting (use the smallest scale if the meter is not auto-ranging). Touch the leads together: the meter should read 0 ohms. If the meter does not respond, check the leads and internal battery and try again. If the display does not read 0 ohms, double-check the range and function settings, and adjust the meter to read 0 ohms (not all digital meters have a 0 adjust, but most analog meters do).

Once the meter has checked out, select the desired function and range, and apply the leads to the circuit under test. Usually, the black lead will be connected to ground, and the red lead will be connected to the various test points in the circuit.

LOGIC PROBE

Meters are typically used for measuring analog signals. *Logic probes* test for the presence or absence of low voltage dc signals that represent digital data. The 0s and 1s are usually electrically defined as 0 and 5 volts, respectively, with TTL ICs. In practice, the actual voltages of the 0 and 1 bits depends entirely on the circuit. You can use a meter to test a logic circuit, but the results aren't

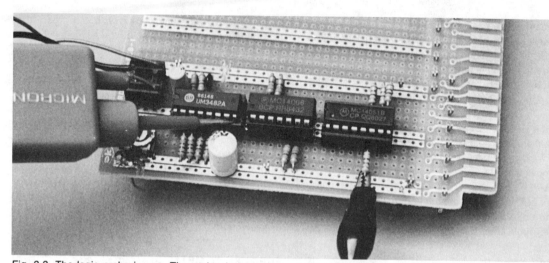

Fig. 3-3. The logic probe in use. The probe derives its power from the circuit under test.

always predictable. Further, many logic circuits change states (pulse) quickly and meters cannot track the voltage switches fast enough.

Logic probes, such as the model in Fig. 3-3, are designed to give a visual and (usually) aural signal of the logic state of a particular circuit line. One LED on the probe lights up if the logic is 0 (or LOW), another LED lights up if the logic is 1 (or HIGH). Most probes have a built-in buzzer, which has a different tone for the two logic levels. That way, you don't need to keep glancing at the probe to see the logic level.

A third LED or tone may indicate a pulsing signal. A good logic probe can detect that a circuit line is pulsing at speeds of up to 10 MHz, which is more than fast enough for robotic applications, even when using computer control. The minimum detectable pulse width (the time the pulse remains at one level) is 50 nanoseconds, again more than sufficient.

Although logic probes may sound complex, they are really simple devices, and their cost reflects this. You can buy a reasonably good logic probe for under $20. Most probes are not battery operated; rather, they obtain operating voltage from the circuit under test. You can also make a logic probe if you wish. A number of project books provide plans.

Using a Logic Probe

The same safety precautions apply when using a logic probe as they do when using a meter. Be wary when working close to high voltages. Cover them to prevent acciden-

tal shock (for obvious reasons, logic probes are not meant for anything but digital circuits, so never apply the leads of the probe to an ac line). Logic probes cannot operate with voltages exceeding about 15 volts dc, so if you are unsure of the voltage level of a particular circuit, test it with a meter first.

Successful use of the logic probe really requires you to have a circuit schematic to refer to. Keep it handy when troubleshooting your projects. It's nearly impossible to blindly use the logic probe on a circuit without knowing what you are testing. And since the probe receives its power from the circuit under test, you need to know where to pick off suitable power. To use the probe, connect the probe's power leads to a voltage source on the board, clip the black ground wire to circuit ground, and touch the tip of the probe against a pin of an integrated circuit or the lead of some other component. Figure 3-3 depicts a probe testing the logic level at an IC pin. For more information on using your probe, consult the manufacturer's instruction sheet.

LOGIC PULSER

A handy troubleshooting accessory when working with digital circuits is the *logic pulser*. This device puts out a timed pulse, letting you can see the effect of the pulse on a digital circuit. Normally, you'd use the pulser with a logic probe or an oscilloscope (discussed below). The pulser is switchable between one pulse and continuous pulsing. You can make your own pulser out of a 555 timer IC. Some of the 555 timer circuits in later chapters may be used as a logic pulser.

Most pulsers obtain their power from the circuit under test. It's important that you remember this. With digital circuits, its generally a bad idea to present an input signal to a device that's greater than the supply voltage for that device. In other words, if a chip is powered by five volts, and you give it a 12 volt pulse, you'll probably ruin the chip. Some circuits work with split (+, −, and ground) power supplies (especially circuits with op amps), so be sure you connect the leads of the pulser to the correct power points.

Also be sure that you do not pulse a line that has an output but no input. Some integrated circuits are sensitive to unloaded pulses at their output stages, and improper application of the pulse can destroy the chip.

OSCILLOSCOPE

An *oscilloscope* is a pricey tool—good ones start at about $500—and only a small number of electronic and robot hobbyists own one. For really serious work, however, an oscilloscope is an invaluable tool, one that will save you hours of time and frustration. I know; I'm a new oscilloscope owner myself. About half way through this project, during a difficult troubleshooting session with a circuit that defied all logic (in both human and electronic terms), I purchased an oscilloscope to find out what the circuit was doing. Within 10 minutes I had located the sources of trouble (about four separate problems contributed to the failure) and I had my circuit working.

Things you can do with a scope include some of the things you can do with other test equipment, but oscilloscopes do it all in one box and generally with greater precision. Among the many applications of an oscilloscope, you can:

- Test dc or ac voltage levels.
- Analyze the waveforms of digital and analog circuits.
- Determine the operating frequency of digital, analog, and rf circuits.
- Test logic levels.
- Visually check the timing of a circuit, to see if things are happening in the correct order and at the prescribed time intervals.

The designs provided in this book don't absolutely require the use of an oscilloscope, but you'll probably want one if you design your own circuits, or want to develop your electronic skills. A basic, no-nonsense model is enough, but don't settle for the cheap, single-trace units.

A dual-trace (two channel) scope with a 20 to 25 MHz maximum input frequency should do the job nicely. The two channels let you monitor two lines at once, so you can easily compare the input signal and output signal at the same time. You do not need a scope with storage or delayed sweep, although if your model has these features, you're sure to find a use for them sooner or later.

Scopes are not particularly easy to use; they have lots of dials and controls that set operation. Thoroughly familiarize yourself with the operation of your oscilloscope before using it for any construction project or for troubleshooting. Knowing how to set the time-per-division knob is as important as knowing how to turn the scope on. As usual, exercise caution when using the scope with or near high voltages.

FREQUENCY METER

A *frequency meter* (or frequency counter) tests the operating frequency of a circuit. Most models, like the one shown in Fig. 3-4, can be used on digital, analog, and rf circuits, for a variety of testing chores—from making sure the crystal in the robot's computer is working properly, to determining the radio frequency of a transmitter. You need only a basic frequency meter—a $100 investment. You can save some money by building a frequency meter kit.

Frequency meters have an upward operating limit, but it's generally well within the region applicable to robotics experiments. A frequency meter with a maximum range of up to 50 MHz is enough. One exception is testing the frequency of some radio control transmitters, which operate at frequencies of 27 MHz to over 75 MHz. A couple of meters are available with an optional prescaler, a device that extends the useful operating frequency to well over 100 MHz.

WIRE-WRAPPING

Making a printed circuit board for a one-shot application is time consuming, though it can be done with the proper kits and supplies. Conventional point-to-point solder wiring is not an acceptable approach when constructing digital circuits, which represent the lion's share of electronics you'll be building for your robots.

The preferred construction method is to use *wire-wrapping*. Wire-wrapping is a point-to-point wiring system that uses a special tool and extra-fine 28 or 30 gauge wrapping wire. When done properly, wire-wrapped circuits are as sturdy as soldered circuits, and you have the added benefit of being able to go back and make modifications and corrections without the hassle of desolder-

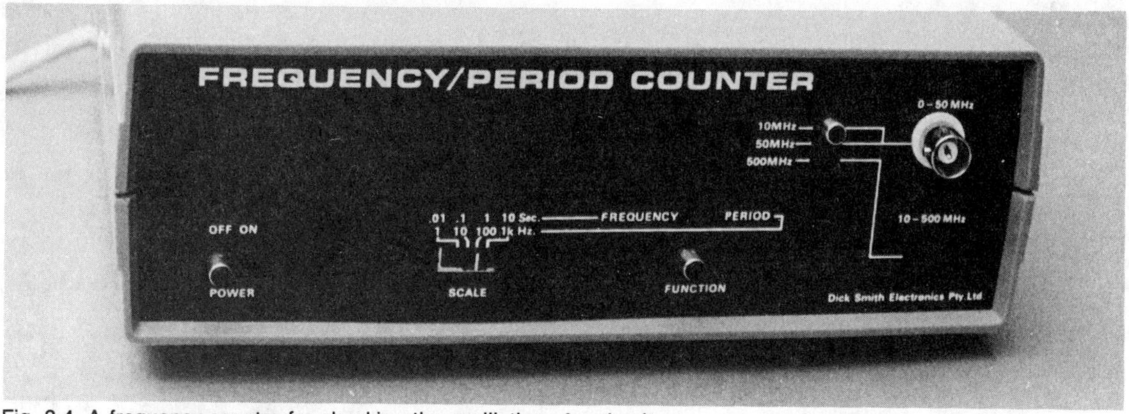

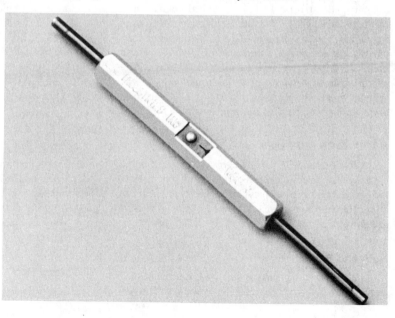

Fig. 3-4. A frequency counter for checking the oscillation of a circuit.

ing and resoldering.

A manual wire-wrapping tool is shown in Fig. 3-5. You insert one end of the stripped wire into a slot in the tool, and place the tool over a square-shaped wrapping post. Give the tool five to ten twirls, and the connection is complete. The edges of the post keep the wire anchored in place. To remove the wire, you use the other end of the tool and undo the wrapping.

A number of different wire-wrapping tools are available. Some are motorized and some automatically strip the wire for you, freeing you of this task and of purchasing the more expensive pre-stripped wire. The basic manual tool is recommended for initial use. You can graduate to other tools as you become proficient in wire-wrapping.

Wrapping wire comes in many forms, lengths, and colors, and you need to use special wire wrapping sockets and posts. See the section on electronic supplies and components for more details.

BREADBOARD

You should test each of the circuits you want to use in your robot (including the ones in this book) on a solderless breadboard before you commit it to wire-wrap or solder. Breadboards consist of a series of holes with internal contacts spaced one-tenth of an inch apart, just the right spacing for ICs. You plug in ICs, resistors, capacitors, transistors, and 20 or 22 gauge wire in the proper contact holes to create your circuit.

Fig. 3-5. A wire-wrapping tool. The long end is for wrapping the wire around the post; the short end for unwrapping (should it be necessary). The blade in the middle is for stripping insulation off the wire.

21

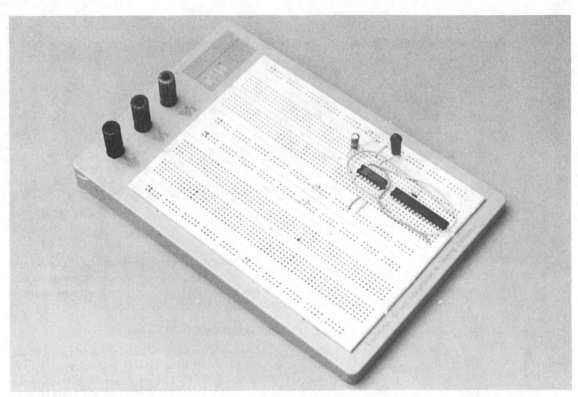

Fig. 3-6. Solderless breadboards are used to ''try out'' a circuit before soldering or wire-wrapping.

Solderless breadboards come in many sizes. For the most flexibility, get a double-width board, one that can accommodate at least 10 ICs. A typical double-width model is shown in Fig. 3-6. Smaller boards can be used for simple projects; circuits with a high number of components require bigger boards. While you're buying a breadboard, purchase a set of prestripped wires. The wires come in a variety of lengths, and are already stripped and bent for use in breadboards. The set costs $5 to $7, but you can bet the price is well worth it.

HARDWARE SUPPLIES

A robot is about 75 percent hardware and 25 percent electronic and electromechanical. Most of your trips to get parts for your robots will be to the local hardware store. Here are some common items you'll want to have around your shop:

Nuts and Bolts

Number 8 and 10 nuts and pan-head stove bolts (8/32 and 10/24, respectively) are good for all around construc-tion. Get a variety of bolts in 1/2-, 3/4-, 1-, 1 1/4-, and 1 1/2-inch lengths. You may also want to get some 2-inch and 3-inch long bolts for special applications.

Motor shafts and other heavy duty applications require 1/4-inch 20 or 5/16-inch hardware. Pan-head stove bolts are the best choice; you don't need hex-head carriage bolts unless you have a specific requirement for them. You can use number 6 (6/32) nuts and bolts for small, lightweight applications.

Washers

While you're at the store, stock up on flat washers, fender washers (large washers with small holes), tooth lockwashers and split lockwashers. Get an assortment for the various sizes of nuts and bolts. Split lockwashers are good for heavy duty applications, because they provide more compression locking power. You usually use them with bolt sizes of 1/4-inch and above.

All-Thread Rod

All-thread is two- to three-foot lengths of threaded

rod stock. It comes in standard thread sizes and pitches. All-thread is good for shafts and linear motion actuators. Get one of each in 8/32, 10/24, and 1/4-inch 20 threads to start.

Special Nuts

Coupling nuts are just like regular nuts but have been stretched out. They are designed to couple two bolts or pieces of all-thread together, end to end. In robotics, you might use them for a variety of tasks, including linear motion actuators and grippers.

Locking nuts have a piece of nylon built into them that provides a locking bite when threaded onto a bolt. Locking nuts are preferred over using two nuts tightened together.

EXTRUDED ALUMINUM

For most of your robot designs, you can take advantage of a rather common hardware item: extruded aluminum stock. It is designed for such things as building bathtub enclosures, picture frames, and other handyman applications and comes in various sizes, thicknesses, and configurations. Length is usually 12 feet, but if you need less, most hardware stores will cut to order (you save when you buy it in full lengths). The stock is available in plain (dull silver) anodized aluminum and gold anodized aluminum. Get the plain stuff: it's 10 to 25 percent cheaper.

Two particularly handy stocks are 41/64- by 1/2- by 1/16-inch channel and 57/64- by 9/16- by 1/16-inch channel. I use these extensively to make the frames, arms, legs, and other parts of my robots. Angle stock measuring 1- by 1- by 1/16-inches is another often-used item, usually employed for attaching cross bars and other structural components. No matter what size you eventually settle on for your own robots, keep several feet handy at all times.

If extruded aluminum is not available, another approach is to use shelving standards, the bar-like channel stock used for wall shelving. It's most often available in steel, but some hardware stores carry it in aluminum (silver, gold, and black anodized).

The biggest problem with using shelving standards is that the slots can cause problems when drilling holes for hardware. The drill bits can slip into the slots, causing the hole to be off-center. Some standards have an extra lip on the inside of the channel, which can interfere with some of the hardware you may use to join the pieces together.

ANGLE BRACKETS

You need a good assortment of 3/8-inch and 1/2-inch galvanized iron brackets to join the extruded stock or shelving standards together. Use 1 1/2-inch by 3/8-inch flat corner irons when joining pieces cut at 45 degree angles to make a frame. The 1-inch by 3/8-inch and 1 1/2-inch by 3/8-inch corner angle irons are helpful when attaching the stock to baseplates, and when securing various components to the robot.

ELECTRONIC SUPPLIES AND COMPONENTS

Most of the electronic projects you'll assemble from this book, and other books with digital and analog circuits, depend on common electronic components. If you do any amount of electronic circuit building, you'll want to stock up on the following standard components. Keeping spares handy prevents you from making repeated trips to the electronics store.

Resistors

Get a good assortment of 1/4 watt resistors. Make sure the assortment includes a variety of common values, and that there are several of each value. Supplement the assortment with individual purchases of the following resistor values: 270 ohm, 330 ohm, 1K ohm, 3.3K ohm, 10K ohm, 100K ohm. The 270 and 330 ohm values are often used with light emitting diodes (LEDs) and the remaining values are common to TTL and CMOS digital circuits.

Variable Resistors

Variable resistors, or potentiometers (pots) are relatively cheap and are a boon when designing and troubleshooting circuits. Buy an assortment of the small PC mount pots (about 80 cents each retail) in the 2.5K, 5K, 10K, 50K, 100K, 250K and 1 Megohm values. You'll find 1 Megohm pots often used in op amp circuits, so buy a couple extra.

Capacitors

Like resistors, you'll find yourself returning to the same standard capacitor values project after project. For a well stocked shop, get a dozen or so each, of the following inexpensive ceramic capacitors: 0.1, 0.01, and 0.001 μF.

Many circuits use in-between values of 0.47, 0.047, and 0.022 μF. You may want to get a couple of these, too. Power supply, timing, and audio circuits often use larger polarized electrolytic or tantalum capacitors. Buy a few each of 1.0, 2.2, 4.7, 10 and 100 μF values. Some projects call for other values (in the picoFarad range and the 1000's of microFarad range). You can buy these as needed unless you find yourself returning to standard values repeatedly.

Transistors

There are thousands of transistors available, and each one has slightly different characteristics than the others. Most applications need nothing more than generic transistors for simple switching and amplifying. Common npn signal transistors are the 2N2222 and the 2N3904 (some transistors are marked with an MPS prefix instead of the "2N" prefix; they are nearly identical). Both kinds are available in bulk packages of 10 for about $1. Common pnp signal transistors are the 2N3906 and the 2N2907. Price is the same or a little higher.

If the circuit you're building specifies another transistor than the generic kind, you may still be able to use one of these if you first look up the specifications of the transistor called for in the schematic. A number of cross-reference guides provide the specifications and replacement-equivalents for popular transistors.

There are common power transistors as well. The npn TIP 31 and TIP 41 are familiar to most anyone who has dealt with power switching or amplification of up to one amp or so. Pnp counterparts are the TIP 32 and TIP 42. These transistors come in the TO-220 style package (see Fig. 3-7).

A larger capacity common npn transistor that can switch 10 amps or more is the 2N3055. It comes in the T0-3 style package and is available everywhere. Price is between 50 cents and $2, depending on the source.

Diodes

Common diodes are the 1N914, for light-duty signal switching applications, and the 1N4000 series (1N4001, 1N4002, 1N4003, and 1N4004). Get several of each and use the proper size to handle the current in the circuit. Refer to a databook on the voltage and power handling capabilities of these diodes.

LEDs

All semiconductors emit light, but light emitting diodes (LEDs) are especially designed for the task. LEDs last longer than regular filament lamps and require less operating current. They are available in a variety of sizes, shapes, and colors. For general applications, the medium size red LED is perfect. Buy a few dozen and use them as needed. Many of the projects in this book call for in-

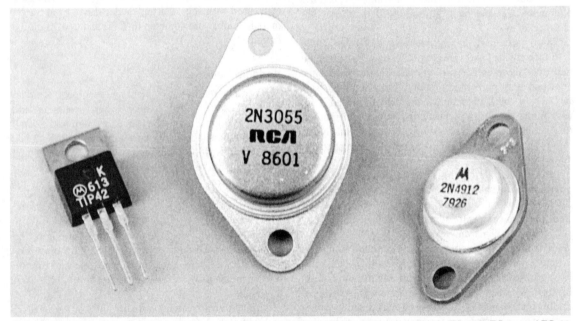

Fig. 3-7. An assortment of power transistors. From left to right, the transistors are packaged in TO-220, TO-3, and TO-66 cases. All three can be mounted on a heatsink for enhanced operation.

frared LEDs. These emit no visible light and are used in conjunction with an infrared-sensitive phototransistor or photodiode.

Integrated Circuits

Integrated circuits let you construct fairly complex circuits from just a couple of components. Although there are literally thousands of different ICs, some with exotic applications, a small handful crops up again and again in hobby projects. You should keep the following ICs in ready stock:

- *555 timer*. This is, by far, the most popular integrated circuit for hobby electronics. With just a couple of resistors and capacitors, the NE555 can be made to act as a pulser, a timer, a time delay, a missing pulse detector, and dozens of other useful things. The chip is usually used as a pulse source for digital circuits. Available in dual versions as the 556 and quadruple versions as the 558. A special CMOS version lets you increase the pulsing rate to 2 MHz.
- *741 op amp*. The LM741 comes in second in popularity to the 555. The 741 can be used for signal amplification, differentiation, integration, sample-and-hold, and a host of other useful applications. The 741 is available in a dual version, the 1458. The chip comes in different package configurations. The schematics in this book, and those usually found elsewhere, specify the pins for the common 8-pin DIP package. If you are using the 14-pin DIP package or the round can package, check the manufacturer's data sheet for the correct pin-outs. Note that there are numerous op amps available and some have design advantages over the 741.
- *TTL chips*. TTL ICs are common in computer circuits and other digital applications. There are many types of TTL packages, but you won't use more than 10 to 15 of them unless you're heavily into electronics experimentation. Specifically, the most common and most useful TTL ICs are the: 7400, 7401, 7402, 7404, 7407, 7408, 7414, 7430, 7432, 7473, 7474, 74154, 74193, and 74244.
- *CMOS chips*. Because CMOS ICs require less power to operate than the TTL variety, you'll often find them specified for use with low-power robotic and remote control applications. Like TTL, there is a relatively small number of common packages: 4001, 4011, 4013, 4016, 4017, 4027, 4040,

4041, 4049, 4060, 4066, 4069, 4071, and 4081.
- *339 quad comparator*. Comparators are used to compare two voltages. The output of the comparator changes depending on the voltage levels at its two inputs. The comparator is similar to the op amp, except that it does not use an external feedback resistor. You can use an op amp as a comparator, but a better approach is to use something like the LM339 chip, which contains four comparators in one package.

Wire

Solid-conductor insulated 22 gauge hookup wire can be used in your finished projects as well as connecting wires in breadboards. Buy a few spools in different colors. Solid-conductor wire can be crimped sharply and it might break when excessively twisted and flexed. If you expect that wiring in your project may be flexed repeatedly, use stranded wire instead. Heavier 12 to 18 gauge hook-up wire is required for connection to heavy-duty batteries, drive motors, and circuit board power supply lines. Table 3-1 lists the power handling capabilities of standard wire gauges.

Wire-wrap wire is available in spool or pre-cut/pre-stripped packages. For ease of use, buy the more expensive pre-cut stuff unless you have a tool that does it for you. Get several of each length. The wire wrapping tool has its own stripper built-in (which you must use instead of a regular wire stripper), so you can always shorten the pre-cut wires as needed. Some special wire wrapping tools require their own wrapping wire. Check the instruction that came with the tool for details.

CIRCUIT BOARDS

Simple projects can be built onto solder breadboards. These are modeled after the solderless breadboard, so you simply transfer the tested circuit from the solderless breadboard to the solder board. You can cut the board with a hacksaw or razor saw if you don't need all of it.

Table 3-1. Wire Gauge Current Ratings.

Current Rating (Amperes)	Minimum Wire Size (AWG)
1	20
3	18
6	16
12	14
20	12

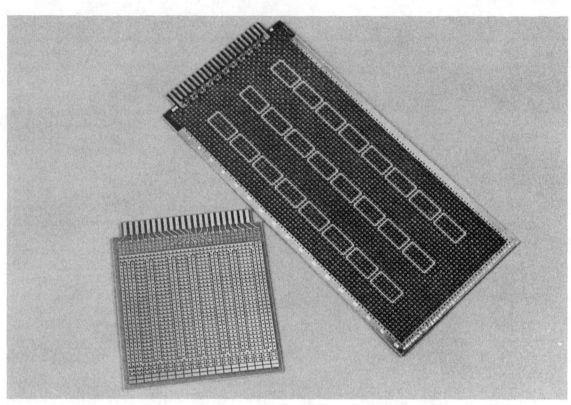

Fig. 3-8. Pre-drilled circuit boards for point-to-point wiring. The boards are most often used with wire-wrapping, and have solder tabs for easy component mounting. Both boards mate with the common 44-pin edge connector (22 pins on each side, 0.056-inch spacing).

Larger projects require perforated boards. Get the kind with solder tabs or solder traces on them. You'll be able to secure the components onto the boards with solder. Most perf boards, like the ones in Fig. 3-8, are designed for wire-wrapping.

Sockets

You should use sockets for ICs whenever possible. Sockets come in sizes ranging from 8-pin to 40-pin. The sockets with extra long square leads are for wire-wrapping, and are for use when assembling a wire-wrapped project.

You can also use wire-wrap IC sockets to hold discrete components like resistors, capacitors, diodes, LEDs, and transistors. You can, if you wish, wire-wrap the leads of these components, but since the leads are not square, the small wire doesn't have anything to bite into, so the connection isn't very strong. After assembly and testing, when you are sure the circuit works, apply a dab of sol-

der to the leads to hold the wires in place.

SETTING UP SHOP

You'll need a work table to construct the mechanisms and electronic circuits of your robots. The garage is an ideal setting because it affords you freedom to cut and drill wood, metal, and plastic, yet not worry about getting the pieces in the carpet.

Electronic assembly can be indoors or out, but I've found that, when working in a carpeted room, it's best to spread another carpet or some protective cover over the floor. When the throw rug fills with solder bits and little pieces of wire and component leads, I can take it outside, beat it with a broom handle, and it's as good as new.

Whatever space you choose for your robot lab, make sure all your tools are within easy reach. Keep special tools and supplies in an inexpensive fishing tackle box. The tackle box has lots of small compartments for plac-

26

ing screws and other parts that you remove.

For best results, your workspace should be an area where the robot-in-progress will not be disturbed if you have to leave it for several hours or several days, as will usually be the case. The work table should also be one that is off limits or inaccessible to young children, or at least an area that can be easily supervised.

Good lighting is a must. Both mechanical and electronic assembly require detail work, and you need good lighting to see everything properly. Supplement overhead lights with a 60 watt desk lamp. You'll be crouched over the worktable for hours at a time, so a comfortable chair or stool is a must. Be sure the seat is adjusted for the height of the worktable.

Chapter 4

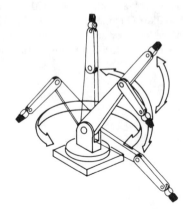

Buying Parts

Building a robot from scratch can be hard or easy. It's up to you. Personally, I go for the easy route; life is too demanding as it is. From experience, I've found that the best way to simplify the construction of a robot is to use standard, off-the-shelf parts, things you can get at the neighborhood hardware store, auto parts store, and electronics store.

Finding parts for your robots is routine, all things considered, and little thought goes into it. Be forewarned, there are some tricks of the trade, shortcuts, and tips that you should consider before you go on a buying spree and stock up for your next project. By shopping carefully and wisely, you can save both time and money in your robot building endeavors.

BUYING AT HARDWARE STORES

The small-town neighborhood hardware store is a great place when you need to buy some nails to fix the back porch, or a rake to clean up the leaves that have scattered around your yard.

It seems that while the neighborhood hardware outlet carries thousands of items, it doesn't have the ones you need at that exact moment. The right size of angle bracket may not be stocked, or may be in short supply.

The store may not carry a wide selection of extruded aluminum, and the exact one you want is no where to be found. Nuts and bolts are confined to pre-packaged sets where they sell four bolts and five nuts—but you need five bolts and six nuts!

More often than not, you'll find it necessary to make trips to a variety of hardware stores, tapping the best that each has to offer. You may find that some stores cater to a specific segment of do-it-yourselfers and professionals. Some stores are designed expressly to please professional painters, while others are for weekend plumbers and electricians. Realize this specialization and you'll have better luck in finding the parts you need.

Warehouse hardware outlets and builder's supply stores (usually open to the public) are the best source for the wide variety of tools and parts you need for robot experimentation. Items like nuts and bolts are generally available in bulk, so you can save a considerable amount of money.

As you tour the hardware stores in your area, keep a notebook handy, and jot down the lines each outlet carries. Then, when you find yourself needing a specific item, you have only to refer to your notes, rather than bash your brain trying to remember where you saw some little dodad. On a regular basis, take an idle stroll through your regular hardware store haunts. Unless the store is very small, you'll always find something new and useful in robot design each time you visit.

BUYING AT ELECTRONICS STORES

As recent as ten years ago, electronic parts stores used to be in plentiful supply. Even some automotive outlets carried a full range of tubes, specialty transistors, and other electronic gadgets. Now, Radio Shack remains as the only national electronics store chain. In many towns across the country, it's the only thing going.

Radio Shack continues to support electronics experimenters, but they stock only the very common components. If your needs extend beyond resistors, capacitors, and a few integrated circuits, you must turn to other sources. Check the local Yellow Pages under Electronics-Retail for a list of electronic parts shops near you.

Radio Shack isn't known for the best prices in electronic parts (although sometimes they have really good bargains), but the neighborhood independent electronics specialty store is even worse. It's not unusual for prepackaged resistors and capacitors to sell for 50 cents to $1 each, and few stores carry cost-saving parts assortments. Unless you need a specific component that isn't available anywhere else near you, stay away from the independent electronics outlet. There are independent stores that don't charge outrageously for parts, of course, but these are the exception, not the rule.

Mail order provides a welcome relief to overpriced electronic components. You can find the ads for these mail order firms in the various electronics magazines, such as Radio-Electronics and Modern Electronics. Both publications are available at newsstands. Several mail order firms are listed in Appendix A, Sources.

SPECIALTY STORES

Specialty stores are those outlets open to the general public that sell items you won't find in a regular hardware or electronic parts store. Specialty stores don't include surplus outlets, which are discussed below.

What specialty stores are of use to robot builders? Consider these:

- *Sewing machine repair shops.* Ideal for small gears, cams, levers, and other precision parts. Some shops will sell broken machines to you. Tear the machine to shreds and use the parts for your robot.
- *Auto parts stores.* The independent stores tend to

stock more goodies than the national chains, but both kinds offer surprises on every aisle. Keep an eye out for things like hoses, pumps, and automotive gadgets.

- *Used battery depot.* Usually a home-based business, where a guy buys old car and motorcycle batteries and refurbishes them. Selling price is usually between $15 and $25, or 50 to 75 percent less than a new battery.
- *Junk yards.* Old cars are good sources for powerful dc motors, used for windshield wipers, electric windows, and automatic adjustable seats. Or how about the hydraulic brake system on a junked 1969 Ford Falcon? Bring tools to salvage the parts you want.
- *Lawn mower sales/service shop.* Lawn mowers use all sorts of control cables, wheel bearings, and assorted odds-and-ends. Pick up new or used parts for a current project or for your own stock.
- *Bicycle sales/service shop.* Not the department store that sells bikes, but a real professional bicycle shop. Items of interest: control cables, chains, brake calipers, wheels, sprockets, brake linings, and more.
- *Industrial parts outlet.* Some places sell gears, bearings, shafts, motors, and other industrial hardware on a one-piece-at-a-time basis. The penalty: high prices.

SHOPPING THE SURPLUS STORE

Surplus is a wonderful thing, but most people shy away from it. Why? If its surplus, as the reasoning goes, it must be worthless junk. That's simply not true. Surplus is exactly what its name implies: extra stock. Because the stock is extra, it's generally priced accordingly—to move it out the door.

Surplus stores that specialize in new and used mechanical and electronic parts (not to be confused with surplus clothing, camping, and government equipment stores) are a pleasure to find. Most areas have at least one such surplus store; some as many as three or four. Get to know each and compare prices. Surplus stores don't have mass market appeal, so finding them is not always easy. Start by looking in the phone book Yellow Pages under Electronics and also under Surplus.

Mail Order Surplus

Some surplus is available through the mail. The number of mail order surplus outfits that cater to the hobbyist is limited, but you can usually find everything you need if you look carefully enough and are patient.

While surplus is a great way to stock up on dc motors, gears, chain, sprockets, and other odds and ends, you must shop wisely. Just because the company calls the stuff surplus doesn't mean that it's cheap. A popular item in a catalog may sell for top dollar. Always compare prices of similar items offered by various surplus outlets before buying. Consider all the variables, such as the added cost of insurance, postage and handling, and COD fees. Also, be sure that the mail order firm has a lenient return policy. You should always be able to return the goods if they are not satisfactory to you.

Shopping surplus can be a tough proposition, because it's hard to know what you'll need before you need it. Still, there are certain items that are almost always in demand by the robotics experimenter. If the price is right (especially on assortments or sets), stock up on:

- *Gears.* Small gears between 1/2-inch and 3-inch are extremely useful. Stick with standard tooth pitches of 24, 32, and 48. Try to get an assortment of sizes with similar pitches. Avoid "grab bag" collections of gears, where there are no mates. Plastic and nylon gears are fine for most jobs, but you should use larger metal gears for the main drive systems of your robots.
- *Chain and sprocket.* Robotics applications generally call for 1/4-inch (#25) chain, which is smaller and lighter than bicycle chain. When you see this stuff, snatch it up, but make sure you have the master links if the chain isn't permanently riveted together. Sprockets come in various sizes, expressed as the number of teeth on the outside of the sprocket. Buy a selection. Plastic and nylon sprockets are fine for general use; steel is preferred for main drives.
- *Bushings.* Bushings can be used as a kind of ball bearing, or they can be used to reduce the hub size of gears and sprockets so they fit smaller shafts. Common motor shaft sizes are 1/8-inch for small motors and 1/4-inch for larger motors. Gears and sprockets generally have 3/8-inch, 1/2-inch and 5/8-inch hubs. Oil-impregnated Oilite bushings are among the best, but cost more than regular bushings.
- *Spacers.* Made of aluminum, brass, or stainless steel. The best kind to get have an inside diameter to accept 10/32-inch and 1/4-inch shafts.
- *Motors.* Particularly the 6 volt and 12 volt dc variety. Most motors turn too fast for robotics applications. Save yourself some hassle by ordering geared motors. Final speeds of 20 to 100 rpm at the output of the gear reduction train are ideal. If gear motors aren't available, be on the lookout for gear boxes that you can attach to your motors. Stepper motors are handy, too, but make sure you know what you are buying. See Chapter 14, "Robot Locomotion with Stepper Motors," for more information.
- *Rechargable batteries.* The sealed lead-acid and gel-cell variety are common in surplus outlets. Test the battery immediately upon arrival to make sure it takes a charge and delivers its full capacity (test it under a load, like a heavy-duty motor). These batteries come in 6 volt dc and 12 volt dc capacity, both of which are ideal for robotics.

Chapter 5

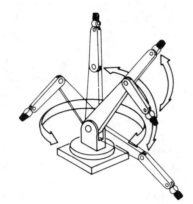

Robots of Plastic

It all started with billiard balls. You see, in the old days, billiard balls were made from elephant tusks. By the 1850's, the supply of tusk ivory was drying up, and its cost skyrocketed. So, in 1863, Phelan & Collender, a major manufacturer of billiard balls, offered a $10,000 prize for anyone who could come up with a suitable substitute for ivory. A New York printer named John Wesley Hyatt was among several who took up the challenge.

Hyatt didn't get the $10,000. His innovation, celluloid, proved to be too brittle for use as billiard balls. But while Hyatt's name won't go down in the poolhall hall of fame, it does go down as the man who started the plastics revolution. Hyatt's celluloid was perfect for such things as gentlemen's collars, ladies' combs, containers, and eventually, even motion picture film.

Since the more than 100 years since the introduction of celluloid, plastics have taken over our lives. Plastic is sometimes the object of ridicule, yet even its critics are quick to point out its many advantages:

- Plastic is cheaper per square inch than wood, metal, and most other construction materials.
- Certain plastics are extremely strong, approaching the tensile strength of such light metals as copper and aluminum.
- Some plastic is unbreakable.

Plastic is an ideal material for use in hobby robotics. Its properties are well suited for numerous robot designs, from simple frame structures to complete assemblies. Read this chapter to learn more about plastic, and how to work with it. Then construct an easy-to-build turtle robot—or Minibot—out of inexpensive and readily available plastic parts.

TYPES OF PLASTICS

Plastics represent a large family. For our application, the exact type of plastic isn't an important consideration—a large sheet of plastic is a large sheet of plastic, whether it carries the tradename of Plexiglas, Lexan, Acrylite, or any of the dozens of other identifiers. Some plastics are better suited for some jobs, however, so a basic understanding of the various types of plastics is beneficial. Here's a short rundown of the plastics you may encounter:

- *ABS.* Short for acrylonitrile butadiene styrene, ABS is most often used in sewer and waste-water plumbing systems. ABS is the large black pipes and fittings you see in the hardware store. Despite its shiny black appearance in plumbing material, ABS is really a glossy, translucent plastic that can take on just about any color and texture. It is tough and hard and relatively easy to cut and drill. Besides plumbing fittings, ABS comes in rods, sheets, and pipes.

- *Acrylic*. Acrylic is clear and strong, the mainstay of the decorative plastics industry. It can be easily scratched, but if the scratches aren't too deep, they can be rubbed out. Acrylic is somewhat tough to cut without cracking and requires careful drilling. The material mostly comes in sheets, but also available in tubing, rods, and the coating in pour-on plastic laminate.
- *Cellulosics*. Lightweight, flimsy, but surprisingly resilient, cellulosic plastics are often used as a sheet covering. They have minor uses in robotics. One useful application: The material softens at low heat, and can be slowly formed around an object. Comes in sheet or film form.
- *Epoxies*. Very durable clear plastic, often used as the binder in fiberglass. Epoxies most often come in liquid form, for pouring over something or onto a fiberglass base. The dried material can be cut, drilled, and sanded.
- *Nylon*. Tough, slippery, self-lubricating stuff most often used as a substitute for twine. Nylon also comes in rods and sheets from plastics distributors. Nylon is flexible, which makes it moderately hard to cut.
- *Phenolics*. An original plastic, phenolics are usually black or brown in color, easy to cut and drill, and smell terrible when heated. The material is usually reinforced with wood or cotton bits, or laminated with paper or cloth. Even with these additives, phenolic plastics are not unbreakable. Comes in rods, sheets, and pour-on coatings. Minor application in robotics except as circuit board material.
- *Polycarbonate*. Polycarbonate plastic is a close cousin to acrylic but more durable and more resistant to breakage. Polycarbonate plastics are slightly cloudy in appearance and easy to mar and scratch. Comes in rods, sheets, and tubing. A common inexpensive window glazing material, polycarbonates are hard to cut and drill without breakage.
- *Polyethylene*. Polyethylene is lightweight and translucent, and is often used to make flexible tubing. It also comes in rod, film, sheet, and pipe form. The material can be reformed with application of low heat, and when in tube form, can be cut with a knife.
- *Polypropylene*. Like polyethylene, above, but harder and more resistant to heat.
- *Polystyrene*. A mainstay in the toy industry. This plastic is hard, clear (can be colored with dyes), and cheap. Although often labeled as high-impact plastic, polystyrene is really brittle and susceptible to damage by low heat and sunlight. Available in rods, sheets, and foam. Moderately hard to cut and drill without cracking and breaking.
- *Polyurethane*. These days, polyurethane is most often used as insulation material, but it's also available in rod and sheet forms. The plastic is durable, flexible, and relatively easy to cut and drill.
- *PVC*. Short for polyvinyl chloride, PVC is an extremely versatile plastic best known as the material used in fresh water plumbing and patio furniture. Usually processed with white pigment, PVC is actually clear and softens in relatively low heat. PVC is extremely easy to cut and drill and almost impervious to breakage. Beside plumbing fixtures and pipes, PVC is supplied in film, sheet, rod, tubing, even nut and bolt form.
- *Silicone*. A large family of plastics all in its own right. Because of their elasticity, silicone plastics are most often used in molding compounds. Silicone is slippery to the touch and comes in resin form for pouring.

HOW TO CUT PLASTIC

Soft plastics may be cut with a sharp utility knife. When cutting, place a sheet of cardboard or artboard on the table. This helps prevent the knife from cutting into the table, ruining the tabletop and dulling the knife. Use a carpenter's square or metal rule when you need to cut a straight line. Use the rule against the knife holder, not the blade. That'll prolong blade life.

Harder plastics can be cut in a variety of ways. When cutting sheet plastic under 1/8-inch, use a utility knife and metal carpenter's square to score a cutting line. If necessary, use clamps to hold down the square. Most sheet plastic comes with a protective peel-off plastic on both sides. Keep it on for cutting.

Carefully repeat the scoring two or three times to deepen the cut. Place a 1/2-inch or 1-inch dowel under the plastic so that the score line is on the top of the dowel. With your fingers or the palms of your hands, carefully push down on both sides of the score line. If the sheet is wide, use a piece of 1-by-2 or 2-by-4 lumber to exert even pressure. Breakage and cracking is most likely to occur on the edges, so press on the edges first, then work your way towards the center. Don't force the break. If you can't get the plastic to break off cleanly, deepen the score line with the utility knife.

Thicker sheet plastic, as well as extruded tubes, pipes, and bars must be cut with a saw. If you have a table saw, outfit it with a plywood/paneling blade. Among other applications, the blade can be used for cutting plastics. You cut through plastic just as you do with wood, but the feed rate—the speed at which the material is sawed—must be slower. Forcing the plastic, or using a dull blade, heats the plastic, deforming it and melting it. A band saw is ideal for cutting plastics less than 1/2-inch thick, especially if you need to cut corners. Again, keep the protective covering on the plastic while cutting. When working with a power saw, use fences or pieces of wood held in place by C-clamps to ensure a straight cut.

You can use a hand saw for cutting smaller pieces of plastic. A hacksaw with a medium- or fine-tooth blade (24 or 32 teeth per inch) is a good choice. You can also use a coping saw (with a fine-tooth blade) or a razor saw. These are good choices when cutting angles and corners, and when doing detail work.

A motorized scroll (or sabre) saw can be used for cutting plastic, but care must be taken to ensure a straight cut. If possible, use a piece of plywood held in place by C-clamps as a guidefence. Routers can be used to cut and score plastic, but unless you are an experienced router user, you should not attempt this method.

HOW TO DRILL PLASTIC

Wood drill bits can be used for cutting plastics, but I've found that bits designed for glass drilling yields better, safer results. If you use wood bits, you should modify them by blunting the tip slightly (otherwise the tip may crack the plastic when it exits the other side) and continue the flute from the cutting lip all the way to the end of the bit (see Fig. 5-1). Blunting the tip of the bit isn't hard, but grinding the flute is a difficult proposition. Best invest in a couple of glass/plastic bits, which are already engineered for the application at hand.

Drilling is best accomplished with a power drill. The drill must either be made for working with metal or have a variable speed control. Reduce the speed of the drill to about 500 to 1,000 rpm when using twist bits, about 1,000 to 2,000 when using spade bits.

When drilling, always back the plastic with a wooden block. Without the block, the plastic is almost guaranteed to crack. When using spade bits or brad-point bits,

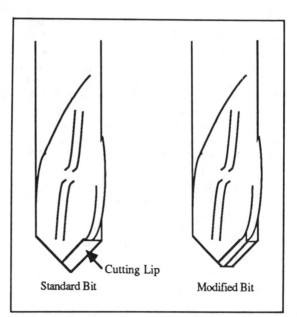

Cutting Lip

Standard Bit   Modified Bit

Fig. 5-1. Suggested modifications for drill bits used with plastic. The end is blunted and the flutes are extended to the end of the cutting lip.

drill partially through from one side, then complete the hole by drilling from the other side. As with cutting, don't force the hole, and always use sharp bits. Too much friction causes the plastic to melt.

Holes larger than 1/4-inch should first be made by drilling a smaller, pilot hole. If the hole is large, over 1/4-inch in diameter, start with a small drill and work your way up several steps. Practice drilling on a piece of scrap until you get the right technique.

HOW TO BEND AND FORM PLASTIC

Most rigid and semi-rigid plastics can be formed by applying low localized heat. A sure way of bending sheet plastic is to use a strip heater. These are available ready-made at some hardware and plastics supply houses, or you can build your own. A narrow element in the heater applies a regulated amount of heat to the plastic. When the plastic is soft enough, you can bend it to just about any angle you want.

There are two important points to remember when using a strip heater. First, be sure that the plastic is pliable before you try to bend it. You may break it otherwise or cause excessive stress at the joint (a stressed joint will looked cracked or crazed).

Second, bend the plastic past the angle you want. The plastic will relax a bit when it cools off and you must anticipate this. The proper amount of overbending comes with experience, and differs depending on the type of plastic and the size of the piece you're working with.

You can mold thinner sheet plastic around shapes by first heating it up with a hair dryer, then using your fingers to form the plastic. Be careful that you don't heat up the plastic too much. You don't want it to melt, just conform to the underlying shape. You can soften an entire sheet or piece by placing it into an oven for 10 or so minutes (remove the protective plastic before baking). Set the thermostat to 300 degrees and be sure to leave the door slightly ajar so that any fumes released during the heating can escape. Ventilate the kitchen and avoid breathing the fumes.

All plastics release gas when they heat up, but the fumes can be downright toxic when the plastic actually ignites. Avoid any overheating where the plastic burns. The dripping, molten plastic can also seriously burn you if it drops on your skin.

HOW TO POLISH THE EDGES OF PLASTIC

Plastic that has been cut or scored usually has rough edges. The edges of cut PVC and ABS can be filed using a wood or metal file. The edges of higher density plastics like acrylics and polycarbonates should be polished by sanding, buffing, or burnishing. Try a fine grit (200 to 300) wet-dry sandpaper, used wet. Buy an assortment of sandpapers and try several until you find the coarseness that works best with the plastic you're using. Jeweler's rouge, available at many hardware stores in large blocks, is applied using a polishing wheel. The wheel can be attached to a grinder or drill motor.

Burnishing uses a very low temperature flame (a match or lighter will do) to slightly melt the plastic. A propane torch kept some distance from the plastic can also be used. Be extremely careful when using a flame to burnish plastic. Don't let the plastic ignite or you'll end up with an ugly blob that will ruin your project, not to mention fill the room with poisonous gas.

HOW TO GLUE PLASTIC

Most plastics aren't really glued together; they are cemented together. The cement contains one or more solvents that actually melt the plastic at the joint. The pieces are then fused together—made one. Household adhesives can be used, of course, but you get better results using specially formulated cements.

Herein lies a problem. The various plastics rarely use the same cement formulations, so to make your project

stick you've got to use the right mixture. That requires you to know the variety of plastic used in the material you are working with. See the discussion above on the various types of plastics. Also refer to Table 5-1, which lists the major types of plastics and how they are used in common household and industrial products, as well as Table 5-2, which indicates the suggested adhesives for cementing various popular plastics.

When using solvent for PVC or ABS plumbing fixtures, apply in the recommended manner by spreading a thin coat on the pieces to be joined. A cotton applicator is included in the can of cement. Plastic sheet, bars, and other items require more careful cementing, especially if you want the end result to look nice.

With the exception of PVC solvent, the cement for plastics is watery thin, and can be applied in a variety of ways. One method is to use a small painter's brush, with a #0 or #1 tip. Join the pieces to be fused together and "paint" the cement on the joint with the brush. Capillary action will draw the cement into the joint, where it will spread out. Another method is to fill a special syringe applicator with cement. With the pieces butted together, squirt a small amount of cement into the joint line.

In all cases, you must be sure that the surfaces at the joint are perfectly flat, and that there are no voids where the cement may not make ample contact. If you are joining pieces where the edges cannot be made flush, apply a thicker type of glue, such as contact cement or white household glue.

After applying the cement, wait at least 15 minutes for the plastic to re-fuse and the joint to harden. Disturbing the joint before it has a time to set up will permanently weaken it. Remember that you cannot apply cement to plastics that have been painted. If you paint the pieces before cementing them, scrape off the paint and refinish the edges to make them smooth again.

HOW TO PAINT PLASTICS

Sheet plastic is available in transparent or opaque colors, and this is the best way to add color to your robot projects. The colors are impregnated in the plastic and can't be scraped or sanded off. You can also add a coat of paint to the plastic, to add color or to make it opaque. Most plastics accept brush or spray painting.

Spray painting is the preferred method for all jobs that don't require extra-fine detail. Carefully select the paint before you use it and always apply a small amount to a scrap piece of plastic before painting the entire project. Some paints contain solvents that may soften the plastic.

Table 5-1. Plastics in Everyday Household Articles.

Household Article	Plastic
Bottles, containers	
Clear	Polyester, PVC
Translucent or opaque	Polyethylene, polypropylene
Buckets, washtubs	Polyethylene, polypropylene
Foam cushions	Polyurethane foam, PVC foam
Electrical circuit boards	Laminated epoxies, phenolics
Fillers	
Caulking compounds	Polyurethane, silicone, PVAC
Grouts	Silicone, PVAC
Patching compounds	Polyester, fiberglass
Putties	Epoxies, polyester, PVAC
Films	
Art film	Cellulosics
Audio tape	Polyester
Food wrap	Polyethylene, polypropylene
Photographic	Cellulosics
Glasses (drinking)	
Clear, hard	Polystyrene
Flexible	Polyethylene
Insulated cups	Styrofoam (polystyrene foam)
Hoses, garden	PVC
Insulation foam	Polystyrene, polyurethane
Lubricants	Silicones
Plumbing pipes	
Fresh water	PVC, polyethylene, ABS
Gray water	ABS
Siding and paneling	PVC
Toys	
Flexible	Polyethylene, polypropylene
Rigid	Polystyrene, ABS
Tubing (clear or translucent)	Polyethylene, PVC

One of the best all-around paints for plastics are the model/hobby cans from Testor. These are specially formulated for styrene model plastic, but I've found that the paint adheres well without softening on most plastics. I've used it successfully on ABS, PVC, acrylic, polycarbonate, and many others. You can purchase this paint in a variety of colors, in either gloss or flat finish. Matching colors are available in bottles and self-contained brush applicators.

If the plastic is clear, you have the choice of painting on the front or back side (or both for that matter). Painting on the front side yields the standard finish of the paint. Gloss colors come out gloss; flat colors come out flat. Flat finish paints tend to scrape off easier, however, so exercise care.

Painting on the back side only yields a glossy appearance, because you look through the clear plastic to the paint on the back side. Painting imperfections are more

Table 5-2. Plastic Bonding Guide.

PLASTIC	CEMENTED TO ITSELF	OTHER PLASTIC	METAL
ABS	ABS-ABS solvent	Rubber adhesive	Epoxy cement
Acrylic	Acrylic solvent	Epoxy cement	Contact cement
Cellulosics	White glue	Rubber adhesive	Contact cement
Polystyrene	Model glue	Epoxy	Super glue
Polystyrene foam	White glue	Contact cement	Contact cement
Polyurethane	Rubber adhesive	Epoxy, contact cement	Contact cement
PVC	PVC-PVC solvent	PVC-ABS (to ABS)	Contact cement
		Other, rubber adhesive	

or less hidden, and external scratches don't mar up the paint job.

BUYING PLASTIC

Some hardware stores carry plastic, but you'll be sorely frustrated at the selection. The best place to look for plastic—in all its styles, shapes, and chemical makeup—is a plastics specialty store. Most larger cities have at least one plastic supply store that's open to the public. Look in the Yellow Pages under Plastics—Retail.

Another useful source is the plastics fabricator. These are actually in greater number than retail plastic stores. They are in business to build merchandise, display racks, and other plastic items. Although they don't usually advertise as selling to the general public, most do. If the fabricator doesn't sell new material, ask to buy the left-over scrap.

The scrap is a valuable source of plastic for your robot designs. One local plastic fabricator practically begged me to take some of his scrap plastic off his hands—for free. I came back with almost 100 pounds of plastic pieces of various sizes, colors, and thicknesses, and hardly made a dent in his scrap pile. Not all fabricators are this generous, but if you look long enough, you may hit the jackpot.

BUILD THE MINIBOT

You can use a small piece of scrap sheet acrylic to build the foundation and frame of the Minibot. The robot is about eight inches square, and scoots around the floor or table on two small rubber tires. The basic version is meant to be wire controlled, although in upcoming chapters, you'll see how to adapt the Minibot to automatic electronic control, even remote control. Power comes from a set of flashlight batteries. I used four "AA" batteries because they are small, lightweight, and provide more driving power than 9-volt transistor batteries.

A parts list for the minibot is in Table 5-3.

Foundation/Base

The foundation is clear or colored Plexiglas, or similar acrylic sheet plastic. The thickness should be at least 1/8-inch; 3/16-inch is even better. The prototype Minibot used 1/8-inch thick acrylic and there was minimum stressing caused by bending or flexing.

Cut as shown in Fig. 5-2. Remember to keep the protective paper cover on the plastic while cutting. File or sand the edges to smooth the cutting and scoring marks. The corners are sharp and can cause some injury if the robot is handled by small children. You can easily fix this by rounding out the corners with a file. Find the center and drill a hole with a #10 bit. Holes are also shown for mounting the drive motors. These holes are spaced for a simple clamp mechanism that secures commonly available hobby motors.

Motor Mount

The small dc motors used in the prototype Minibot were surplus gearmotors with an output speed of about 30 rpm. The motors for your Minibot should have a similar speed, because even with fairly large wheels, 30 rpm makes the robot scoot around the floor or a table at about four to six inches a second. Choose motors small enough so that they don't crowd the base of the robot and add unnecessary weight. Remember that you have other items to add, such as batteries and control electronics.

Use 3/8-inch wide metal mending braces to secure the motor (the prototype used plastic pieces from a Fastech toy construction kit; you can use these or something similar). You must add spacers or extra nuts to balance the motor in the brace. Drill holes for 8/32 bolts (#19 bit), spaced to match the holes in the mending plate. Another method is to use U-bolts. Drill the holes for the U-

Table 5-3. Minibot Parts List.

Minibot:

1	8 1/2-inch by 8 1/2-inch acrylic plastic (1/16-inch or 1/8-inch thickness)
2	Small hobby motors, with gear reduction
2	Model airplane wheels
1	6 1/4-inch (approx) 10/24 all-thread rod
1	8- to 9-inch diameter clear plastic dome
1	Four-cell "AA" battery holder

Assorted   1/2" by 8/32 bolts
               8/32 nuts, lockwashers
               1/2-inch by 10/25 bolts
               10/24 nuts, lockwashers, capnuts

Motor Control Switch:

1	Small electronic project enclosure
2	DPDT momentary switches, with center off return
Misc.	Hookup wire

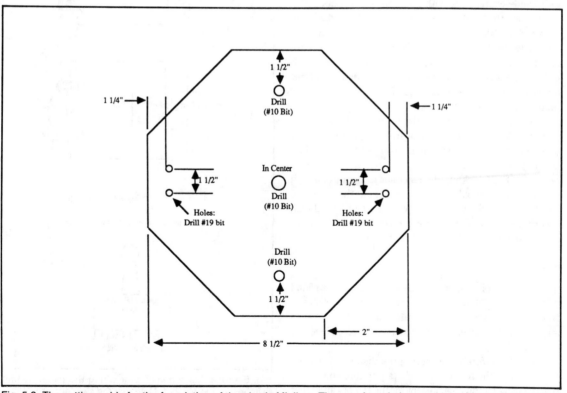

Fig. 5-2. The cutting guide for the foundation of the plastic Minibot. The set of two holes on either side are for the motor mount, and should be spaced according to the specific mount you are using.

Fig. 5-3. How the drive motors of the Minibot should look. The wheels are threaded directly onto the motor shaft. Note the gear-reduction system built onto the hobby motor.

bolts and secure them with a double set of nuts.

Attach the tires to the motor shafts. Tires designed for a radio control airplane or race car are good choices. The tires are well made and the hubs are threaded in standard screw sizes (the threads may be metric, so watch out!). I threaded the motor shaft and attached a 4-40 nut on each side of the wheel. Figure 5-3 shows a mounted motor with a tire attached.

Finishing the foundation/base plate are the counterbalances. These keep the robot from tipping backward and forward along its drive axis. You can use small ball bearings, tiny casters, or—as I did in the prototype—the head of a 10/24 locknut. The locknut is smooth enough to act as a kind of ball bearing and is about the right size for the job. Attach the locknut with a 10/24 by 1/2-inch bolt (if the bolt you have is too long to fit in the locknut, add washers or a 10/24 nut as a spacer).

Top Shell

The top shell is optional. The prototype used a round display bowl 8 1/2 inches in diameter. It was purchased from a plastics specialty store. Alternatively, you can use any suitable half sphere for your robot, such as an inverted salad bowl.

Attach the top by measuring the distance from the foundation to the top of the shell, taking into consideration the gap that must be present for the motors and other bulky internal components. Cut a length of 10/24 all-thread rod to size. The length of the prototype shaft was 6 1/4 inches.

Secure the center shaft to the base using a pair of 10/24 nuts and a tooth lockwasher. Secure the center shaft to the top shell with a 10/24 nut and a 10/24 locknut. Use a tooth lockwasher on the inside or outside of the shell to keep the shell from spinning loose.

Battery Holder

Battery holders are available that hold from one to six dry cells, in any of the popular battery sizes. The Minibot motors, like most all small hobby motors, run off 1.5 to 6 volts. A four cell, AA battery holder does the

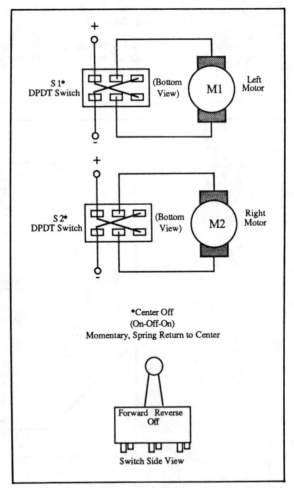

Fig. 5-4. Use this schematic for wiring the motor control switches for the Minibot. Note that the switches are DPDT, with spring return to center off.

job nicely. The wiring in the holder connects the batteries in series, so the output is six volts.

Secure the battery holder to the base with 8/32 nuts and bolts. Drill holes to accommodate the hardware. Be sure the nuts and bolts don't extend too far below the base or they may drag when the robot is moving. Likewise, be sure the hardware doesn't interfere with the batteries.

Wiring Diagram

The wiring diagram in Fig. 5-4 allows you to control the movement of the Minibot in all directions. This simple two-switch system, used in many other projects in the book, uses double-pole, double-throw (DPDT) switches. The switches called for in the circuit are spring loaded so that they return to a center-off position when you let go of them.

Chapter 6

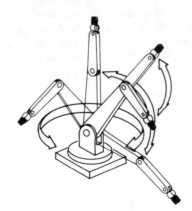

Building a Basic Wooden Platform

Wood may not be hi-tech, but it's an ideal building material for hobby robots. Wood is available just about everywhere. It's relatively inexpensive, it's easy to work with, and mistakes can be readily covered up or painted over. Here, in this chapter, we'll take a look at using wood in robots, and how to apply simple woodworking skills to construct a basic wooden robot platform. This platform can serve as the foundation for a number of robot designs you may care to explore.

CHOOSING THE RIGHT WOOD

There is good wood and there is bad wood. Obviously, you want the good stuff, but you have to be willing to pay for it. For reasons you'll soon discover, you should buy only the best stock you can get your hands on. The better woods are available at specialty wood stores, particularly the ones that sell mostly hardwoods and exotic woods. The local lumber and hardware store may have great buys on rough-hewn redwood planking, but it's hardly the stuff of robots.

Best Woods

The best overall wood for robotics use, especially foundation platforms, is plywood. In case you are a little unfamiliar with plywood, this common building material comes in many grades and is made by laminating thin sheets of wood together. The cheapest is called shop grade, and is the kind often used for flooring and projects where looks aren't too important. The board is full of knots and knotholes and there may be considerable voids inside the board.

The remaining grades specify the quality of both sides of the plywood. Grade N is the best and signifies natural finish veneer. The surface quality of grade N really isn't important to us, so we can settle for grade A. Since we want the board to be in good shape on both sides, a plywood with a grade of A-A (grade A on both sides) is desired. Grades B and C are acceptable, but only if better plywoods aren't around. Depending on availability, for example, you may have to settle on A-C grade plywood (grade A on one side; grade C on the other).

Most plywoods you purchase at the lumber stores are made of softwoods—usually fir and pine. You can get hardwood plywood as well. In fact, these are more desirable because they are denser and less likely to become chipped. Don't confuse hardwood plywood with hardboard. The latter is made up of sawdust epoxied together under high pressure. Hardboard has a smooth finish; its close cousin particle board does not. Both types are wholly unsuitable for robotics because they are too heavy and too brittle.

Plywood comes in various thicknesses starting at about 5/16-inch to over 1-inch. Something in the middle of the range is perfect for a robot platform.

Plywood generally comes in 4- by 8-foot panels. You don't need a piece of plywood this large to make the platform, so if you don't already have a scrap of it laying around the workshop, get a pre-cut panel instead. These typically come in sizes ranging from 1- by 1-feet to 2- by 4-feet.

Planking

An alternative to working with plywood is *planking*. Use ash, birch, or some other solid hardwood; stay away from the less meaty softwoods such as fir, pine, and hemlock. Most hardwood planks are available in widths no more than 12 or 15 inches, so you must take this into consideration when designing the platform. You can butt two smaller widths together if absolutely required. Use a router to fashion a secure joint, or attach metal mending plates to mate the two pieces together.

When choosing wood, be especially wary of warpage and moisture content. Take along a carpenter's square and check the squareness and levelness of the lumber in every possible direction. Reject any piece that isn't perfectly square; you'll regret it otherwise. Defects in milled wood goes by a variety of colorful names, such as crook, bow, cup, twist, wane, split, shake and check, but they all mean headache to you.

Wood with excessive moisture may bow and bend as it dries, causing cracks and warpage. These can be devastating in a robot you've just completed and perfected. Buy only seasoned lumber stored inside the lumberyard, not outside. Watch for green specks or grains—these indicate trapped moisture. If the wood is marked, look for an "MC" specification. An "MC-15" rating means that the moisture content doesn't exceed 15 percent. Good plywoods and hardwood planks meet or exceed this requirement. Don't get anything marked MC-20 or higher, or marked S-GREEN.

Dowels

Wood *dowels* come in every conceivable diameter from about 1/16-inch to over 1 1/2-inch. Length is three or four feet. Most dowels are made of high quality hardwood, such as birch or ash. The dowel is always cut with the grain lengthwise to increase strength. Other than choosing the proper dimension, there is little to buying dowels. You should, however, inspect the dowel to make sure that it is straight. At the store, roll the dowel on the floor. The dowel should lie flat and should roll easily.

THE WOODCUTTER'S ART

You've cut a piece of wood in two before, haven't you? Sure you have; everyone has. You don't need any special tools or techniques for cutting wood to make a robot platform. The basic shop cutting tools will suffice: a hand saw, a back saw, a circular saw, a jig saw (if the wood is thin enough for the blade), a table saw, a radial arm saw, or—you name it.

Whatever cutting tool you use, make sure the blade is the right one for the wood. The combination blade that probably came with your power saw isn't the right choice for plywood and hardwood. Outfit the saw with a cutoff blade or a plywood/paneling blade. Both have many more teeth per inch. Handsaws generally come in two versions: crosscut and ripsaw. You need the crosscut kind.

DEEP DRILLING

You can use a hand or motor drill for drilling through wood. Electric drills are great and do the job fast, but I prefer the old fashioned hand drill for applications that require precision. Either way, it's important that you use only sharp drill bits. If your bits are dull, replace them or have them sharpened.

It's important that the holes you drill are straight, or your robot may not go together properly. If your drill press is large enough, you can use it to drill perfectly straight holes in plywood and other large stock. Otherwise, use a portable drill stand. These attach to the drill or work in such a way that you're guaranteed a straight hole.

FINISHING

You can easily shape wood with rasps and files. If the shaping is extensive—like a circle in the middle of a large plank—you may want to consider getting a rasp for your drill motor. They come in various sizes and shapes. You'll want the project to be as smooth as possible, for both appearance and ease in attaching other wood pieces or components of the robot to it. A medium grit sandpaper is fine for the job.

Add a coat of paint to the wood when you're done. Silver, gloss black, or some other hi-tech opaque color are good choices. The paint also serves to protect the wood against the effects of moisture and aging.

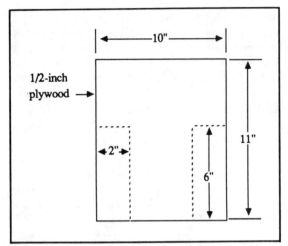

Fig. 6-1. Cutting plan for the plywood base.

BUILDING A WOODEN MOTORIZED PLATFORM

Figures 6-1 and 6-2 present two approaches for constructing the basic wooden platform. As shown in Fig. 6-1, cut a piece of 1/2-inch plywood to 11 by 10 inches. Make sure it is square. Notch the wood as shown to make a T-shape. Attached to the two ends of the T will be the drive motors; attached to the stem of the T will be the swivel caster.

Alternatively, as shown in Fig 6-2, cut two lengths

of 4-inch wide by 1/2-inch thick hardwood plank to 10 inches and 7 inches. Glue the pieces together as shown into a T-shape, using 1/2-inch mending plates on the top and bottom.

Drill the 10/32 holes for the mending plate first. Glue the pieces then attach the mending plates with 10/32 by 1-inch bolts. Secure with 10/32 nuts and tooth lockwashers. Let dry until the glue hardens and the joint is rigid.

Attaching the Motors

The wooden platform constructed so far is perfect for a fairly sturdy robot, so the choice of motors should reflect this. Use heavy-duty motors, geared down to a top speed of no more than about 75 rpm. Anything faster will cause the robot to dash about at speeds exceeding a few miles per hour, which is unacceptable unless you plan on entering your creation in the Robot-olympics.

Note: You can use electronic control (see Chapter 13) to reduce the speed of the gearmotor by 15 or 20 percent without losing much torque, but you can't slow a fast motor by more than 30 percent. The closer the motor operates at its rated speed, the better results you'll have.

Attach the motors using corner brackets, if the motors have mounting flanges and holes on them. Some motors do not have mounting holes or hardware, and you must fashion a hold-down plate for them. You can make

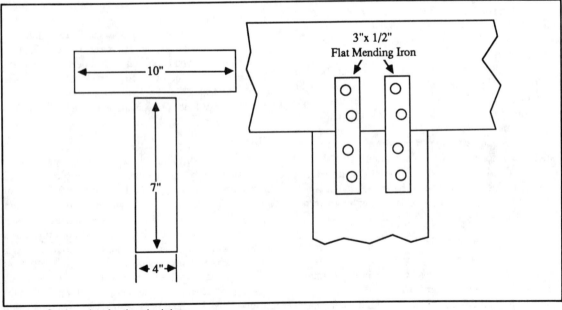

Fig. 6-2. Cutting plan for the plank base.

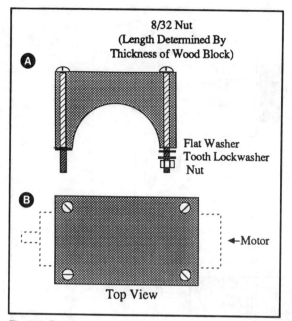

8/32 Nut
(Length Determined By
Thickness of Wood Block)

A

Flat Washer
Tooth Lockwasher
Nut

B

←Motor

Top View

Fig. 6-3. One way to secure the motors to the base is to use a wood block hollowed out to the shape of the motor casing. A. Side view; B. Top view.

an effective hold-down plate, as shown in Fig. 6-3, out of wood. Round out the plate to match the cylindrical body of the motor casing. Secure the plate to the platform.

Finish by attaching the wheels to the motor shafts. You may need to thread the shafts with a die so you can secure the wheels. Use the proper size nuts and washers on either side of the hub to keep the wheel in place. Life is made much easier if the wheels have a set screw. Once attached to the shaft, tighten the setscrew. The complete parts list is in Table 6-1.

Stabilizer Caster

Using two motors and a centered caster, as depicted in Fig. 6-4, allows you to have full control over the direction of travel of your robot. You turn by stopping or reversing one motor while the other continues turning. Attach the caster using four 8/32 by 1-inch bolts. Secure with tooth lockwashers and 8/32 nuts.

Alternatively, you can add a steering motor/wheel in place of the caster. This is generally a more difficult design approach, but it does allow some interesting possibilities. For example, you can couple the drive motor to the steering motor. For additional information on using

Table 6-1. Wood Robot Parts List.

Robot Base:

1	10-inch by 11-inch plywood (1/2-inch thickness)
or	
1	10-inch by 4-inch hardwood plank (1/2-inch thickness)
1	7-inch by 4-inch hardwood plank (1/2-inch thickness)
4	3-inch by 1/2-inch flat mending plates
4	1-inch by 10/24 stove bolts, 10/24 nuts, flat washers

Motor/drive:

2	dc gear motors
2	5- to 7-inch rubber wheels
1	1 1/4-inch caster
2	2- by 4-inch lumber (cut to length to fit motor)
8	3″ by 10/24 stove bolts, 10/24 nuts, flat washers (for motor mount)
4	1 1/4-inch by 8/32 stove bolts, 8/32 nuts, flat washers, lock washers (for caster mount)
1	Four-cell ''D'' battery holder

See the parts list in Chapter 5 for motor control switch

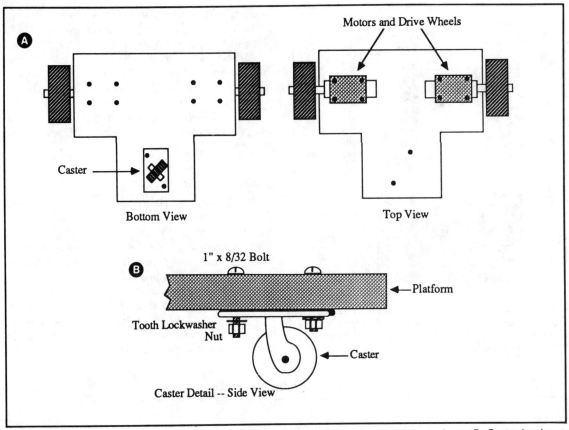

Fig. 6-4. Attaching the caster to the platform (either plywood or plank). A. Top and bottom views; B. Caster hardware assembly detail.

a steering wheel motor, see Chapter 17, "Advanced Locomotion Systems."

Note that the caster must be level with the drive motors. You may, if necessary, use spacers to increase the distance from the baseplate of the caster to the bottom of the platform, but if the caster is already lower than the wheels, you'll have trouble. You can take care of this by adding spacers when mounting the drive motors. You can also rectify the problem by simply using larger diameter wheels.

Battery Holder

Battery holders are available that contain from one to six dry cells in any of the popular battery sizes. When using six volt motors, you can use a four cell "D" battery holder. You can also use a single six volt lantern or rechargeable battery. Motors that require 12 volts will need two battery holders, two six volt batteries, or one 12 volt battery (the later is somewhat hard to find unless you're looking for a car battery, which you are not). For the prototype, I used six volt motors and a 4-cell "D" battery holder.

Secure the battery holder(s) to the base with 8/32 nuts and bolts. Drill holes to accommodate the hardware. Be sure the nuts and bolts don't extend too far below the base or they may drag when the robot is moving. Likewise, be sure the hardware doesn't interfere with the batteries.

Wire the batteries and wheels to the DPDT through control switches, as shown at the end of Chapter 5, "Robots of Plastic." One switch controls the left motor; the other switch controls the right motor.

Chapter 7

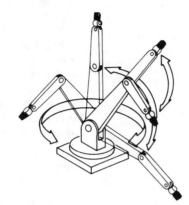

Building a Metal Platform

Metal is perhaps the best all-around material for build-ing robots because it offers extra strength where other materials, such as plastic and wood, cannot. If you've never worked with metal before, you shouldn't worry; there is really nothing to it. The designs outlined in this chapter, as well as the chapters that follow, show how to construct robots both large and small out of readily-available metal stock, without resorting to welding or cus-tom machining.

METAL STOCKS

Metal stock is available from a variety of sources. The local hardware store is the best place to start. Some stock may be available only at the neighborhood sheet metal shop. Look around and you're sure to find what you need.

Shelving Standards

You've seen those shelving materials where you nail two metal rails on the wall, then attach brackets and shelves to them. The rails are referred to as standards, and are perfectly suited as girders in robot frames. The standards come in either aluminum or steel and measure 41/64- by 1/2- by 1/16-inch. The steel stock is cheaper

and easier to find, but considerably heavier. Limit its use to structural points in your robot that need extra strength. Otherwise, if you can find it, use the aluminum stock. Not all hardware stores carry the aluminum variety.

Extruded Aluminum

Extruded stock is made by pushing molten metal out of a shaped orifice. The metal cools as it exits the ori-fice, and is shaped exactly like it. Extruded aluminum stock is readily available at most hardware stores. It generally comes in 12 foot sections, but most hardware stores let you buy cut pieces if you don't need all 12 feet.

Extruded aluminum is available in more than two dozen common styles, from thin bars to pipes to square posts. Although you can use any of it, as you see fit, a couple of styles may prove to be particularly beneficial in your robot building endeavors.

- 1- by 1- by 1/16-inch angle stock.
- 57/64- by 9/16- by 1/16-inch channel stock.
- 41/64- by 1/2- by 1/16-inch channel stock.
- Bar stock, widths from one to three inches; thick-nesses 1/16-inch to 1/4-inch.

Mending Plates

Galvanized mending plates are designed to strengthen the joint of two or more pieces of lumber. Much of it is pre-formed, in all sorts of weird shapes, and is pretty much unusable for building robots. But flat plates are available in a number of widths and lengths. You can use the plates as is or cut to size. The plates are made of metal, and have numerous pre-drilled holes in it to facilitate hammering with nails. The material is soft enough that you can easily drill new holes.

The plates are available in lengths of about 4, 6, and 12 inches. Widths are not as standardized, but 2, 4, 6, and 12 inches seem common. You can usually find mending plates near the rain gutter/roofing section in the hardware store.

Rods and Squares

Most hardware stores carry a limited amount of short, extruded steel or zinc rods and squares. The material is solid and somewhat heavy, and is perfect for use in some advanced projects, like robotic arms. Lengths are typically limited to 12 or 24 inches, and thicknesses ranging from 1/16-inch to about 1/2-inch.

Iron Angle Brackets

You need a way to connect all the metal pieces together. The easiest way is to use galvanized iron brackets. These come in a variety of sizes and shapes, and have pre-drilled holes to facilitate construction. The 3/8-inch wide brackets fit easily into the two sizes of channel stock mentioned above. You need only to drill a corresponding hole in the channel stock and attach the pieces together with nuts and bolts. The result is a very sturdy and clean-looking frame. You'll find the flat corner angle iron, corner angle iron, and flat mending iron to be particularly useful.

WORKING WITH METAL

If you have the right tools, working with metal is only slightly harder than working with wood or plastic. You'll have better than average results if you always use sharpened, well made tools. Dull, bargain-basement tools can't effectively cut through aluminum or steel stock. Instead of the tool doing most of the work, you do.

Cutting

Use a hacksaw outfitted with a fine-tooth blade, something on the order of 24 or 32 teeth per inch, to cut metal.

Coping saws, keyhole saws, and other hand saws are generally engineered for wood cutting, and their blades aren't fine enough for metal work. You can use a power saw, like a table saw or reciprocating saw, but again, make sure that you use the right blade.

You'll probably do most of your cutting by hand. You can help guarantee straight cuts with an inexpensive miter box. You don't need anything fancy, but try to stay away from the wooden boxes. They wear out too fast. The hardened plastic and metal boxes are the best buys. Be sure to get a miter box that lets you cut at 45 degrees both vertically and horizontally. Firmly attach the miter box to your workbench using hardware or a large clamp.

Drilling

Metal requires a slower drilling speed than wood, and you need a power drill that either runs at a low speed or lets you adjust the speed to match the work. Variable speed power drills are available for under $30 these days, and it's a good investment. Be sure to use only sharp drill bits. If your bits are dull, replace them or have them sharpened. Quite often, buying a new set is cheaper than professional resharpening. It's up to you.

You'll find that when cutting metal, the bit will skate all over the surface until the hole is started. You can eliminate or reduce this skating by using a punch prior to drilling. Use a hammer to gently tap a small indentation into the metal with the punch. Hold the smaller pieces in a vise while you drill.

When it comes to working with metal, particularly channel and pipe stock, you'll find a drill press a godsend. Accuracy is improved, and you'll find the work goes much faster. Always use a proper vise when working with a drill press. Never hold the work with your hands. Especially with metal, the bit can snag as it's cutting and yank the piece out of your hands. If you can't place the work in the vise, use a pair of Vice Grips or other suitable locking pliers.

Finishing

Cutting and drilling often leave rough edges, called flashing, in the metal. These must be filed down using a medium or fine pitch metal file, or the pieces won't fit together properly. Aluminum flash comes off quickly and easily; you need to work a little harder when removing the flash in steel or zinc stock.

BUILD THE SCOOTERBOT

The Scooterbot is a small robot built from channel

Table 7-1. Scooterbot Parts List.

Frame:

2	11-inch length aluminum or steel shelving standards
2	5 3/4-inch length aluminum or steel shelving standards
4	1 1/2- by 3/8-inch flat corner irons
8	1/2-inch by 8/32 stove bolts, nuts, lockwashers

Motor and Mount:

1	Surplus Big Trak motor (or two dc gear reduction motors)
1	5 3/4-inch length 1 1/4-inch wide galvanized nail mending plate
2	1-inch by 8/32 stove bolts, nuts, flat washers, tooth lockwashers
1	Four-cell "D" battery holder

Support Caster:

1	5 3/4-inch length 1 1/4-inch wide galvanized nail mending plate
1	1 1/4-inch swivel caster
2	1/2" by 8/32 stove bolts, nuts, tooth lockwashers, flat washers (as spacers)

See parts list in Chapter 5 for motor control switch

---

aluminum, mending plates, nuts and bolts, and a few other odds and ends. You can use the Scooterbot as the foundation and running gear for a very sophisticated pet-like robot. As in the previous chapters that dealt with robots of plastic and wood, the basic design of the all-metal Scooterbot can be enhanced just about any way you see fit. This chapter details the construction of the framework, locomotion, and power systems for a wired remote control robot. Future chapters deal with adding more sophisticated features, like wireless remote control, automatic navigation, and collision avoidance/detection. All the parts needed are in Table 7-1.

Framework

Build the frame of the Scooterbot from a single three foot length of channel aluminum or steel stock. The prototype used aluminum shelving standards; you can use steel standards or extruded aluminum channel. Cut the pieces, using a hacksaw and miter box, as shown in Fig. 7-1. Be sure to cut precise 45 degree angles, and that the pieces are as close to the specified length as possible. A deviation of as little as 1/8-inch will cause the frame to be off-square, and the robot may not roll in a straight line.

Using 1 1/2-inch by 3/8-inch flat corner irons, as shown in Fig. 7-2, attach the pieces in picture frame style. The flat corner has two holes on each leg; drill only one matching hole in the channel or standard stock to match. Assemble using 8/32 by 1/2-inch pan-head stove bolts, and secure with tooth lockwashers and nuts. Do not

tighten the nuts until the entire frame is assembled. Using a carpenter's square, align the frame as close to square as possible. If a corner will not square up, try reversing one of the pieces, or file down the ends for a better match. Tighten the nuts and bolts when you are satisfied with the alignment.

Motor Mount

The prototype Scooterbot made use of a dandy surplus geared motor system. This all-in-one drive system, shown in Fig. 7-3, was originally designed for the Milton-Bradley Big Trak toy, and contains two motors and two gear reduction systems, not to mention a novel magnetic clutch arrangement to assure that the motors turn at the same speed when the toy is propelled directly forward.

Thousands of these motors appeared in the surplus market a few years ago, and as of this writing, they are still in plentiful supply. The motors are available from a variety of sources, including Jerryco, Edmund Scientific, and H&R. The best part: cost is reasonable—I've never seen it over $5. See Appendix A, "Sources," for a list of these surplus outlets. The motor is probably available elsewhere, too.

As usual when buying surplus, check first before you order because when the stock of these motors is gone, it's gone forever. Of course, you're always free to use another motor system. Look around and you are sure to find something suitable.

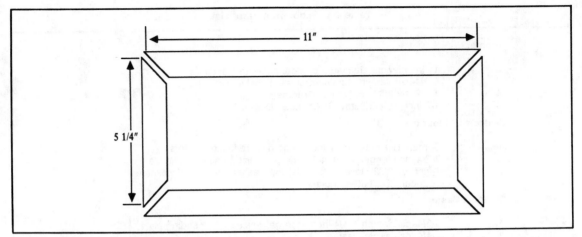

Fig. 7-1. Cutting diagram for the Scooterbot frame. Note the orientation of the mitered ends.

The motor attaches to the frame with a 1 1/4-inch by 5 3/4-inch mending plate, as shown in Fig. 7-4. The plate was cut to size from one measuring 12 inches long. Secure the motor, at the center flanges of the unit, with two 8/32 by 1-inch stove bolts and 8/32 nuts. Use washers as needed. Secure the plate to the frame using 8/32 by 1/2-inch bolts and nuts. Attach 5- or 6-inch rubber wheels to the motor shafts. The shafts of the motors are notched,

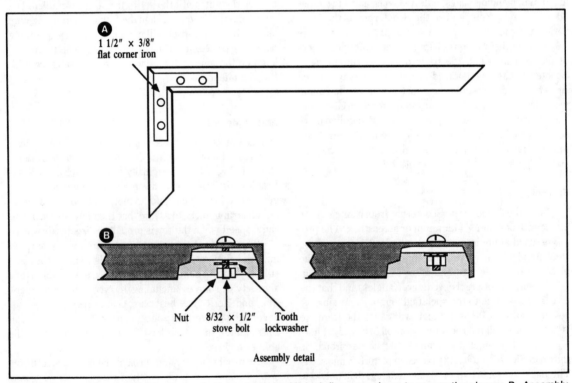

Fig. 7-2. Construction detail of the frame. A. Using 1 1/2- by 3/8-inch flat corner irons to secure the pieces; B. Assembly detail using 1/2-inch by 8/32 hardware.

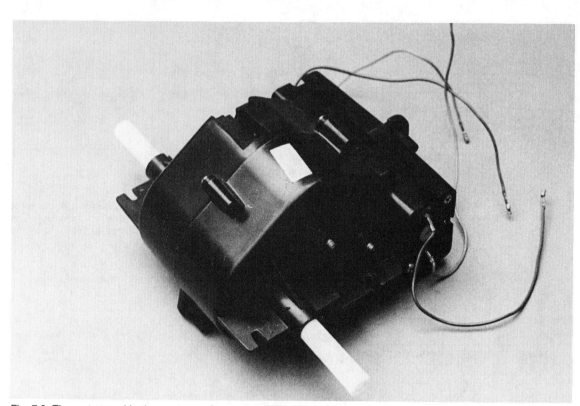

Fig. 7-3. The motor used in the prototype Scooterbot. This dual motor is widely available surplus, originally used in the Milton-Bradley Big Trak toy.

and you can easily slot the hubs of the wheels to match.

Support Caster

The Scooterbot uses the two-wheel drive tripod arrangement. You need a caster on the other end of the frame to balance the robot and provide a steering swivel. The 1 1/4-inch swivel caster is not driven and doesn't do the actual steering. Driving and steering are taken care of by the drive motors.

Refer to Fig. 7-5. Attach the caster using another piece of 1 1/4-inch by 6-inch mending plate. Secure the plate to the frame with 8/32 by 1/2-inch bolts and 8/32 nuts. Using the baseplate of the caster as a drilling guide, drill two holes in the mending plate. Secure the caster with 8/32 by 1/2-inch bolts and nuts. Be sure to add a few washers or another nut on the top side of the mending plate, as depicted in the figure. If you don't, the end of the bolt will interfere with the swivel motion of the caster. Alternatively, you can cut the bolt to length.

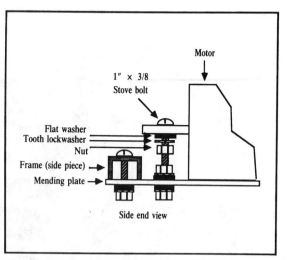

Fig. 7-4. Motor mounting detail. Nuts and lockwashers are used as a spacer to secure the bolt against the motor flange and the mending plate.

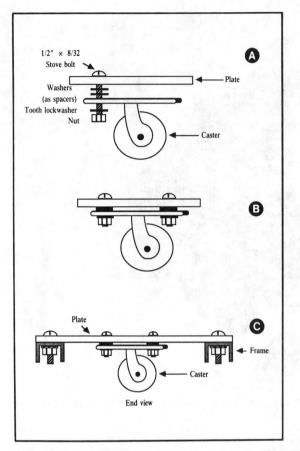

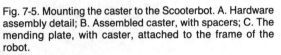

Fig. 7-5. Mounting the caster to the Scooterbot. A. Hardware assembly detail; B. Assembled caster, with spacers; C. The mending plate, with caster, attached to the frame of the robot.

Fig. 7-6. The completed Scooterbot, with batteries and wheels. Note the whip used to keep the control wires from tangling with the robot.

Battery Holder

The motors require an appreciable amount of current, so the Scooterbot really should be powered by heavy-duty "C" or "D" size cells. The smaller "AA" cells just won't cut it. The prototype Scooterbot used a four-cell "D" battery holder nearly six inches wide, and fit nicely on the top of the frame. Drill holes in the corners of the holder and secure it to the base using 6/32 by 1/2-inch pan-head stove bolts and 6/32 nuts. Be sure the head of the bolts do not interfere with any of the batteries.

Wiring Diagram

The basic Scooterbot uses a manual wired switch control. The control is the same one used in the plastic Minibot detailed in Chapter 5, "Robots of Plastic." Refer to the wiring diagram given at the end of that chapter for information on powering the Scooterbot.

To prevent the control wire from interfering with the operation of the robot, attach a piece of heavy wire (the bottom rail of a coat hanger will do) to the caster plate and lead the wire up it. Use nylon wire ties to secure the wire. The finished Scooterbot is shown in Fig. 7-6. Note that a small piece of perforated circuit board was used to solder the motor, battery, and control wires together. Alternatively, you can use wire nuts to bundle the wires together.

Test Run

You'll find that the Scooterbot is an amazingly agile robot. It turns in a distance a little longer than its length (about 14 inches in the prototype) and has plenty of power to spare. There is room on the front and back of the robot to mount additional control circuitry. You can also add control circuits and other enhancements over the battery holder. Just be sure that you can remove the circuit(s) when it comes time to change or recharge the batteries.

Chapter 8

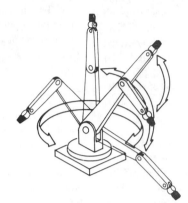

Converting Toys
Into Working Robots

Ready-made toys can be used as the basis for more complex homebrew hobby robots. The toy industry is robot crazy, and you can buy a basic motorized or unmotorized robot for parts, building on it and adding sophistication and features. Snap or screw-together kits, such as the venerable Erector Set, let you use pre-machined parts for your own creations. And some kits, like the Milton-Bradley Robotix sets, are even designed to create futuristic motorized robots and vehicles. You can use the parts in the kits as is, or cannibalize them, modifying them in any way you see fit. Because the parts already come in the exact or approximate shape you need, construction of your own robots is greatly simplified.

About the only disadvantage to using toys as the basis for more advanced robots is that the plastic and lightweight metal used in the kits and finished products are not suitable for a homemade robot of any size or strength. You are pretty much confined to building small Minibot or Scooterbot-type robots from toy parts. Even so, you can sometimes apply toy parts to robot subsystems, such as a light-duty arm/gripper mechanism installed on a larger automaton.

Let's take a closer look at using toys in your robot designs, and examine a couple of simple, cost-effective designs using readily available toy construction kits.

ERECTOR SET

Erector Set, sold by Ideal, has been around since the Dawn of Time—or so it seems, anyway. The basic Erector Set hasn't changed in the over 70 years since it came out, although instead of all-metal parts, some of the components in the latest Erector Set kits are made of plastic. The kits come in various sizes, and are generally designed to build a number of different projects. Some kits, like the Maxx Steele robot, are engineered for a specific design with, perhaps, provisions for moderate variations. I've found the general purpose sets to be the best bets.

Useful components of the kits include pre-punched metal girders, plastic and metal mending plates, tires, wheels, shafts, and plastic mounting panels. You can use any as you see fit, assembling your robots with the hardware supplied with the kit, or with 6/32 or 8/32 nuts and bolts.

Erector Set Project: Motorized Platform

One specialized Erector Set kit, the Wreck & Rescue Crane, comes with assorted parts to construct a giant construction crane. The crane is mounted on a wheeled base. That base, and some of the other parts, can be easily used to construct a motorized robotic platform. Motors are not included in this, and most other, Erector Set kits; you have to supply your own. In the following, only basic construction details will be provided. It's up to you to adapt the design to the parts you have on hand.

The pre-punched metal girders make excellent motor mounts. They are lightweight enough that they can be bent, using a vise, into a U-shaped motor holder. Bend the girder at the ends to create tabs for the bolts, or use the angle stock provided in an Erector Set kit. The basic platform is designed for four or more wheels, but the wheel arrangement makes steering the robot difficult. The design presented in Fig. 8-1 uses only two wheels, as in previous projects. The platform is stabilized using

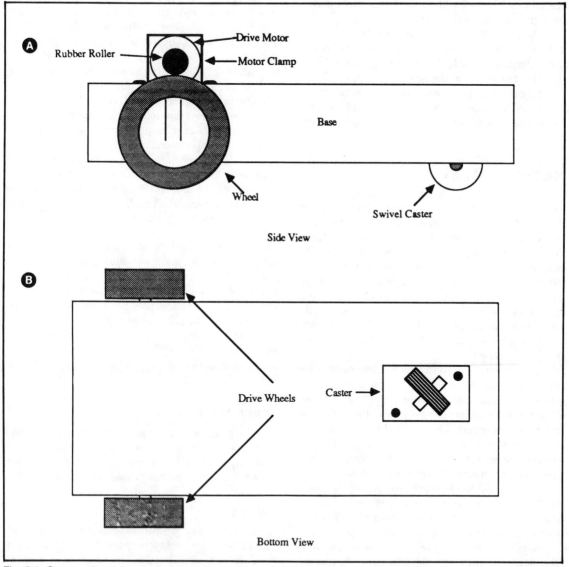

Fig. 8-1. Constructing the motorized base for a robot out of Erector Set parts. A. Attaching the motor and drive roller over the wheel; B. Drive wheel/caster arrangement.

a swivel caster at the other end. The Crane set doesn't come with the caster. Purchase a small one at the hardware store.

Note that the shafts of the motors are not directly linked to the wheels. The shaft of the wheels connect to the baseplate as originally designed in the kit. The drive motors are equipped with rollers, which engage against the top of the wheels for traction. You can use a metal or rubber roller, but rubber is better. The pinch roller from a discarded cassette tape player is a good choice, as is a 3/8-inch beveled bibb washer, found in the plumbing section of the hardware store.

A battery holder can be easily mounted on the top of the platform. Position the battery holder in the center of the platform, towards the caster end. This will help distribute the weight of the robot.

The basic platform is now complete. You can attach a dual-switch remote control, as described in Chapter 5, "Robots of Plastic," or connect automatic control circuitry.

ROBOTIX

The Milton-Bradley Robotix kits are specially designed to make snap-together walking and rolling robots. The complete kits come with two or more gear motor assemblies, and you can buy additional motors separately. You control the motors using a central switch pad. The motor control panel in the deluxe set, the R-2000, has provisions for five motors. Pushing the switch forward turns the motor in one direction; pushing the switch back turns the motor in the other direction. The output speed of the motors is about six rpm, which makes them a bit slow for moving a robot across the room, but perfect for arm/gripper designs.

The structural components in the Robotix kits are molded high-impact plastic. You can connect pieces together to form just about anything. One useful project is building a robotic arm using several of the motors and structural components. The arm can be used by itself as a robotic trainer, or attached to a larger robot. It can lift a reasonable eight ounces or so, and its pincher claw is strong enough to firmly grasp most small objects.

While the Robotix kit allows you to snap the pieces apart when you're doing experimenting, the design presented here is meant to be permanent. Glue the pieces together using plastic model cement or contact cement. Cementing is optional of course, and you're free to try other, less permanent methods for securing the parts together, such as small nuts and bolts, screws, or allen set screws.

When cemented, the pieces hold together much better and the arm is considerably stronger. Remember that once cemented, the parts cannot be easily disassembled, so make sure that your design works properly before you commit to it. When used as a stand-alone arm, you can plug the shoulder motor into the battery holder/base. You don't need to cement this joint.

Building the Arm

Refer to Fig. 8-2 as you build the arm. Temporarily attach a motor (we'll call it "Motor 1") to the Robotix battery holder/baseplate. Position the motor so that the drive spindle points straight up. Attach a double plug to the drive spindle and the end connector of another motor, "Motor 2." Position this motor so that the drive spindle is on one side. Next, attach another double plug and an elbow to the drive spindle of Motor 2. Attach the other end of the elbow connector to a beam arm.

Connect a third motor, "Motor 3," to the large connector on the opposite end of the beam arm. Position this motor so that the drive spindle is on the other end of the beam arm. Attach a double plug and an elbow between the drive spindle of Motor 3 and the connector opposite the drive spindle of the fourth motor, "Motor 4." The two claw levers directly attach to the drive spindle of Motor 4.

Motorize the joints by plugging in the yellow power cables between the power switch box and the motor connectors. Try each joint and note the various degrees of freedom. Experiment with picking up various objects with the claw. Make changes now before disassembling the arm and cementing the pieces together.

After the arm is assembled, route the wires around the components, making sure there is sufficient slack to permit free movement. Attach the wires to the arm using nylon wire ties.

OTHER TOYS

Toy stores are full of plastic put-together kits and ready-made robot toys that seem to beg you to tear them apart and use them in your own robot designs. Here are some toys you may want to consider for your next project.

Capsula

Capsula, from Japanese manufacturing giant Mitsubishi, is another popular snap-together motorized parts kit. Capsula kits come in different sizes and have one or more gearmotors that can be attached to various components. The kits contain unique parts that other put-

Fig. 8-2. A robot arm constructed with parts from a Robotix construction kit.

together toys don't, such as plastic chain and chain sprockets/gears. Advanced kits come with remote control and computer circuits. All the parts from the various kits are interchangeable.

The links of the chain snap apart, so you can make any length chain you want. Combine the links from many kits and you can make an impressive drive system for an experimental lightweight robot.

Fastech

Fastech construction kits are sold by Shaper, and is one of the best assortment of parts you can buy. All the parts are plastic, and the kits come with a plastic temporary riveting system that you probably won't use in your designs.

The plastic parts are pre-drilled and come in a variety of shapes and styles. The components can be used on their own to make a small, light-duty robot frame and body, or they can be used as parts in a larger robot (the prototype plastic Minibot described in Chapter 5 used some plastic plates from a Fastech kit, as does an experimental human-like hand in Chapter 21).

Armatron

The Tomy Armatron, now sold exclusively by Radio Shack, is a perennial gadgeteer's favorite. The Armatron is a remotely controlled arm that you can use to manipulate small lightweight objects. You control the Armatron (Fig. 8-3) by moving two joysticks.

A rewarding but demanding project is converting the Armatron to computer control. Because of the way the Armatron works, using a single motor and clutched drive shafts, it's necessary to rebuild it with separate motors so that a computer can individually control its various joints. A number of magazines have published conversion projects for the Armatron. See Appendix B, "Further Reading," for a bibliography.

One approach is to connect small 12 volt dc solenoids to the joysticks. Each joystick requires six solenoids: four for the up, down, left, and right movement of the joystick, and two for turning the post of the joystick clockwise or counter clockwise (or use rotary solenoids). A basic diagram is shown in Fig. 8-4. A computer controls the action of the solenoids.

Mobile Armatron

The Mobile Armatron, shown in Fig. 8-5, is another Tomy classic, again sold exclusively by Radio Shack. The Mobile Armatron is different from the regular Armatron. Besides providing a movable base, the Mobile Armatron is controlled by a wired keypad controller. The controller makes it easy to convert the toy to computer control. Refer to Chapter 32, "Build a Robot Interface Card," to learn how to interface a computer to dc drive motors. Unfortunately, despite its motorized base, the Mobile Arma-

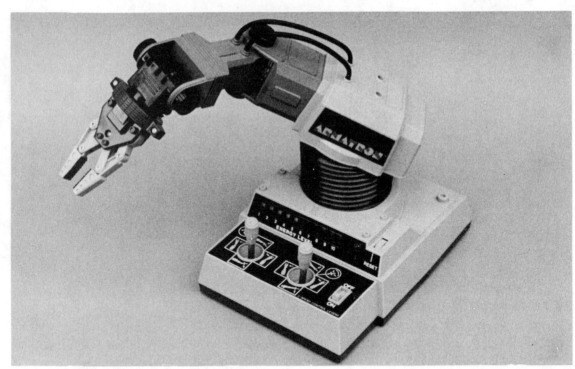

Fig. 8-3. The indefatigable Armatron motorized robotic arm.

Fig. 8-4. One approach to converting the Armatron to computer or remote control.

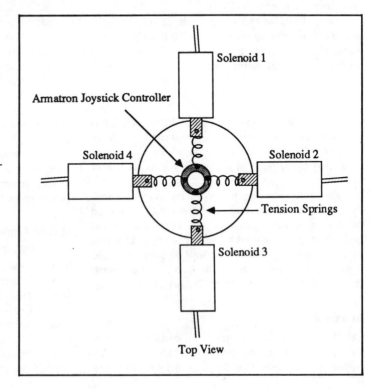

Solenoid 1

Armatron Joystick Controller

Solenoid 4

Solenoid 2

Tension Springs

Solenoid 3

Top View

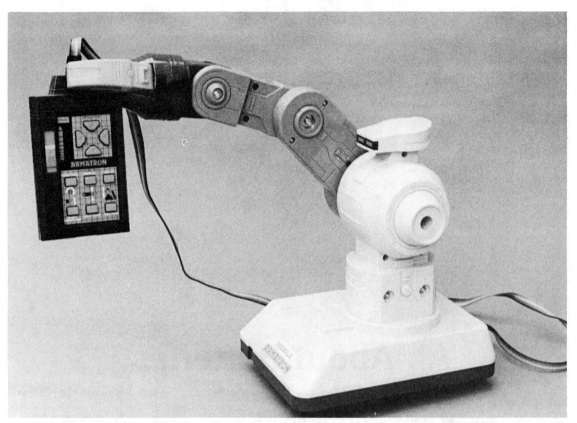

Fig. 8-5. The Mobile Armatron has fewer functions than the regular Armatron but has a switch-type operating panel that makes it easy to convert the toy to computer control.

tron doesn't have all the functions that the original Armatron has.

Fischertechnik

The Fischertechnik kits, made in West Germany by Fischer and sold in the U.S. through Heath/Zenith Computers and Electronics Centers, are the Rolls-Royces of construction toys. Actually, toy isn't the proper term, because the Fischertechnik kits are not designed for use by small children. They are primarily designed for high school and college industrial engineering students, and offer a snap-together approach to making working electromagnetic, hydraulic, pneumatic, static, and robotic mechanisms.

One somewhat pricey kit is designed to teach the fundamentals of arm-based robotics. The kit comes with various molded plastic gears, chains, levers, and other assorted parts, plus motors, solenoids, and an RS-232C serial interface for operating the contraption by computer control. Operating software comes on disk so you can program the robot.

All the Fischertechnik parts are interchangeable, and attach to a common plastic baseplate. You can extend the lengths of the baseplate to just about any size you want, and the baseplate can serve as the foundation for your robot. You can use the motors supplied with the kits, or use your own motors with the parts provided. Because of the cost of the Fischertechnik kits, you won't be cannibalizing them for robot components. That's reserved for the $25 Erector Sets. But if you are interested in learning more about mechanical theory and design, the Fischertechnik kits, used as is, provide a thorough and programmed method for jumping in with both feet.

Chapter 9

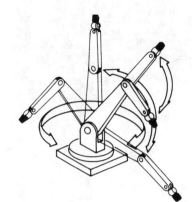

All About Batteries

The robots in science fiction films are seldom like the robots in real life. Take the robot power supply. In the movies, robots almost always have some type of advanced nuclear drive, or perhaps a space-age solar cell that can soak up the sun's energy, then slowly release it for a two or three day period.

Nuclear power plants are out of the question for all but some top-secret robotic experiment conducted by the U.S. Army. Solar cells are too inefficient to be used on anything but a monstrous robot, and as yet, solar cells have no power storage capabilities.

Self-contained real-life robots are powered by batteries, the same kind of batteries used to provide juice to a flashlight, cassette radio, portable television, or other electrical device. Batteries are an integral part of robot design, as important as the frame, motor, and electronic brain—the components we most often think about when the subject turns to robots. To robots, batteries are the elixir of life, and without them, the robot ceases to function.

While great strides have been taken in electronics during the past 20 years—with entire computers that fit on a chip—battery technology is behind the times. On the whole, today's batteries don't pack much wallop for their size and weight, and the rechargeable ones take hours

to come back to life. The hi-tech batteries you may have heard about exist, but they are largely confined to the laboratories and a few high-priced applications, such as space or medical science. That leaves us with the old, run-of-the-mill batteries used in everyday applications.

With judicious planning and use, however, along with learning how to make do with the limitations, these common everyday batteries can provide more than adequate power to all of your robot creations.

TYPES OF BATTERIES

There are five main types of batteries. These types come in a variety of shapes, sizes, and configurations.

Zinc

Zinc batteries are the staple of the battery industry, and are often referred to simply as flashlight cells. The chemical makeup of zinc batteries comes in two forms: carbon zinc and zinc chloride. Carbon zinc, or regular duty, batteries die out the quickest and are unsuited for robotic applications.

Zinc chloride, or heavy duty, batteries provide a little more power than regular carbon zinc cells and last 25 to 50 percent longer. Despite the increase in energy, zinc chloride batteries are also unsuitable for most robotics applications.

Both carbon zinc and zinc chloride batteries can be rejuvenated a few times after being drained. See the section on Battery Recharging for more information on recharging batteries. Zinc batteries are available in all the standard flashlight ("D," "C," "A," "AA," and "AAA") and lantern battery sizes.

Alkaline

Alkaline cells use a special alkaline manganese dioxide formula that lasts up to 800 percent longer than carbon zinc batteries. The actual increase in life expectancy ranges from about 300 percent to 800 percent, depending on the application. In robotics, where the batteries are driving motors, solenoids, and electronics, the average increase is a reasonable 450 to 550 percent.

Alkaline cells, which come in all the standard sizes, (as well as 6 and 12 volt lantern cells), cost about twice as much as zinc batteries. But the increase in power and service life is worth the cost. Unlike zinc batteries, however, alkaline batteries cannot be rejuvenated by recharging. In fact, recharging an alkaline battery is not recommended, as the battery may heat up and explode. When the battery is dead, throw it away.

Nickel-Cadmium

When you think "rechargeable battery," you undoubtedly think nickel-cadmium—or "Ni-Cad" for short. Ni-Cad's aren't the only battery specifically engineered to be recharged, but they are among the least expensive and easiest to get. With the exception of driving large motors, Ni-Cad's are ideal for most robotics applications.

The cells are available in all standard sizes, plus special-purpose "sub" sizes, for use in sealed battery packs (as in rechargeable handheld vacuum cleaners, photoflash equipment, and so forth). Most sub-size batteries have solder tabs on them, so you can directly attach wires to the battery instead of placing the cells in a battery holder.

Ni-Cad's don't last nearly as long as zinc or alkaline batteries, but you can easily recharge them when they wear out. Most battery manufacturers claim their Ni-Cad cells last for 1,000 or more recharges. I have yet to wear out a set of Ni-Cad batteries.

A new, higher capacity Ni-Cad battery recently became available that offers two to three times the service life of regular Ni-Cad's. More importantly, the hi-capacity

cells provide considerably more power, and are ideally suited for robotics work. Of course, cost is higher.

Lead-Acid

The battery in your car is a lead-acid battery. Its makeup is not much more than lead plates crammed in a container that's filled with an acid-based electrolyte. These brutes pack a wallop and have an admirable between-charge life. When the battery goes dead, recharge it, just like a Ni-Cad.

Not all lead-acid batteries are as big as the one in your car. You can also get—new or surplus—6 volt lead-acid batteries that are about the size of a small radio. The battery is sealed, so the acid doesn't spill out (most automotive batteries are now sealed as well). The sealing isn't complete, though: during charging, gases develop inside the battery and these are vented out through very small pores. Without proper venting, the battery would be ruined after discharging and recharging.

Lead-acid batteries typically come in self-contained packs. Six volt packs are the most common, but you can also get 12 and 24 volt packs. The packs are actually made by combining several, smaller cells. The cells are wired together to provide the rated voltage of the entire pack. Each cell typically provides two volts, so three cells are required to make a 6 volt pack. You can, if you wish, take the pack apart, unsolder the cells, and use them separately.

Although lead-acid batteries are powerful, they are heavy. A single 6 volt pack can weigh four or five pounds. If you need a 12 volt supply, you must add two batteries, for a combined weight of eight to 10 pounds. You can easily see how the weight of your robot can quickly increase just by adding the batteries.

Lead-acid batteries are often used as a backup or emergency power supply for computers, lights, and telephone equipment. The cells are commonly available on the surplus market, and although used, still have many more years of productive life. Retail price for new lead-acid cells is about $25 for a 6 volt pack. Surplus prices are 50 to 80 percent less.

Motorcycle batteries make good power cells for robots. They are easy to get, compact, and relatively lightweight. The batteries come in various amp-hour capacities (discussed below), so you can choose the best one for your application. Motorcycle batteries are somewhat pricey, however. Car batteries can also be used, as long as the robot is large enough and sturdy enough to support it. It's not unusual for a car battery to weigh 20 pounds.

Fig. 9-1. A lead-acid (left) and gel-cell rechargeable battery. The lead-acid battery is rated 6 volts at 8 Ah; the gel-cell is rated 6 volts at 9 Ah.

Gel-Cell

Gel-cell batteries use a special gelled electrolyte. They are rechargeable and provide high current for a reasonable period of time, making them perfect for robots. As shown in Fig. 9-1, gel-cells are physically similar to lead-acid batteries, and you may not be able to tell that a battery is the gel-cell type unless it is labeled as such on the outside of the pack. Gel-cells are the "Cadillac" of batteries, in both cost and quality of service. A new

12 volt gel-cell costs about $35, but pound for pound, it will provide more power than any of the other types of batteries, including lead-acid. Gel-cells are typically available only in 6 or 12 volt packs. Some specially designed cells (available through surplus) may come with different voltage ratings.

A rough comparison of the service life of the main types of batteries for robot work is shown in Fig. 9-2. Note that Ni-Cad batteries last the least between charges.

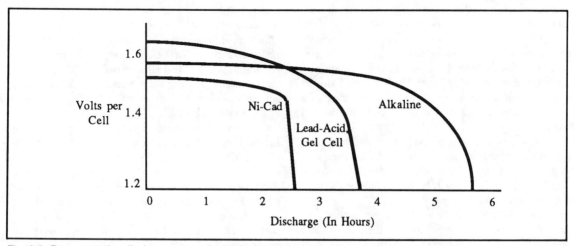

Fig. 9-2. Representative discharge curves for Ni-Cad, lead-acid/gel-cell, and alkaline batteries.

BATTERY RATINGS

Batteries carry all sorts of ratings and specifications. The two most important specifications are voltage and amp-hour current.

Voltage

The voltage rating of a battery is fairly straightforward. If the cell is rated for 1.5 volts, it puts out 1.5 volts, give or take. That "give or take" is more important than you may think, because few batteries actually deliver their rated voltage throughout their life span. Most rechargeable batteries are recharged 20 to 30 percent higher than their specified rating. For example, the 12 volt battery in your car, a type of lead-acid battery, is charged to about 13.8 volts.

Standard zinc and alkaline flashlight batteries are rated at 1.5 volts per cell. Assuming a well-made battery in the first place, the voltage may actually be 1.65 volts when the cell is fresh, to 1.2 volts, when the battery is considered dead. The circuit or motor you are powering with the battery must be able to operate sufficiently throughout this range.

Most batteries are considered dead when their power level reaches 80 percent of their rated voltage. That is, if the cell is rated at 6 volts, it's considered dead when it puts out only 4.8 volts. Some equipment may still function at levels below 80 percent, but the efficiency of the battery is greatly diminished. Past the 80 percent mark, the battery no longer provides the rated current (see below) and if it is the rechargeable type, the cell is likely to be damaged and not take a new charge.

When experimenting with your robot systems, keep a volt-ohm meter handy and periodically test the output of the batteries. Perform the test while the battery is in use. The test results may be erroneous if the battery is not tested under load.

It is often helpful to know the battery condition when the robot is in use. Using a volt-ohm meter to periodically test the robots power plant is inconvenient. You can build a number of battery monitors into your robot that sense voltage level. The output of the monitor can be a light emitting diode (LED), allowing you to see the relative voltage level, or the output can be connected to a circuit that instructs the robot to seek a recharge, or turn off. Several such circuits are given later in this chapter.

If your robot has an on-board computer, you want to avoid running out of juice mid-way through some task. Not only will you lose the operating program and have to rekey or reload it, the robot may damage itself or its surroundings if the power to the computer is suddenly turned off.

Amp-hour Current

The amp-hour current is the amount of power, in amps or milliamps, the battery can deliver over a specified period of time. The amp-hour current rating is a little like the current rating of an ac power line, but with a twist. Ac power is considered never-ending, available night and day, always in the same quantity. But a battery can only store so much energy before it poops out, so the useful service life must be taken into account.

The current rating of a battery is at least as important as the voltage rating, because a battery that can't provide enough juice won't be able to turn a motor, or sufficiently power all the electronic junk you've stuck onto your robot.

What exactly does the term "amp-hour" mean? Basically, the battery will be able to provide the rated current for one hour before failing. If a battery has a rating of 5 amp-hours (expressed as "Ah"), it can provide up to five amps continuously for one hour, one amp for five hours, and so forth, as shown in Fig. 9-3. So far, so good, but the amp-hour rating is not that simple.

You may or may not have been around when service stations decided to price their gas in 9/10ths of a gallon increments. A price of 29 cents per 9/10ths-gallon seems better than 32 cents per gallon, yet the 32 cent price is actually a better buy (not by much, but it is better).

In a way, battery companies came up with something similar when they settled on the rating system for batteries. The 5 Ah rating is actually taken at a 10 or 20 hour discharge interval.

That is, the battery is used for 10 or 20 hours, at a low or medium discharge rate. After the specified time, the battery is tested to see how much juice it has left. The rating of the battery is then calculated taking the difference between the discharge rate and the reserve power, and multiplying it by the number of hours under test.

What all this means is that it's an unusual battery that provides the stated amps in the one hour period. The battery is much more likely to fail after 30 or 45 minutes of heavy-duty use, and won't be able to supply the specified current for more than about 15 to 20 minutes. Discharging at or above the amp-hour rating may actually cause damage to the battery. This is especially true of Ni-Cad cells.

The lesson to learn is that you should always choose a battery that has an amp-hour rating 20 to 40 percent

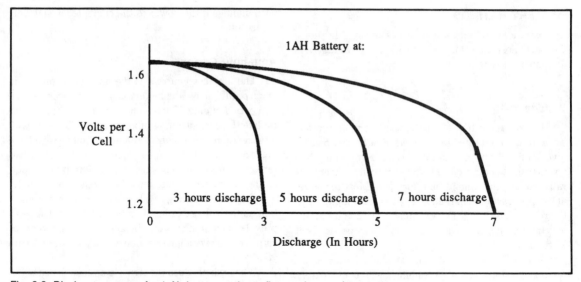

Fig. 9-3. Discharge curves of a 1 Ah battery at three, five, and seven hour rates.

more than what you need to power your robot. Figuring the desired capacity is nearly impossible until the entire robot is designed and built (or unless you are very good at computing current consumption). The best advice is to design the robot with the largest battery you think practical. If you find that the battery is way too large for the application, you can always swap it out for a smaller one. It's not so easy when doing the reverse.

Note that some components in your robot may draw excessive current when they are first switched on, then settle down to a more reasonable level. Motors are a good example of this. A motor that draws one amp under load may actually require four or five amps at start up. The period is very brief, on the order of 100 to 200 milliseconds. No matter; the battery should be able to accommodate the surge, which means that the 20 to 40 percent overhead in using the larger battery is a necessity, not just a design suggestion.

With this in mind, it's easy to see how a battery will wear down faster if the motors in the robot are constantly stopping and starting. You can help increase the life expectancy of the battery by reducing the repetitive start-stop of the motors.

Recharge Rate

Batteries must be recharged slowly, over a 12 to 24 hour period. The battery can't take on too much current without breaking down and destroying itself, so the current from the battery charger must be kept at a safe level.

A good rule of thumb to follow when recharging any battery is to limit the recharging level to 1/10th the amp-hour rating of the cell. For a 5 Ah battery, then, a safe recharge level is 500 milliamps. Current limiting is extremely important when recharging Ni-Cad's, which can be permanently damaged by charging too quickly. Lead-acid and gel-cell batteries can take an occasional fast-charge—a quickie at 25 to 50 percent of the rated amp-hour capacity of the battery. Repeated quick-charging will warp the plates and disturb the electrolyte action in the battery, and is not recommended.

The recharge period, the number of hours the battery is recharged, varies depending on the type of cell. The recharge rate is usually a function of the discharge rate. You can't recharge a battery any faster than it is discharged. If your robot discharges its battery continually over a cumulative five hour period, (a reasonable average), it will take at least five hours to recharge it. In practice, a recharge interval of two to ten times that of the discharge rate is recommended. Figure 9-4 shows a typical discharge/recharge curve for a lead-acid or gel-cell battery.

Most manufacturers specify the recharge time for their batteries. If so, follow it. If no recharge time is specified, assume a three- or four-to-one discharge/recharge ratio, and place the battery under charge for that period of time. Continue experimenting until you find an optimum recharge interval.

INCREASING BATTERY RATINGS

You can obtain higher voltages and current by con-

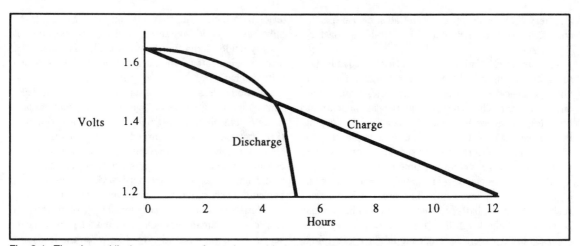

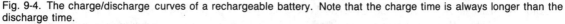

Fig. 9-4. The charge/discharge curves of a rechargeable battery. Note that the charge time is always longer than the discharge time.

necting several cells together, as shown in Fig. 9-5.

- To increase voltage, connect the batteries in series. The resultant voltage is the sum of the voltage outputs of all the cells combined.
- To increase current, connect the batteries in parallel. The resultant current is the sum of the current capacities of all the cells combined.

Take note that when connecting cells together, not all cells may be discharged or recharged at the same rate. This is particularly true if you combine two half-used batteries with two new ones. The new ones will do the lion's share of the work, and won't last as long. Therefore, you should always replace or recharge all the cells at once. Similarly, if one or more of the cells in a battery pack is permanently damaged, and can't deliver or take on a charge like the others, it should be replaced.

BATTERY RECHARGING

Most lead-acid and gel-cell batteries can be recharged using a 200 to 800 mA battery charger. The charger can even be an ac adapter for a video game or other electronics device (the output of the adapter must be dc, of course). You must exercise care when recharging lead-acid and gel-cell batteries. They can be overcharged and damaged if left charging for periods longer than 24 hours. Buy a 12-hour timer if you are the forgetful type.

Standard Ni-Cad batteries can't withstand recharge rates exceeding 50 to 100 mA, and using a charger that supplies too much current will destroy the cell. Use only

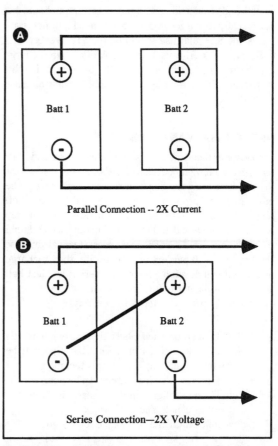

Fig. 9-5. Wiring batteries to increase ratings. A. Parallel connection increases current; B. Series connection increase voltage.

63

a battery charger designed for Ni-Cad's, or use the circuit described in Chapter 10, "Build a Battery Recharger."

Hi-capacity Ni-Cad batteries can be charged at higher rates, and are used with a recharger designed for them. Unlike lead-acid and gel-cell batteries, Ni-Cad's have self-limiting charging characteristics (but they are not current-limiting), and generally cannot be damaged by keeping them on charge for extended periods. Still, to prevent possible internal deterioration, you should remove the batteries from charge when their time is up.

Zinc batteries can be rejuvenated by placing them in a recharger for a few hours. The process is not true recharging, as the battery is not restored to its original power or voltage level. The rejuvenated battery lasts about 20 to 30 percent as long as it did during its initial use. Most well-built zinc batteries can be rejuvenated two or three times before they are completely depleted.

Ni-Cad, lead-acid, and gel-cell batteries should be periodically recharged, whether they need it or not. Batteries not in regular use should be recharged every two to four months. Always observe polarity when recharging batteries. Inserting the cells backward in the recharger will destroy the batteries and possibly damage the recharger.

NI-CAD DISADVANTAGES

Despite their numerous advantages, Ni-Cad batteries have a few odd peculiarities about them you'll want to consider when designing your robot power system. The most annoying problem is what's termed "memory effect." Ni-Cad's tend to remember the rate at which they have been discharged and recharged in the past, and tend to follow this curve in subsequent uses. If you always use the battery to 50 percent of its capacity, then recharge it, the cell will develop a memory where it will last half as long as it should before giving out.

Memory effect is altered in two ways:

- The dangerous way. Short the battery until it's dead. Recharge it as usual. Some batteries may be permanently damaged with this technique.
- The safe way. Use the battery in a low-current circuit, like a flashlight, until it is dead. Recharge the battery as usual. You must repeat this process a few times until the memory effect is gone.

The best way to combat memory effect is to avoid it in the first place. Always fully discharge Ni-Cad batteries before charging them. If you don't have a flashlight handy, build yourself a discharge circuit using a battery holder and a flashlight bulb. The bulb acts as a discharge indicator. When it goes out, the batteries are fully discharged.

The other disadvantage is that the polarity of Ni-Cad's can change—positive becomes negative and vice versa—under certain circumstances. Polarity reversal is common if the battery is left discharged too long, or if it is discharged below 75 or 80 percent capacity. Excessive discharging can occur if one or more cells in a battery pack wears out. The adjacent cells must work overtime to compensate, discharging themselves too fast and too far.

You can test for polarity reversal by hooking the battery to a volt-ohm meter (remove it from the pack if necessary). If you get a negative reading when the leads are connected properly, the polarity of the cell is reversed.

Polarity reversal can sometimes be corrected by fully charging the battery (connecting it in the recharger in reverse), then shorting it out. Repeat the process a couple of times, if necessary. You have about a 50-50 chance that the battery will survive this. The alternative is to throw the battery out, so you actually stand to lose very little.

Recharging in the Robot

You'll probably want to recharge the batteries while they are inside the robot. This is no problem as long as you provide a connector for the charger terminals on the outside of the robot. When the robot is ready for a charge, connect it to the charger. The robot should be turned off during the charge period, otherwise the batteries may never recharge.

Turning off the robot during recharging may not be desirable, however. There are several schemes you can employ that will continue to supply current to the electronics of the robot, yet allow the batteries to charge. One way is to use a relay switchout. In this system, the external power plug on your robot consists of four terminals: two for the battery and two for the electronics. When the recharger is plugged in, the batteries are disconnected from the robot. You can use relays to control the changeover, or heavy-duty open-circuit jacks and plugs (the ones for audio applications may work). While the batteries are switched out and being recharged, a separate power supply provides operating juice to the robot.

BATTERY CARE

Batteries are rather sturdy little creatures, but you should follow some simple guidelines when using them.

You'll find that your batteries will last much longer, so you'll save yourself some money.

- Store new batteries in the fresh food compartment of your refrigerator (not the freezer). Put them in a plastic bag to prevent contamination from possible leakage. Warm up the batteries for several hours before using them.
- Avoid using or storing batteries in temperatures above 75 or 80 degrees F. The life of the battery will be severely shortened otherwise. Using a battery above 100 to 125 degrees causes rapid deterioration.
- Unless you're repairing a misbehaving Ni-Cad, avoid shorting out the terminals of the battery. Besides possibly igniting fumes exhausted by the battery, the sudden and intense current output shortens the life of the cell.
- Keep rechargeable batteries charged. Make a note when the battery was last charged.
- Fully discharge Ni-Cads before charging them again. This prevents memory effect.
- Given the right circumstances, all batteries will leak, even the sealed variety. When not in use, keep batteries in a safe place where leaked electrolyte will not cause damage. Remove batteries from their holder when not in use.

POWER DISTRIBUTION

Now that you know about batteries, you can start using them in your robot designs. The most simple and straightforward arrangement is to use a commercial-made battery holder. Holders are available that contain from two to eight "AA," "C," or "D" batteries. Wiring in the holders connect the batteries in series, so a four-cell holder puts out 6 volts (1.5 times 4). You attach the leads of the holder (red for positive and black for ground or negative) to the main power supply rail in your robot. You'd follow a similar procedure if using a gel-cell or lead-acid battery.

Fuse Protection

Flashlight batteries don't deliver extraordinary current, so fuse protection is not required on the most basic robot designs. Gel-cell, lead-acid, and hi-capacity Ni-Cad batteries can deliver a most shocking amount of current, enough that if the leads of the battery accidentally touch each other or there is a short in the circuit, the wires may melt and a fire could erupt.

Fuse protection helps eliminate the calamity of a short circuit or power overload in your robot. As illustrated in Fig. 9-6, connect the fuse in-line with the positive rail of the battery, as near to the battery as possible. You can purchase fuse holders that connect directly to the wire, or mount on a panel or printed circuit board.

Choosing the right value of fuse can be a little tricky, but not impossible. It does require that you know how much current your robot draws from the battery during normal and stalled motor operation. You can figure the value of the fuse by adding up the current draw of each separate sub-system, then tacking on 20 to 25 percent overhead.

Let's say that the two drive motors in the robot draw two amps each, the main circuit board one amp, and other small motors 0.5 amp each (for a a total of, perhaps two amps). Add all these up and you get seven amps. Installing a fuse with a rating of at least seven amps at 125 or 220 volts will help assure that the fuse won't burn out prematurely during normal operation. Adding that 20 to 25 percent margin calls for an eight to 10 amp fuse.

Recall that motors draw excessive current when first started. You can still use that eight to 10 amp fuse, but make sure it is the slow-blow type. Otherwise, the fuse will burn out every time one of the heavy-duty motors kick in.

Fuses don't come in every conceivable size. For the sake of standardization, choose the regular 1 1/4-inch long by 1/4-inch diameter buss fuses. You'll have an easier job finding fuse holders for them, and you have a greater selection of values. Even with a standard fuse size, there is not much to choose from past eight amps, other than 10, 15, and 20 amps, but this is acceptable for most robot applications. For values over eight amps, you may have to go with ceramic fuses, used mainly for microwave and kitchen appliances.

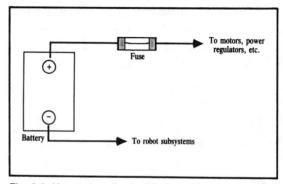

Fig. 9-6. How to install a fuse in-line with the battery and robot electronics.

Multiple Voltage Requirements

Some advanced robot designs require several voltages for proper operation. The drive motors may require 12 volts, at perhaps two to four amps whereas, the electronics require +5, −5 and +12 volts.

Multiple voltages can be handled in several ways. The easiest and most straightforward is to use a different set of batteries for each main sub-section. The motors operate off one set of large lead-acid or gel-cell batteries; the electronics are driven by smaller capacity Ni-Cad's.

This approach is actually desirable when the motors used in the robot draw a lot of current. Motors naturally distribute a lot of electrical noise throughout the power lines, noise that electronic circuitry is extremely sensitive to. The electrical isolation provided by using different batteries nearly eliminates problems caused by noise (the remainder of the problems occur when arcing of the motor commutators cause rf interference).

In addition, the excessive current draw from the motors when they are first started may zap all the juice from the electronics. This can cause failed or erratic behavior, and could cause your robot to lose control.

The other approach is to use one main battery source and "step" it down for use with the various components in the system. One 12 volt battery can be regulated (see below) to just about any voltage under 12 volts. The battery can directly drive the 12 volt motors, and with proper regulation, supply the +5 volt power to the circuit boards.

Judicious connection of the batteries can also yield multiple voltage outputs. By connecting two 6-volt batteries in series, as shown in Fig. 9-7, you get +12 volts, +6 volts, and −6 volts. This system isn't nearly as foolproof as it seems, however. More than likely, the two batteries will not be discharged at the same rate. This causes extra current to be drawn from one to the other, and the batteries may not last as long as they might otherwise.

If all of the subsystems in your robot use the same batteries, be sure to add sufficient filtering capacitors across the positive and negative power rails (see the voltage regulator hookup below). The capacitors help soak up excessive current spikes and noise, most often contributed by motors. Place the capacitors as near to the batteries and the noise source as possible. Exact values are not critical, but should be over 100 μF. Be certain the capacitors you use are rated at the proper voltage. Using an underrated capacitor will burn it out and possibly cause a short circuit.

Smaller value capacitors, such as 0.1 μF, should be placed across the positive and negative power rails wherever power enters or exits a circuit board. As a general rule, you should add these "decoupling" capacitors beside clocked logic ICs, particularly flip-flops and counters. A few linear ICs, such as the NE555 timer, need decoupling capacitors, or the noise they generate through the power lines can ripple through to other circuits. If many ICs are on the board, you can usually get by with adding one 0.1 μF decoupling capacitor for every three or four packages.

Voltage Regulation

Some electronic circuits require a precise voltage or they may be damaged or act erratically. Generally, you provide voltage regulation only to those components and circuit boards in your robot that require it. It is impractical, not to mention bad design, to regulate the voltage for the entire robot as it exits the battery.

Solid-state voltage regulators can be easily added to all your electronic circuits. They are easy to obtain and you can choose from among several styles and output capacities. Two of the most popular voltage regulators, the 7805 and 7812, provide +5 volts and +12 volts, respectively. You connect them to the "+" and "−" rails of your robot, as shown in Fig. 9-8.

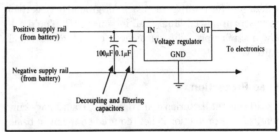

Fig. 9-8. Filtering and decoupling capacitors should be added to the voltage regulator circuit to provide cleaner power. The exact values of the capacitors are not important, but components must be rated to handle at least (and preferably double) the supply voltage. A good rule of thumb: use capacitors rated at 35 volts or higher.

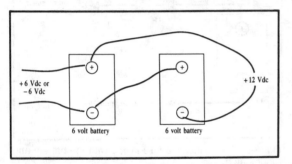

Fig. 9-7. Various voltage tap-offs from two six volt batteries.

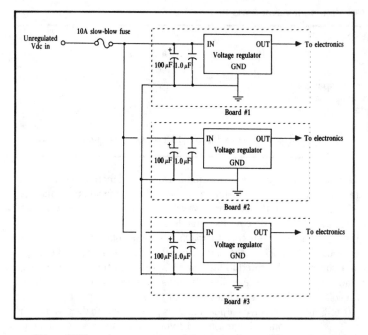

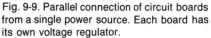

Fig. 9-9. Parallel connection of circuit boards from a single power source. Each board has its own voltage regulator.

Other 7800 series power regulators are for +15, +18, +20, and +24 volts. The 7900 series provide negative power supply voltages in similar increments. The current capacity of the 7800 and 7900 series that come in the TO-220 style transistor packages (no suffix or "T" in the part number), is limited to less than one amp, so they must be used in circuits that do not draw in excess of this.

Other regulators are available in a more traditional TO-3 style transistor package ("K" suffix) offering current output to several amps. The "L" series regulators come in the small TO-92 transistor packages and are designed for applications that require less than about 500 mA.

Other regulators:

- The 328K provides an adjustable output to 5 volts, with a maximum current of 5A.
- The 78H05K offers a 5 volt output at 5A.
- The 78H12K offers a 12 volt output at 5A.
- The 78P05K delivers 5 volts at 10 amps.

These regulators come in TO-3 packages and are very expensive ($10 to $15 for the 78P05K). Fortunately, there's not much call for them in hobby robotics.

Power Distribution

You may choose to place all or most of your robot's electronic components on a single board. You can mount the regulator(s) directly on the board. You can also have several smaller boards share one regulator, as long as the boards together don't pull power in excess of what the regulator can supply.

Figure 9-9 shows how to distribute the power from a single battery source to many separate circuit boards. The individual regulators provide power for one or two large boards or a half dozen or so smaller ones.

Voltage regulators are great devices, but they are somewhat wasteful. To work properly, the regulator must be provided with several volts more than the desired output voltage. For example, the 7812 +12 volt regulator

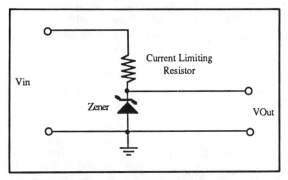

Fig. 9-10. Using a zener diode to regulate voltage. Use a zener and current-limiting resistor large enough to handle the current consumption of the circuit.

67

Table 9-1. Voltage Inverter Parts List.

U1	NE555 timer IC
R1	1.2KΩ resistor
R2	3.9KΩ resistor
R3	1KΩ resistor
C1	0.05 μF ceramic capacitor
C2,C3	220 μF electrolytic capacitor
D1,D2	1N4002 diode

All resistors 5 or 10 percent tolerance, 1/4-watt; all capacitors 10 percent tolerance.

needs at least 13 to 15 volts to deliver the full voltage and current specified for the device. Regulated 12 volt robotic systems may require you to use an 18 volt supply.

Voltage regulators do an extremely good job at keeping just the right amount of juice flowing to the circuit. Sometimes, such tight regulation isn't absolutely required. On circuits that can stand a fluctuation of half a volt or so, use zener diodes to provide a reasonably stable voltage output. Zeners are rated by their breakdown voltage. A 5.1 volt zener (common for TTL circuits) acts to limit voltage to the circuit to about 5.1 volts.

Zeners are diodes, so they must be installed in the circuit with the proper orientation. They also are rated by the amount of current they can withstand. Small circuits can use the small 1 watt zeners, but higher current applications may need the 5+ watt types.

In most circuits, zeners must be used in conjunction with a current limiting resistor (also with a wattage that can handle the current flowing through the circuit). Figure 9-10 shows a typical zener volt application. There is a 0.6 volt drop across the zener, so for a 5.1 volt system, provide the circuit with at least 6.0 volts.

IC VOLTAGE DOUBLERS AND INVERTERS

If your robot is equipped with 12 volt motors and uses circuitry that requires only +5 and/or +12 volts, then your work is made easy for you. But if you require negative supply voltages for some of the circuits, you're faced

with a design dilemma. Do you add more batteries to provide the negative supply? That's one solution, and maybe the only one available to you if the current demand of the circuits is moderate to high.

Another approach is to use a polarity reversal circuit, such as the one in Fig. 9-11. The current at the output is limited to less than 200 mA, but this is often enough for devices like op amps, CMOS analog switches, and other small devices that require a −5 or −12 Vdc. The negative output voltage is proportional to the positive output voltage. So, +12 volts in, roughly −12 volts out.

A few microprocessors use negative supply voltages as well, but these chips routinely draw in excess of 200 mA. If your robot has an on-board computer, and the microprocessor requires a negative supply voltage, you're better off using a separate battery.

Some computer components, such as microprocessors and electronically erasable programmable read only memory chips (EEPROM), require a "program" voltage of 24 volts or higher. You can use the circuit in Fig. 9-12 to double the supply voltage to provide the desired program voltage. The current level at the output is very low, but it should be enough for use as a high voltage pulse. The voltage inverter parts list is in Table 9-1, and the doublers are in Table 9-2.

BATTERY MONITORS

Quick! What's the condition of the battery in your

Table 9-2. Voltage Doubler Parts List.

U1	NE555 timer IC
R1	2.2KΩ resistor
R2	15KΩ resistor
C1	0.01 μF ceramic capacitor
C2,C3	220 μF electrolytic capacitor
C4	470 μF electrolytic capacitor
D1,D2	1N4002 diode

All resistors 5 or 10 percent tolerance, 1/4-watt; all capacitors 10 percent tolerance.

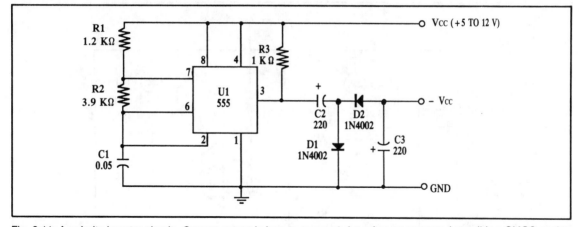

Fig. 9-11. A polarity inverter circuit. Current output is low, but enough for a few op-amps and possibly a CMOS analog switch or two.

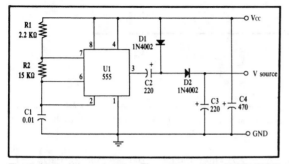

Fig. 9-12. A voltage doubler. The voltage at the V SOURCE terminal is roughly double the voltage applied to the Vcc terminal.

robot? With a battery monitor, you'd know in a flash. A battery monitor continually samples the output voltage of the battery during operation of the robot (the best time to test the battery), and provides a visual or logic output.

Figure 9-13 shows a simple battery monitor with a one LED output. Potentiometer R1 sets the trip point of the circuit, or the minimum voltage level before the circuit is activated. The LED lights when the battery drops below the trip point. When you see the LED light up, you can stop the robot and place its batteries on charge.

A similar circuit is shown in Fig. 9-14. Once again, a variable resistor, R1, sets the trip point of the circuit. When the input voltage drops below the trip point, the LED turns on. This circuit can be used with other sup-

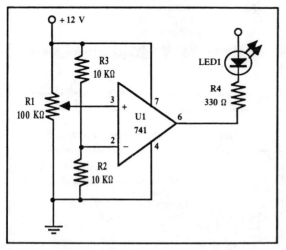

Fig. 9-13. A simple "high-low" battery monitor. For reverse logic indication, connect the LED to ground through the curent-limiting resistor, R4.

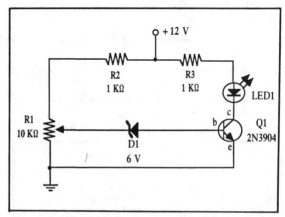

Fig. 9-14. Another battery monitor. The circuit is designed for use with a 12 volt battery; for 6 volt operation, readjust R1 and install a 3.3 volt zener in place of D1. R1 must be calibrated with a discharged battery prior to putting the circuit into use.

69

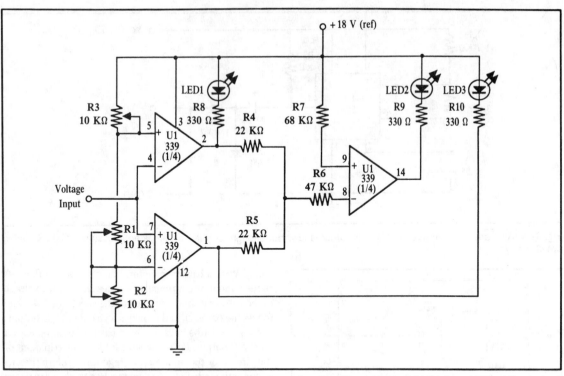

Fig. 9-15. A three-state "window comparator" battery monitor. Adjust R1, R2, and R3 to light the three LEDs at the desired voltage levels throughout the discharge life of the battery.

Table 9-3. Basic Comparator Circuit Parts List.

U1	LM741 op amp IC
R1	100 kΩ potentiometer
R2,R3	10 kΩ resistor
R4	330 Ω resistor
LED1	Light Emitting Diode

All resistors 5 or 10 percent tolerance, 1/4-watt.

Table 9-5. Three-Indicator Battery Monitor Parts List.

U1	LM339 Quad Comparator
R1-R3	10K potentiometer
R4,R5	22KΩ resistor
R6	47KΩ resistor
R7	68KΩ resistor
R8-R10	330 Ω resistor
LED1-3	Light Emitting Diode

All resistors 5 or 10 percent tolerance, 1/4-watt.

Table 9-4. Twelve-Volt Battery Monitor Parts List.

R1	10KΩ resistor
R2,R3	1KΩ resistor
D1	6 volt zener, 1 watt
Q1	2N3904 npn transistor
LED1	Light Emitting Diode

All resistors 5 or 10 percent tolerance, 1/4-watt.

Table 9-6. Bar-Dot Battery Monitor.

U1	LM3914 Bar-Dot Generator IC
R1	100K potentiometer
R2	1K resistor
LED1-10	Light Emitting Diode

All resistors 5 or 10 percent tolerance, 1/4-watt.

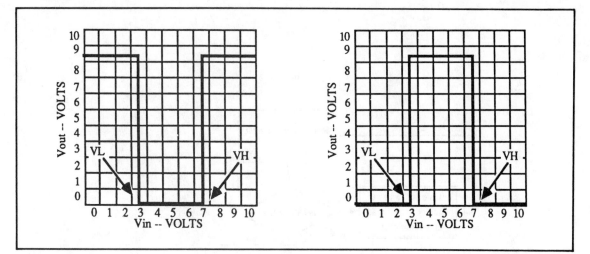

Fig. 9-16. A graphic representation of the window comparator. The state of the comparators change when the input voltage exceeds the high and low trip points.

ply voltages by changing the value of the zener diode and by readjusting R1.

The last two circuits provide a high/low indication of the battery level. The circuit in Fig. 9-15 allows you to set high, normal, and low voltages and monitor the bat-

tery as it passes through these phases during its discharge cycle. The circuit should be operated by a separate 12 volt supply, perhaps from a long-life lithium or mercury battery. Connect the negative terminal of this battery to the common ground rail shared by all the other bat-

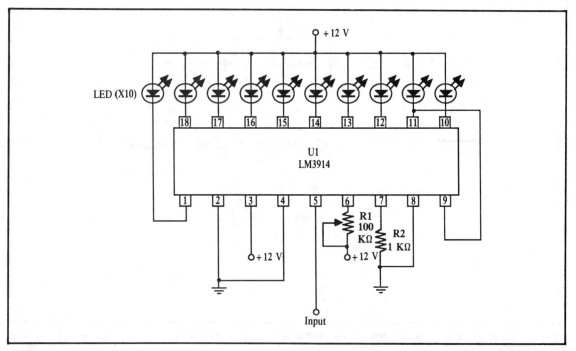

Fig. 9-17. A sophisticated bar-graph battery monitor. The LEDs light up and extinguish as the voltage output of the battery changes.

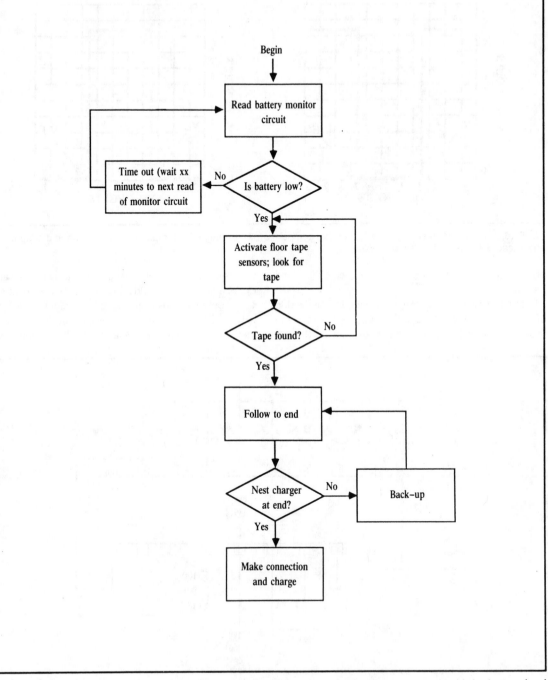

Fig. 9-18. Software can be used to command the robot to return to its battery recharger nest should the battery level exceed a certain low point.

teries. The input voltage is tapped directly off the main supply battery for the robot.

Three potentiometers determine the trip points for the three LED indicators. Here is one way to calibrate the circuit. Set R3 for the new charge level. LED 1 will glow when the battery has been freshly recharged and is at its peak. Set R1 so that LED2 glows during the useful discharge cycle of the battery. This will be a fairly wide range. Set R2 so that LED3 lights up when the battery dips below the 80 percent voltage level, indicating the need for a new charge. Figure 9-16 shows a graph that depicts the trip-points and the changes in state of the comparators as the input voltage changes.

There is one minor design flaw in this circuit, and you can correct it if you wish by adding a 1 K resistor between ground and R2. Note that when the wipers of the potentiometers are rotated to their extremes in a certain way, there is a direct path between the positive supply rail and ground. If you were to allow this to happen, the circuit would short, and the battery would quickly burn out.

Figure 9-17 shows another way to implement the *window comparator* circuit described above. This version uses the LM3914 bar/dot generator. As with the previous circuit, the chip should be powered by a separate battery, otherwise the battery level indication may not be accurate. The voltage from the battery is applied to pin 5 of the chip. Adjust R1 so that at full charge, the second or third to the last LED (the ones connected to pins 16 or 17) lights up. As the supply battery is discharged, the LEDs down

the scale will light up one after the other. By experimenting, you'll be able to identify which LED is lit when the battery reaches its fully discharged level.

With all the circuits, you can attach the LED outputs to a gate or microprocessor/computer port to provide the robot with internal monitoring. In the window comparator schematic in Fig. 9-17, you could connect LED3 to an interrupt port on a microprocessor or computer system bus (be sure to shift the voltage level to 5 volts—the output of the comparator is too high; see Appendix C for details). When the LED is triggered, the interrupt signals the microprocessor.

Software running on the computer interprets the interrupt as "low battery; quick get a recharge." The robot can then place itself into nest mode, where it seeks out its own battery charger. If the charger terminals are constructed properly, it's possible for the robot to plug itself in. The robot won't release itself from charge until the "fully recharged" LED, LED1 in the example, is activated.

You don't need a computer to activate a nest search routine. When the "battery low" LED comes on, it could trigger a track following circuit, such as the ones described in Chapter 29, "Navigating Through Space." The robot then follows a white track taped to the floor. The track leads to—what else—battery food! An operating flowchart for the nest-search process is shown in Fig. 9-18. Parts lists for these comparators and monitors are in Tables 9-3, 9-4, 9-5, and 9-6.

Chapter 10

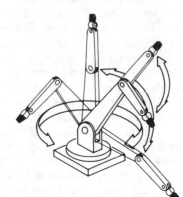

Build an
Experimenter's Power Supply

Using a battery while testing or experimenting with new robot designs is both inconvenient and anti-productive. Just when you get a circuit perfected, the battery goes dead and must be recharged. A stand-alone power supply, which operates off 117 volt house current, can supply your robot designs with regulated dc power, without the need to install, replace, or recharge batteries. You can buy a ready-made power supply (they are common in the surplus market) or make your own.

This chapter presents two power supply circuits. The single-voltage supply can be engineered to provide +5 volts or +12 volts, depending on the transformer and power regulator you use. You'd likely use the +5 volt version to power TTL ICs and computer boards; the 12 volt version to operate motors and CMOS circuits. As an alternative, you can install a variable voltage regulator and choose the voltage you need by turning a dial.

The multi-voltage supply is designed to provide the four voltages common in robot systems: +5 volt, +12 volt, −5 volt, and −12 volt. These voltages are used by most motors and TTL, CMOS, linear, and microprocessor ICs.

SINGLE-VOLTAGE POWER SUPPLY

Refer to Figs. 10-1 and 10-2 for schematics of the single-voltage power supplies. Figure 10-1 shows the circuit for a +5 volt supply; Fig. 10-2 shows the circuit for a +12 volt supply. There are few differences between them so the following discussion applies to both. For the sake of simplicity, we'll refer just to the +5 volt circuit.

For safety, the power supply must be enclosed in a plastic or metal chassis (plastic is better because there is less chance of a short circuit). Use a perforated board to secure the components and solder them together using 18 or 16 gauge insulated wire. Alternatively, you can make your own circuit board using a home etching kit. Before constructing the board, collect all the parts and design the board to fit the specific parts you have. There is little size standardization when it comes to power supply components and large value electrolytic capacitors, so pre-sizing is a must.

The incoming ac is routed to the ac terminals on a 12.6 volt transformer. The hot side of the ac is connected through a 2 amp slow-blow fuse and a single-pole single-throw (SPST) toggle switch. With the switch in the off (open) position, the transformer receives no power so the supply is off.

The 117 Vac is stepped down to 12.6 volts. The transformer specified here is rated at 1.2 amps, sufficient for the task at hand. Remember that the power supply is

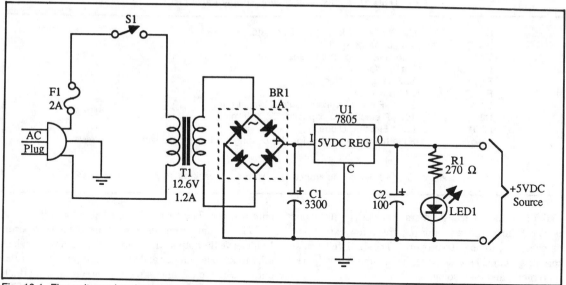

Fig. 10-1. Five-volt regulated power supply.

limited to delivering the capacity of the transformer (and later, the voltage regulator), no more. A bridge rectifier, BR1, converts the ac to dc (shown schematically in the dotted box). You can also construct the rectifier using discrete diodes, and connect them as shown within the dotted box.

When using the bridge rectifier, be sure to connect the leads to the proper terminals. The two terminals marked with a "~" connect to the transformer. The "+" and "–" terminals are the output and must connect as shown in the schematic. A 5 volt, 1 amp regulator, a 7805, is used to maintain the voltage output at a steady 5 volts.

Note that the transformer supplies a great deal more voltage than is necessary. This is for two reasons. First, lower voltage 6.3 or 9 volt transformers are available, but most do not deliver more than .5 amp. It is far easier to

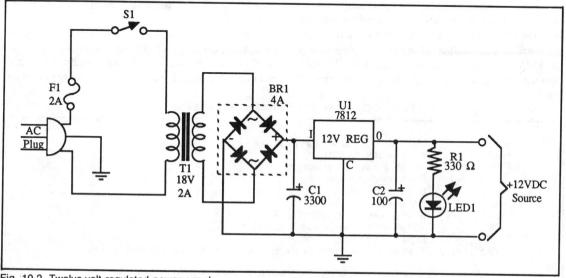

Fig. 10-2. Twelve-volt regulated power supply.

Table 10-1. Five Volt Power Supply Parts List.

U1	LM7805 +5 Vdc Voltage Regulator
BR1	4 amp bridge rectifier
T1	12.6 volt, 1.2 amp ac transformer
F1	2 amp slow-blow fuse
S1	SPST toggle switch
R1	270 Ω resistor
C1	3000 μF electrolytic capacitor, 35 volt min.
C2	100 μF electrolytic capacitor, 35 volt min.
LED1	Light Emitting Diode
Misc.	Fuse holder, heat sink for U1, binding posts, ac cord with plug, chassis

All resistors 5 or 10 percent tolerance, 1/4-watt; all capacitors 10 percent tolerance.

find 12 or 15 volt transformers that deliver sufficient power. Second, the regulator requires a few extra volts as overhead, to operate properly. The 12.6 volt transformer specified here delivers the minimum voltage requirement, and then some.

The 7805 regulator comes in different styles (see Chapter 9, "All About Batteries"). The 7805T, which comes in a TO-220 style transistor case, delivers up to one amp of current. You can increase the current rating, as long as the transformer will comply, with a 7805K regulator. This regulator comes in a TO-3 style transistor case, and mounted on a heat sink, can deliver in excess of three amps. Unless you plan on using the power supply for testing only small circuits, why not opt for the larger capacity regulator. You never know when the extra current will come in handy.

Capacitors C1 and C2 filter the ripple inherent in the rectified dc found at the outputs of the bridge rectifier. With the capacitors installed as shown (note the polarity), the ripple at the output of the power supply is negligible. LED1 and R1 form a simple indicator. The LED will glow when the power supply is on. Remember the 270

ohm resistor. The LED will burn up without it.

The output terminals are insulated binding posts. Don't leave the output wires bare. The wires may accidentally touch one another and short the supply. Solder the output wires to the lug on the binding posts, and attach the posts to the front of the power supply chassis. The posts accept bare wires, alligator clips, even banana plugs. Tables 10-1 and 10-2 contain the parts lists for both the 5 and 12 volt power supplies.

Differences In the 12 Volt Version

The 5 and 12 volt versions of the power supply are basically the same, with but a few important changes. First, the transformer is rated for 18 volts at 2 amps. The 18 volt output is more than enough for the overhead required by the 12 volt regulator, and is commonly available.

The regulator is a 7812, which is the same as the 7805, except that it puts out a regulated +12 volts instead of +5 volts. Use the T series regulator (TO-220 case) for low-current applications; the K series (TO-3) for

Table 10-2. Twelve Volt Power Supply Parts List.

U1	LM7812 +12 Vdc Voltage Regulator
BR1	4 amp bridge rectifier
T1	18 volt, 2 amp ac transformer
F1	2 amp slow-blow fuse
S1	SPST toggle switch
R1	330 Ω resistor
C1	3000 μF electrolytic capacitor, 35 volt min.
C2	100 μF electrolytic capacitor, 35 volt min.
LED1	Light Emitting Diode
Misc.	Fuse holder, heat sink for U1, binding posts, ac cord with plug, chassis

All resistors 5 or 10 percent tolerance, 1/4-watt; all capacitors 10 percent tolerance.

higher capacity applications. Lastly, R1 is increased to 330 ohms.

MULTI-VOLTAGE POWER SUPPLY

The multi-voltage power supply is like four power supplies in one. Rather than using four bulky transformers, this circuit uses just two. The project could have been constructed using one five or 10 amp transformer, tapping the voltage at the proper locations to operate the +5, +12, -5, and -12 regulators. In researching parts availability, however, I found heavy-duty center-tapped transformers in short supply and unusually expensive. This project calls for two commonly available and reasonably priced transformers. If you happen to locate a new or surplus extra heavy-duty 25.2 volt center tapped transformer, by all means, buy and use it.

The circuit, as shown in Fig. 10-3, is really two power supplies in one. One half of the supply provides +12 and -12 volts; the other half provides +5 and -5 volts. Each side is connected to a common fuse, switch, and wall plug.

The basic difference between the multi-voltage supply and the single-voltage supplies described earlier in this chapter is the addition of negative power regulators and the additional filtering capacitors. Circuit ground is the center tap of the transformer.

The supply provides approximately one amp for each of the outputs. You can substitute heavier duty regulators for the two positive voltages (use heat sinks) and draw approximately 2 amps at +12 and +5 volts, and .5 amps at -12 and -5 volts.

Large robot drive motors routinely draw in excess of 2 amps at +12 volts, and the supply may not provide adequate power. Rather than build a heavy-duty regulated power supply for just the motors, use a dedicated set of lead-acid or gel-cell batteries. During testing, the motors are on only intermittently, and you can usually get through the afternoon and evening on a single charge of the battery.

Use nylon binding posts for the five outputs (ground, +5, +12, -5, -12). Clearly label each post so you don't mix them up when using the supply. Check for proper operation with your volt-ohm meter. Table 10-3 contains the parts list for this power supply.

SAFETY PRECAUTIONS

The circuits described in this chapter, as well as Chapter 11, are the only ones where you are required to work with 117 Vac. Be extra careful when wiring the power supply circuits and triple check your work. Never operate the supply when the top of the cabinet is off, unless you are testing it. Even then, stay away from the incoming ac. Touching it can cause a serious shock, and depending on the circumstances, it could kill you. Be extra certain that no wires touch the chassis or front panel. Use an all-plastic enclosure whenever possible.

Do not operate the power supply if it has gotten wet,

Table 10-3. Multiple Voltage Power Supply Parts List.

U1	LM7812 +12 Vdc Voltage Regulator
U2	LM7912 -12 Vdc Voltage Regulator
U3	LM7805 +5 Vdc Voltage Regulator
U4	LM7905 -5 Vdc Voltage Regulator
BR1,BR2	4 amp bridge rectifier
T1	25.2 volt, 3 amp center-tapped ac transformer
T2	18 volt, 2 amp center-tapped ac transformer
F1	5 amp slow-blow fuse
S1	SPST toggle switch
C1,C5, C9,C13	2000 μF electrolytic capacitor, 35 volt min.
C2,C3, C6,C7 C10,C11 C14,C15	1 μF tantalum capacitor, 35 volt min.
C4,C8 C12,C16	100 μF electrolytic capacitor, 35 volt min.
Misc.	Fuse holder, heat sink for voltage regulators, binding posts, ac cord with plug, chassis

All capacitors 10 percent tolerance.

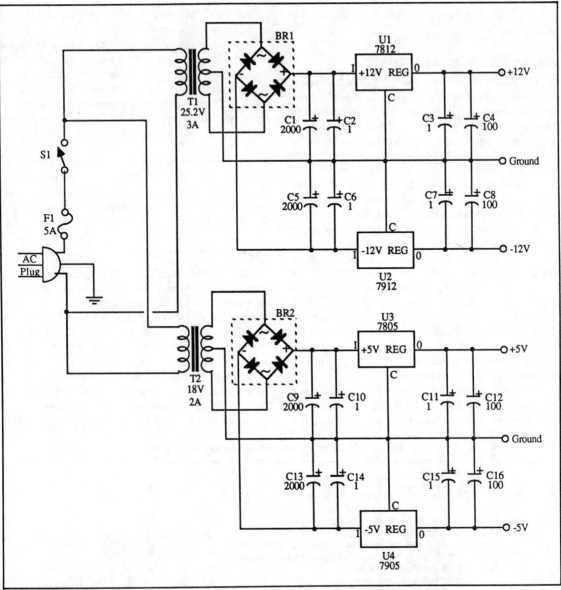

Fig. 10-3. Multiple voltage power supply—output is +12, +5, −12, and −5 volts.

has been dropped, or shows signs of visible damage. Fix any problems before plugging it in. Do not assemble the supply without a fuse, and don't use a fuse rated much higher than rated in the schematic. Defeating the fuse protection diminishes or eliminates the only true safety net in the circuit.

Chapter 11

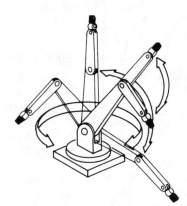

Build a Battery Recharger

Robots are power hungry and if you haven't discovered it by now, you'll soon see the advantages of using only rechargeable batteries in your robot designs (exceptions: small robots and constant-voltage reference sources). With a rechargeable battery, whether it be Ni-Cad, gel-cell, or lead-acid, you can use it once, zap new life into it, use it again, zap some more life into it, and repeat the process again and again.

In fact, most rechargeable batteries let you zap them with new life several hundred—even thousand times before wearing out. The higher initial cost of rechargeable batteries more than pays for itself after the third or fourth recharging.

Rechargeable batteries can't be revived simply by connecting them to a dc power supply, like the kind discussed in Chapter 10, "Build an Experimenter's Power Supply Station." The supply delivers too much current, and so tries to charge the battery too quickly. The result: broken battery.

Effective charging requires a special recharger. Inexpensive battery rechargers for flashlight and transistor Ni-Cad batteries are in plentiful supply, and they're cheap. Buy one. But rechargers for the larger 6 and 12 volt gel cell, lead-acid, and Ni-Cad packs are not as easy to find—or as cheap.

This chapter presents an easy-to-build recharger that delivers just the right amount of current to the battery. It has an auto shut-off circuit that prevents the battery from being overcharged. You can use the recharger for 6 or 12 volt batteries and adjust it to deliver just the right amount of current for the battery you are recharging.

ABOUT BATTERY RECHARGE RATES

In Chapter 9, "All About Batteries," you learned that the output current of batteries is stated in amp-hours. The amp-hour rating is the amount of current that the battery can deliver in a certain period of time. Theoretically, a battery with a five amp-hour (AH) rating can deliver five amps for one hour, or one amp for five hours, or any combination thereof. Figure 11-1 shows a graph that makes this a little easier to follow.

In actuality the battery can't deliver the full five amps for the 60 minute period. The 5Ah rating is arrived at by testing the battery under average load conditions for a period of 10 to 20 hours (usually 20 hours). Therefore, that 5Ah battery is theoretically designed to deliver 250 milliamps for a period of 20 hours.

The amp-hour rating is important because it also determines the amount of time and current required to prop-

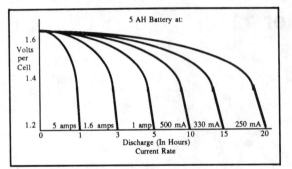

Fig. 11-1. The discharge curve of a 5 Ah battery taken over various discharge rates.

erly recharge the batteries. Unless the battery is designed otherwise, it can't be recharged faster than it was discharged. A battery that is in use for a three hour period before going dead needs at least three hours to recharge. The actual recharge time is usually two to 10 times the discharge period, depending on the battery.

Selecting the amount of current delivered to the battery is also important. Most manufacturers suggest that the current not exceed 10 percent of the amp-hour rating of the battery. A 5Ah battery, therefore, must not be charged with a current exceeding 500 mA. Some batteries, particularly car and motorcycle lead acid types, can take a quick charge, but this effectively shortens the life of the battery. It shouldn't be done often.

The voltage output of the individual cells in a battery pack differs depending on the chemical makeup of the battery. When fully recharged, Ni-Cad batteries have an output of about 1.5 volts per cell. But the nominal voltage output of the Ni-Cad taken over its useful between-charge life, is approximately 1.2 volts. When the cell reaches about 1.1 volts, or 80 percent of its rated voltage, it is considered dead and must be recharged.

Gel cell and lead-acid cells deliver about 2.3 volts when fully charged, dropping down to about 2 volts during its discharge interval (2 volts is the nominal rated capacity of lead-acid and gel cells). The cells are considered dead when the output reaches about 1.6 volts—80 percent of the rated voltage.

A SIMPLE APPROACH

If you are recharging gel-cell or lead-acid batteries, you may be able to get away with using an ac power adapter, the kind designed for video games, portable tape recorders, and other battery-operated equipment (the output must be dc). By design, these adapters limit themselves to their maximum current specifications, usually 300 to 600 mA. A 300 mA recharger can be effectively used on batteries with capacities of 2.5Ah to 5Ah. A 400 mA or 500 mA ac adapter can be used on batteries with capacities of 3.5Ah to 6.5Ah.

One problem though: you must be careful that the battery doesn't stay on charge much longer than about 12 to 16 hours. Leave it on for a day or two and the battery may be ruined. This is especially true of lead-acid batteries.

BUILDING THE CHARGER

The universal battery recharger shown in Fig. 11-2 is built around the LM317 adjustable voltage regulator IC. This IC comes in a TO-3 transistor case and should be used with a heat sink to provide for cool operation. The heat sink is absolutely necessary when recharging batteries at 500 mA or more.

The circuit works by monitoring the voltage level at the battery. During recharging, the circuit supplies a constant current output; the voltage level gradually rises as the battery is charged. When the battery nears full-charge, the circuit removes the constant current source

Table 11-1. Battery Charger Parts List.

U1	LM317 Adjustable Voltage Regulator
R1	See Text; Table 11-2
R2	220 Ω resistor
R3	470 Ω resistor
R4,R5	5K 10-turn precision potentiometers
R6	330 Ω resistor
D1	1N4004 diode
SCR1	200 volt silicon controlled rectifier (1 amp or up)
LED1	Light Emitting Diode
S1	SPST toggle switch
Misc.	Heat sink for voltage regulator, binding posts or contacts for battery under charge, dc source (use power supply presented in Chapter 10).

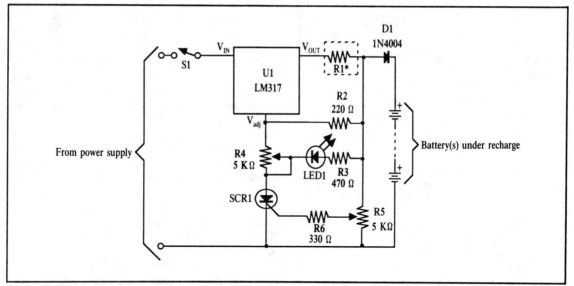

Fig. 11-2. Battery charger circuit. See text for the value or R1.

and maintains a regulated voltage to complete or maintain charging. By switching to constant voltage output, the battery can be left on charge for periods longer than recommended by the manufacturer.

Before you build the circuit you should consider the batteries you want recharged. You'll have to consider whether you will be recharging 6 volt or 12 volt batteries (or both), and the maximum current output that can be safely delivered to the battery (use the 10 percent rule or follow the manufacturer's recommendations).

Resistor R1 determines the current flow to the battery. Its value can be found by using this formula:

$$R1 = 1.25/Icc$$

where Icc is the desired charging current in mA. For example, to recharge a battery at 500 mA (0.5 amp), the calculation for R1 is 1.25/0.5 or 2.5 ohms. Table 11-2 lists common currents for recharging and the calculated values of R1. For currents under 400 mA, you can use a one watt resistor. With currents between 400 mA and 1 amp, use a two watt resistor. If the resistor you need isn't a standard value, choose the closest one to it, as long as the value is within 10 percent. If not, use two standard value resistors, in parallel or series, to equal R1.

If you'd like to make the charger selectable, wire a handful of resistors to a one-pole multi-position rotary switch, as shown in Fig. 11-3. Dial in the current setting you want.

The output terminals can be banana jacks, alligator clips, or any other hardware you desire. If you are building the circuit as a recharger nest—a place that your robot returns to when it needs a charge—you'll want to devise a touch-plate terminal assembly so your robot has only to butt itself against the charger to make contact. Table 11-1 lists parts, while Table 11-2 gives resistor values for R1.

CALIBRATING THE CIRCUIT

The circuit must be calibrated before use. You must first set R4, voltage adjust. This potentiometer sets the end-of-charge voltage. Then you must set the trip-point, which is adjusted by R5.

1. Before attaching a battery to the terminals and turning the circuit on, set variable resistors R4 and R5 to their mid ranges. With the recharger off, use a volt-ohm meter to calibrate R4, referring to Table 11-3. Adjust R4 until the ohm-meter

Table 11-2. Values For R1.

Milliamperes	Ohms
50	25.00
100	12.50
200	6.25
400	3.13
500	2.50
700	1.79
800	1.56
1 amp	1.25

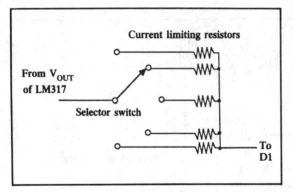

Current limiting resistors

From V$_{OUT}$ of LM317

Selector switch

To D1

Fig. 11-3. Wire up a series of resistors with a rotary switch to select a current output for the battery charger.

displays the proper resistance for the current setting you've chosen for the charger.

2. Connect a 4.7K, five watt resistor across the output terminals of the charger (this simulates a battery). Apply power to the circuit. Measure the output across the resistor. For 12 volt operation when using gel cells and lead-acid batteries, the output should be approximately 13.8 volts; for 6 volt operation, the output should be approximately 6.9 volts. If you don't get a reading, or if it is low, adjust R5. If you still don't get a reading, or if it is off the described mark, turn R4 a couple of times in either direction.

3. Connect the volt-ohm meter to between ground and the wiper of R5, the trip-point potentiometer. Turn R5 until the meter reads zero. Turn the charger off.

4. Remove the 4.7K resistor, and in its place, connect a partially discharged battery to the output terminals (be sure to use a discharged battery). Observe the correct polarity. Turn the charger on and watch the LED. It should not light.

5. Connect the volt-ohm meter across the battery terminals and measure the output voltage. Monitor the voltage until the desired output is reached—13.8 volts for 12 volt lead-acid and gel-cell batteries, and 6.9 volts for 6 volt batteries.

6. When the output is reached, adjust R5 so that the LED glows. At this point, the constant current source is removed from the output, and the battery float charges at the set voltage.

If you have both 6 and 12 volt batteries to charge, you may find yourself readjusting the potentiometers each time. A better way is to construct two battery rechargers

(the components are inexpensive) and use one at 6 volts and the other at 12 volts. Alternatively, you can wire up a selector switch that chooses between two sets of voltage adjustment and trip point pots.

At least one manufacturer of the LM317, National Semiconductor, provides extensive application notes on this and other voltage regulators. Refer to the National Semiconductor *Voltage Regulator Handbook* if you need to recharge batteries with unusual supply voltages and currents.

Depending on your battery and the tolerances of the components you use, you may need to experiment with the values of two other resistors. If the output voltage cannot be adjusted to the point you want (either high or low), increase or decrease the value of R2. If the LED never glows, or glows constantly, adjust the value of R6. Be careful not to go under about 200 ohms for R6, or the SCR may be damaged.

When recharging a battery, you know it has reached full charge when the LED goes on. To be on the safe side, turn the charger off and wait five to 10 seconds for the SCR to unlatch. Reapply power. If the LED remains lit, the battery is charged. If the LED goes out again, keep the battery on charge a little longer.

You can use the activation of the LED to signal the robot that it is finished charging. One approach is to use an infrared LED in addition to the visible red LED. Construct the charger and position the infrared LED in such a way that the robot comes in close proximity to the LED.

On the body of the robot, mount an infrared phototransistor (see schematics provided). When the phototransistor is triggered by the LED, the robot knows it has a full tank, and it can go on its merry way. Chapter 9, "All About Batteries," provides additional self-triggering circuits.

Table 11-3. Values For R4.

R1 =	6 volt (in Ω)	12 volt (in Ω)
25	1578	2950
12.5	1497	2799
6.25	1457	2724
3.13	1437	2686
2.5	1433	2679
1.79	1428	2670
1.56	1427	2668
1.25	1425	2664

Chapter 12

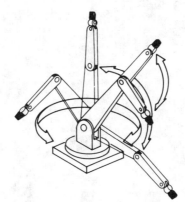

Choosing the
Right Motor for the Job

Motors are the muscles of robots. Attach a motor to a set of wheels and your robot can scoot around the floor. Attach a motor to a lever, and the shoulder joint for your robot can move up and down. Attach a motor to a roller and the head of your robot can turn back and forth, scanning its environment.

There are many kinds of motors; only a select few are truly suitable for homebrew robotics. In this chapter, we'll examine the various types of motors and how they are used.

AC OR DC?

Direct current—dc—dominates robotics. It's used as the main power source, for operating the on-board electronics, for opening and closing solenoids, and—yes, running motors. Few robots use motors designed to operate from ac, even those automatons, used as factory workers or training systems. Such plug-in systems convert the ac to dc, then distribute the dc to various subsystems in the robot. From now on in this book, we'll assume the use of dc motors.

DC motors may be the motors of choice. But that doesn't mean that all dc motors should or can be used in your robot designs. When looking for suitable motors,

be sure the ones you buy are reversible. Few robotic applications call for just unidirectional (one direction) motors. You must be able to operate the motor in one direction, stop it, and change its direction. Dc motors are inherently bidirectional, but some design limitations may prevent reversibility.

The most important factor is the commutator brushes. If the brushes are slanted, the motor probably can't be reversed. In addition, the internal wiring of some dc motors prevent them from going in any but one direction. Spotting the unusual wiring scheme is difficult, at best, even for a seasoned motor user.

The best and easiest test is to try the motor with a suitable battery or dc power supply. Apply the power leads from the motor to the terminals of the battery or supply. Note the direction of rotation of the motor shaft. Now, reverse the power leads from the motor. The motor shaft should rotate in reverse.

CONTINUOUS OR STEPPING

DC motors can be either continuous or stepping. Here is the difference:

With a *continuous motor*, like the ones in Fig. 12-1, application of power causes the shaft to rotate continu-

Fig. 12-1. Dc motor assortme

ally. The shaft stops only when the power is removed, or if the motor is stalled because it can no longer drive the load attached to it.

There are many types of continuous motors, such as permanent magnet, brushless, and variable reluctance. The differences between them aren't germain to the discussion here, but broadly speaking, permanent magnet motors are the type you'll most often use. Readers interested in the dynamics and design of various motors are referred to Appendix B, "Further Reading."

With *stepping motors*, shown in Fig. 12-2, application of power causes the shaft to rotate a few degrees, then stop. Continuous rotation of the shaft requires that the power be pulsed to the motor. As with continuous dc mo-

Fig. 12-2. Stepper motor assortment.

tors, there are sub-types of stepping motors. Permanent magnet steppers are the ones you'll likely encounter, and they are also the easiest to use.

The design differences between continuous and stepping dc motors require in-depth attention. Chapter 13, "Robot Locomotion Using DC Motors," focuses entirely on continuous motors. Chapter 14, "Robot Locomotion Using Stepper Motors," focuses entirely on the stepping variety. Although the text of these two chapters are geared toward the main drive motors of your robot, the information can be applied to motors used for other purposes, as well.

MOTOR SPECIFICATIONS

Motors carry with them numerous specifications. The meaning and purpose of some of the specifications are obvious; others aren't. Let's take a look at the primary specifications of motors—voltage, current draw, speed, and torque—and see how they relate to your robot designs.

Voltage

All motors are rated by their *operating voltage*. With small dc hobby motors, the rating is actually a range, usually 1.5 to 6 volts. Some high-quality dc motors are designed for a specific voltage, such as 12 or 24 volts. The kinds of motors of most interest to robot builders are the low-voltage variety—those that operate at 1.5 to 12 volts.

Most motors can be operated satisfactorily at voltages higher or lower than that specified. A 12 volt motor is likely to run at 8 volts, but it may not be as powerful as it could be, and it will run slower (an exception to this is stepper motors; see Chapter 14 for details). You'll find that most motors will refuse to run at voltages under 50 percent of the specified rating.

Similarly, a 12 volt motor is likely to run at 16 volts. As you may expect, the speed of the shaft rotation is increased, and the motor will exhibit greater power. Running a motor continuously at more than 30 or 40 percent its rated voltage is not recommended, however. The windings will overheat which may cause permanent damage. Motors designed for high speed operation may turn faster than their ball bearing construction allows.

If you don't know the voltage rating of a motor, you can take a guess at it by trying various voltages and seeing which one provides the greatest power with the least amount of heat dissipated through the windings—and felt on the outside of the case. You can also listen to the motor. It should not seem as if it is straining under the stress of high speeds.

Current Draw

Current draw is the amount of current, in milliamps or amps, that the motor requires from the power supply. Current draw is more important when the specification considers motor loading; that is, when the motor is turning something or doing some work. The current draw of a free-running (no load) motor can be quite low. But have the same motor spin a wheel, which in turn moves a robot across the floor, and the current draw jumps 300, 500, even 1,000 percent.

With most permanent magnet motors, which are the most popular kind, current draw increases with load. You can see this visually in Fig. 12-3. The more the motor has to work to turn the shaft, the more current is required. The load used by the manufacturer when testing the motor isn't standardized, so in your application, the current draw may be more or less than specified.

A point is reached when the motor does all the work it can do, and no more current will flow through it. The shaft stops rotating; the motor has stalled. Some motors, but not many, are rated (by the manufacturer) by the amount of current it draws when stalled. This is considered the worst case condition: the motor will never draw more than this current, unless it is shorted out, so if the system is designed to handle the stall current, it can handle anything. Motors rated by their stall current will be labeled as such. Motors designed for the military, and available through surplus, are usually rated by their stall current.

When providing motors for your robots, you should always know the approximate current draw under load. The only way you can test this is with a meter. Most volt-ohm meters can test current. Some special purpose amp meters are made just for the job.

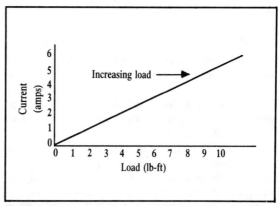

Fig. 12-3. The current draw of a motor increases proportionately to the load on the motor shaft.

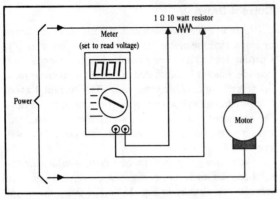

1 Ω 10 watt resistor

Meter
(set to read voltage)

Power

Motor

Fig. 12-4. How to test the current draw of a motor by measuring the voltage developed across an in-line resistor. The actual value of the resistor can vary, but should be under about 20 ohms.

Beware that some volt-ohm meters can't handle the kind of current pulled through a motor. Most digital meters, as discussed more completely in Chapter 3, "Tools and Supplies," can't deal with more than 200 to 400 milliamps of current. Even small hobby motors can draw in excess of this. Be sure your meter can accommodate current up to five or 10 amps.

If your meter cannot register this high without popping fuses or burning up, insert a 1 to 10 ohm power resistor (10 to 20 watts) between one of the motor terminals and the positive supply rail, as shown in Fig. 12-4. With the meter set on dc voltage, measure the voltage developed across the resistor.

A bit of Ohm's Law, $I = E/R$ (I is current, E is voltage, R is resistance) reveals the current draw through the motor. For example, if the resistance is 10 ohms and the voltage is 2.86 volts, the current draw is 286 mA. You can watch the voltage go up by loading the shaft of the motor.

Speed

The *rotational speed* of a motor is given in revolutions per minute (rpm). Most continuous dc motors have a normal operating speed of 4,000 to 7,000 rpm, although some special purpose motors, such as those used in tape recorders and computer disk drives, operate as slow as 2,000 to 3,000 rpm.

For just about all robotic applications, these speeds are much too high. The speed must be reduced to no more than 150 rpm (even less for motors driving arms and grippers) with the use of a gear train. Some reduction can be obtained using electronic control, as described in Chapter 13, but such control is designed to make fine-tune

speed adjustments, not reduce the rotation of the motor from 5,000 rpm to 50 rpm. See the later sections in this chapter for more details on gear trains, and how they are used.

Note that the speed of stepping motors is not rated in rpm, but pulses (or steps) per second. The speed of a stepper motor is a function of the number of steps required to make one full revolution plus the number of steps applied to the motor each second. As a comparitors operate at the equivalent of a minimum of 100 to 140 rpm. See Chapter 14, "Robot Locomotion with Stepper Motors," for more information.

Torque

Torque is the force the motor exerts upon its load. The higher the torque, the larger the load can be, and the faster the motor will spin under that load. Reduce the torque, and the motor slows down, straining under the workload. Reduce the torque even more, and the load may prove too demanding for the motor. The motor will stall to a grinding halt, eating up current (and putting out a lot of heat) in doing so.

Torque is perhaps the most confusing design aspect of motors, not because there is anything inherently difficult about it, but motor manufacturers have yet to settle on a standard means of measurement. Motors made for industry are rated one way; motors for the military in another way.

At its most basic level, torque is measured by attaching a lever to the end of the motor shaft, and a weight or gauge on the end of that lever, as depicted in Fig. 12-5.

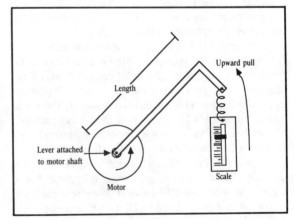

Upward pull

Length

Lever attached
to motor shaft

Scale

Motor

Fig. 12-5. The torque of a motor is measured by attaching a weight or scale to the end of a lever, and mounting the lever to the motor shaft.

That lever can be any number of lengths: one centimeter, one inch, or one foot. Remember this because it plays an important role in torque measurement. The weight can either be a hunk of lead, or more commonly, a spring-loaded scale (as shown in the figure). Turn the motor on and it turns the lever. The amount of weight it lifts is the torque of the motor. There is more to motor testing than this, of course, but it'll do for the moment.

Now for the ratings game. Remember the length of the lever? That length is used in the torque specification. If the lever is one inch long, and the weight successfully lifted is two ounces, then the motor is said to have a torque of two ounce-inches (or oz-in). Some people reverse the "ounce" and "inches" and come up with "inch-ounces." It makes no difference; the measurement is the same.

The length of the lever usually depends on the unit of measurement given for the weight. When the weight is in grams, the lever is in centimeters (gm-cm). When the weight is in ounces, as already seen, the lever used is in inches (oz-in). Finally, when the weight is in pounds, the lever used is in feet (lb-ft). Like the ounce-inch measurement, gram-centimeter and pound-foot specifications can be reversed—"centimeter-gram" or "foot-pound." Once again, it really makes no difference other than to confuse you!

Note that these easy-to-follow conventions aren't always used. Some motors may be rated by a mixture of the standards—ounces and feet, or pounds and inches. Here's a test that proves the point. Is there any difference between a 24 lb-in motor and a two lb-ft motor? No. The motor has the same torque. To arrive at the 24 lb-in specification, the two pound load was multiplied by the number of inches in a foot. You can use similar math to convert grams into ounces, ounces into pounds, centimeters into inches, and so forth. How would you convert 300 oz-in to pound-feet?

More on Torque

Most motors are rated by their *running torque*, or the force they exert as long as the shaft continues to rotate. For robotic applications, it's the most important rating, because it determines how large the load can be and still guarantee that the motor turns. Conducting running torque tests vary from one motor manufacturer to another, so results can differ. The tests are impractical to duplicate in the home shop, unless you have an elaborate slip-clutch test stand, precision scale, and sundry other test jigs.

If the motor(s) you are looking at don't have running torque ratings, you must estimate their relative strength by mounting them on a makeshift wood or metal platform, attaching wheels to them, and having them scoot around the floor. If the motor supports the platform, start piling on weights. If the motor continues to operate with, say, 40 or 50 pounds of junk on the platform, you've got an excellent motor for driving your robot (such a robot would probably have a torque of 6 to 15 lb-ft).

Some motors you may test aren't designed for hauling heavy loads, but may be suitable for operating arms, grippers, and other mechanical components. You can test the relative strength of these motors by securing them in a vise, then attaching a large pair of Vise Grips or other lockable pliers to them. Use your own hand as a test jig, or rig one up with fishing weights. Determine the rotational power of the motor by applying juice to the motor and seeing how many weights it can successfully handle.

Such crude tests make more sense if you have a "standard" by which to judge others. If you've designed a robotic arm before, for example, and are making another one, test the motors successfully used in your prototype. If subsequent motors fail to match or exceed the test results of the standard, you know they are unsuitable for the test.

Another torque specification, *stall torque*, is sometimes provided by the manufacturer instead of, or additional to, running torque (this is especially true of stepping motors). Stall torque is the force exerted by the motor when the shaft is clamped tight. There is an indirect relationship between stall torque and running torque, and although it varies from motor to motor, you can use the stall torque rating when selecting candidate motors for your robot designs. In most cases, the stall torque will not deviate more than 10 or 20 percent from the running torque specification.

GEARS AND GEAR REDUCTION

We've already examined how the normal running speed of motors is far too fast for most robotics applications. Locomotion systems need motors with running speeds of 75 to 150 rpm. Any faster and the robot will skim across the floor and bash into walls and people. Arms, gripper mechanisms, and most other mechanical subsystems need even slower motors. The motor to position the shoulder joint of an arm needs to have a speed of less than 20 rpm; five to eight rpm is even better.

There are two general ways to greatly decrease motor speed: build a bigger motor (impractical) or add gear reduction. Gear reduction is used in your car, in your record player, on your bicycle, in the washing machine and dryer, and in countless other motor-operated mechanisms.

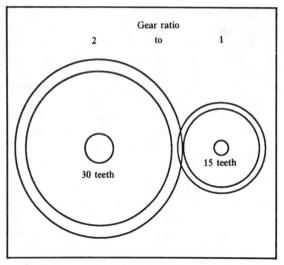

Fig. 12-6. A representation of a 2:1 gear reduction ratio.

Gears 101

Gears perform two important duties. First, they can make the distance of travel applied to one gear greater or lesser than the distance of travel of another gear connected to it. They also increase or decrease power, depending on how the gears are oriented. Gears can also serve to simply transfer force from one place to another, but this is a minor application in comparison to the others.

Gears are actually round levers, and it may help to explain the function of gears by first examining the basic mechanical lever. Place a lever on a fulcrum so that the majority of the lever is on one side. Push up on the long side and the short side moves in proportion. Although you may move the lever several feet, the short side is moved only a few inches. Also note that the force available on the short end is proportionately larger than the force applied on the long end. You use this small wonder of physics-fact when digging a rock out of the ground with your shovel, or when jacking up your car to replace a tire.

Now back to gears. Attach a small gear to a large gear, as shown in Fig. 12-6. The small gear is directly driven by a motor. For each revolution of the small gear, the large gear turns one half a revolution. Expressed another way, if the motor and small gear turn at 1,000 rpm, the large gear turns at 500 rpm. The gear ratio is said to be 2:1.

Note that another important thing happens, just as it did with the lever and fulcrum. Decreasing the speed of the motor also increased its torque. The power output is approximately twice the input. Some power is lost in the reduction process due to the friction of the gears.

If the drive and driven gears are the same size, the rotation speed is neither increased nor decreased, and the torque is not affected (apart from small frictional losses). You'd use same-size gears in robotics design to transfer motive power from one shaft to another, such as driving a set of wheels at the same speed and in the same direction.

Establishing Gear Reduction

Gears are an old invention, going back to ancient Greece. Today's gears are more refined, and available in all sorts of styles and materials, but are based on the

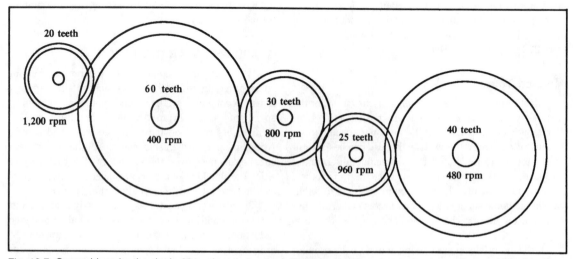

Fig. 12-7. Gears driven by the single 20 tooth gear (on the left) rotate at different speeds, depending on their diameter.

88

old Greek design of teeth from the two mating gears meshing into one another. The teeth provide an active physical connection between the two gears, and force is transferred from one gear to another.

Gears with the same size teeth are usually expressed not by their physical size, but by the number of teeth around the circumference. In the example, the small gear contains 15 teeth; the large gear, 30 teeth.

You can string a number of gears one after the other, all with varying numbers of teeth (see Fig. 12-7). Attach a tachometer to the hub of each gear, and you can measure its speed. You'll find that:

- The speed always *decreases* when going from small to large gear.
- The speed always *increases* when going from large to small gear.

There are plenty of times when you need to reduce the speed of a motor from 5,000 rpm to 50 rpm. That kind of speed reduction requires a reduction ratio of 100:1. That would require a drive gear of, say, 10 teeth and a driven gear of 1,000 teeth. That 1,000 tooth gear would be quite large, bigger than the drive motor itself.

You can reduce the speed of a motor in steps by using the arrangement shown in Fig. 12-8. Here, the drive gear turns a larger gear, which in turn has a smaller gear permanently attached to its shaft. The small hub gear turns the driven gear for the final output speed. You can repeat this process over and over again until the output speed is but a tiny fraction of the input speed. This is the arrangement most often used in motor gear reduction systems.

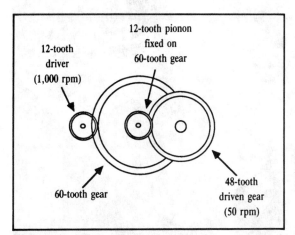

Fig. 12-8. True gear reduction is achieved by ganging gears on the same shaft.

Using Motors with Gear Reduction

It's always easiest to use dc motors that already have a gear reduction box built onto them, such as the motor in Fig. 12-9. That saves you from finding a gear reducer that fits the motor and application, and attaching it yourself. When selecting gear motors, you'll be most interested in the output speed of the gear box, not the actual running speed of the motor. Note, as well, that the running and stall torque of the motor will be greatly increased. Make sure that the torque specification on the motor is for the output of the gear box, not the motor itself.

With most gear reduction systems, the output shaft is opposite (but usually off center) to the input shaft. With other boxes, the output and input are on the same side of the box. When the shafts are offset to one another, the reduction box is said to be a right-angle drive. If you have the selection available to you, choose the kind of gear reduction that best suits the design of your robot. I have found that the shafts on opposite sides is the all-around best choice. Right-angle drives also come in handy, but they usually carry high pricetags.

You'll need to add reduction boxes, such as the model in Fig. 12-10, or make your own, when using motors without built-in gear reduction. Although it is possible to do these things, there are many pitfalls.

- Shaft diameters of motors and ready-made gear boxes may differ, so you must be sure that the motor and gear box mate.
- Separate gear reduction boxes are hard to find. Most must be cannibalized from salvage motors. Old ac motors are one source of surplus boxes.
- When designing your own gear reduction box, care must be taken to assure the same hub size for all the gears, and that meshing gears exactly match each other. This is not always easy, as discussed later in this chapter.
- Machining the gear box takes precision, since even a small error can cause the gears to mesh improperly.

Anatomy of a Gear

Gears are made up of teeth, but the teeth can come in any number of styles, sizes, and orientations. *Spur gears* are most common. The teeth surround the outside edge of the gear, as shown in Fig. 12-11. Spur gears are used when the drive and driven shafts are parallel. *Bevel* gears have teeth on the surface of the circle, rather than the

Fig. 12-9. A motor with an enclosed gear box.

edge. They are used to transmit power to perpendicular shafts. *Miter* gears serve a similar function but are designed so that no reduction takes place. Spur, bevel, and miter gears are reversible. That is, unless the gear ratio is very large, you can drive the gears from either end of

the gear system, thus increasing or decreasing the input speed.

Worm gears transmit power perpendicularly, like bevel and miter gears, but their design is unique. The worm (or lead screw) resembles a threaded rod. The rod

Fig. 12-10. A gear-reduction box, originally removed from an open-frame ac motor. The input and output shafts are on the same side.

90

Fig. 12-11. Spur gears. These gears are made of nylon and have aluminum hubs. Metal hubs are better when the gear is secured to the shaft with a set screw.

provides the power. As it turns, the threads engage a modified spur gear (the modification takes into consideration the cylindrical shape of the worm).

Worm gear systems are specifically designed for large-scale reduction. The gearing is not usually reversible; you can't drive the worm by turning the spur gear. This is an important point, because it gives worm gear systems a kind of automatic locking capability. Worm gears are particularly well suited for arm mechanisms where you want the joints to remain where they are. With another gear system, the arm may droop or sink back due to gravity once the power from the drive motor is removed.

Rack gears are like spur gears unrolled into a flat rod. They are primarily intended to transmit rotational motion to linear motion. Racks have a kind of self-locking characteristic, as well, but it's not as strong as that found in worm gears.

The size of gear teeth is expressed as *pitch*, which is calculated by counting the number of teeth on the gear and dividing it by the diameter of the gear. For example, a gear that measures two inches and has 48 teeth has a tooth pitch of 24.

Common pitches are 12 (large), 24, 32, 48. Some gears have extra fine 64 pitch teeth, but these are usually confined to miniature mechanical systems, such as radio controlled models. Odd-size pitches exist of course, as do metric sizes, so you must be careful when matching gears that the pitches are exactly the same. Otherwise, the gears will not mesh properly and may cause excessive wear.

The degree of slope of the face of each tooth is called the *pressure angle*. The most common pressure angle is 20 degrees, although some gears, particularly high-quality worms and racks, have a 14 1/2-degree pressure angle. Textbooks claim that you should not mix two gears with different pressure angles, even if the pitch is the same, but it can be done. Some excessive wear may result, because the teeth aren't meshing fully.

The orientation of the teeth on the gear can differ. The teeth on most spur gears is perpendicular to the edges of the gear. But the teeth can also be angled, as shown in Fig. 12-12, in which case it is called a helical gear.

There are a number of other unusual tooth geometries in use, including double-teeth, where there are two rows of teeth offset to one another, and herringbone, where there are two sets of helical gears at opposite angles. These gears are designed to reduce the backlash phenomena: The space (or "play") between the teeth when meshing can cause the gears to rock back and forth.

MOUNTING THE MOTOR

Every motor requires a different mounting arrangement. Things are made easier for you when the motor has its own mounting hardware or holes. You can use

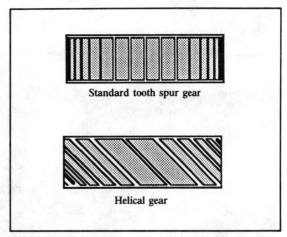

Fig. 12-12. Standard tooth spur gear versus diagonal helical spur gear. The latter is used to decrease backlash—the play inherent when two gears mesh. Some helical gears are also made for diverting the motion at right angles.

these to mount the motor in your robot. Remember that Japanese-made motors often have metric threads, so be sure to use the proper-size bolt.

Other motors may not be as cooperative. Either the mounting holes are in a position where they don't do you much good, or the motor is completely devoid of any way to secure it to your robot. You can still successfully mount these motors by using an assortment of clamps, brackets, wood blocks, and homemade angles.

For example, to secure the motor in Fig. 12-13, mounting brackets where fashioned using 6-inch galvanized iron mending "T" plates. A large hole was drilled for the drive shaft and gear to poke through, and the two halves of the mounting bracket were joined together with nuts, bolts, and spacers. The bracket was then attached to the frame of the robot using angle irons and standard hardware. This motor arrangement was made a little more difficult with the addition of a drive gear and sprocket. Construction time for each motor bracket was about 90 minutes. You can read more about this particular design in Chapter 16, "Build a Six-Legged Walking Robot."

Another example is shown in Fig. 12-14. Here, the motor has mounting holes on the end, by the shaft, but these holes are in the wrong position for the design of

the robot. Two commonly available flat corner irons were used to mount the motor. This is just one approach; a number of other mounting schemes might have worked satisfactorily as well.

If the motor lacks mounting holes, you'll need to use clamps to hold it in place. U-bolts, available at the hardware store, are excellent solutions. Choose a U-bolt large enough to fit around the motor. The rounded shape of the bolt is perfect for motors, which typically have round casings. If desired, you can make a holding block out of wood to keep the bottom of the motor from sliding. Cut the wood to size and round it out with a router, rasp, or file, to match the shape of the motor casing.

CONNECTING TO THE SHAFT

Connecting the shaft of the motor to a gear, gear box, wheel, lever, or other mechanical part is probably the most difficult task of all. Motor shafts come in many different sizes, and because most—if not all—of the motors you'll use will come from surplus outlets, the shaft design may be peculiar to the specific application for which the motor was designed.

Common shaft sizes are 1/16-inch and 1/8-inch for

Fig. 12-13. One approach to mounting a motor to a robot. The motor is sandwiched inside two large hardware plates, and secured to the frame of the motor with angle irons.

Fig. 12-14. Another approach to mounting a motor to a robot. Flat corner irons secure the motor flange to the frame.

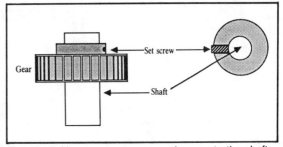

Fig. 12-15. Use a screw to secure the gear to the shaft.

small hobby motors, and 1/4-inch or 5/16-inch for larger motors and gear boxes. Gear hubs are generally 1/2-inch or 5/8-inch, so you'll need to find reducing bushings at an industrial supply store. Surplus is also a good source. The same goes for wheels, sprockets (for chains and timing pulleys), and bearings.

Attaching things like gears and sprockets usually requires that the gear or sprocket be physically secured to the shaft by way of a set screw, as depicted in Fig. 12-15. Most better-made gears and sprockets have the set screws in them, or have provisions for them. If the gear or sprocket has no set screw, and there is no hole for one, you'll have to drill and tap the hole for the screw.

There are two alternatives when you can't use a set screw. The first method is to add a spline, or key, to secure the gear or sprocket to the shaft. This requires some careful machining, as you must make a slot for the spline in the shaft, as well as the hub of the gear/sprocket.

Another method is to thread the gear shaft and mount the gear or sprocket using nuts and split lockwashers (the split in the washer provides compression that keeps the

assembly from working loose). Shaft threading is also sometimes necessary when attaching wheels. I've successfully used both alternative methods, but have found that threading the shaft is easier. Threading requires that the shaft be locked so it won't turn, which can be a problem with some motors. Also, exercise care that the shavings from the threading die do not fall into the motor.

Attaching two shafts to one another is another common, but not insurmountable, problem. The best approach is to use a coupler. You tighten the coupler to the shaft using set screws. Couplers are available from industrial supply houses and can be expensive, so shop carefully. Some couplers are flexible; that is, they give if the two shafts aren't perfectly aligned. These are the best, considering the not-too-close tolerances inherent in home-built robots. Some couplers are available that accept two shafts of different sizes.

You can fashion your own couplers, if need be, out of aluminum or rigid tube stock brass. A piece of 1/4-inch I.D - 1/2-inch O.D. tube stock is a good choice. Use a tube with a smaller inside diameter when attaching smaller shafts. To make a coupler, cut the tube to 1/2-inch or 1-inch lengths. If the hole isn't the right size, drill it out using a drill press or drill attachment to assure a perfect, centered hole. If anything, make the hole on the small size. Eliminate slop where possible.

Drill two smaller holes on either end of the tube with a #43 drill (the hole need go through only one wall). Using a 4/40 tap, tap the hole, and insert a 4/40 set screw. You can use a smaller tap and set screw if desired. When attaching two shafts of different sizes, drill one end of the coupler to fit one shaft, and the other end of the coupler to fit the other shaft.

Chapter 13

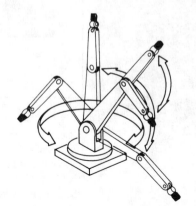

Robot Locomotion with DC Motors

Continuous dc motors are the mainstay of robotics. A surprisingly small motor, when connected to wheels through a gear reduction system, can power a 50—even 100—pound robot seemingly with ease. A flick of a switch, a click of a relay, or a tick of a transistor, the motor stops in its tracks and turns the other way. A simple electronic circuit enables quick and easy control over speed—from a slow crawl to a fast sprint.

This chapter shows you how to apply continuous (as opposed to stepping) dc motors to power your robots. The emphasis is on using motors to propel a robot across your living room floor, but you can use the same control techniques for any motor application, including gripper closure, elbow flexion, and sensor positioning.

FUNDAMENTALS

Next to the batteries, the drive motors are probably the heaviest component in your robot. You'll want to carefully consider where the drive motor(s) are located, and how the weight is distributed throughout the base. Most robot designs use two identical motors to spin two wheels. These wheels provide forward and backward momentum, as shown in Fig. 13-1, as well as left and right steering. By stopping one motor, the robot turns in that direction.

By reversing the motors relative to one another, the robot turns by spinning on its wheel axis. You use the forward-reverse movement to make hard or sharp right and left turns.

The wheels—and hence the motors—can be placed just about anywhere along the length of the platform. If placed in the middle, as shown in Fig. 13-2, two casters must be added to either end of the platform to provide stability. Since the motors are in the center of the platform, the weight is more evenly distributed across it. The battery or batteries can be placed above the center line of the wheel axis, maintaining the even distribution.

The wheels can also be positioned on one end of the platform. In this case, one swivel caster is added on the other end to provide stability and a pivot for turning. Obviously, the weight is now more concentrated on the motor side of the platform. The battery or batteries can be placed on the caster side, helping to redistribute the weight. Should the weight of the platform be concentrated at the wheels, maneuverability and stability may be diminished.

Travel Speed

The speed of the drive motors is one element that

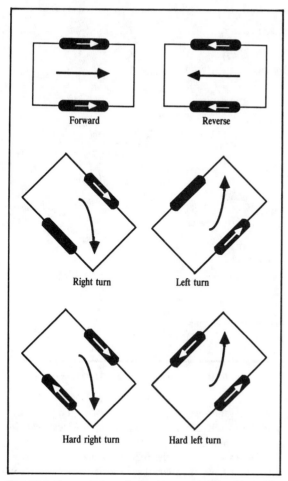

Fig. 13-1. Two motors and wheels, mounted on either side of the robot, can move the automaton forward and backward, and left or right.

partly determines the travel speed of your robot. The other element is the diameter of the wheels. For most applications, the speed of the drive motors should be under 130 rpm (under load). With average size wheels, the resultant travel speed will be approximately four feet per second. That's actually pretty fast. A better travel speed is one to two feet per second, which requires smaller diameter wheels, a slower motor, or both.

How do you calculate travel speed of your robot? Follow these steps:

1. Divide the rpm speed of the motor by 60. The result is the revolutions of the motor per second (rps). A 100 rpm motor runs at 1.66 rps.
2. Multiply the diameter of the drive wheel by pi,

or approximately 3.14. That yields the circumference of the wheel. A seven-inch wheel has a circumference of about 21.98 inches.
3. Multiply the speed of the motor (in rps) by the circumference of the wheel. The result is the number of inches covered by the wheel in one second.

With a 100 rpm motor and seven-inch wheel, the robot will travel at a top speed of 35.168 inches per second, or just under three feet. That's about two miles per hour!

You can readily see that you can slow down a robot by decreasing the size of the wheel. By reducing the wheel to five inches instead of eight, the same 100 rpm motor will propel the robot at about 25 inches per second. By reducing the speed to, say, 75 rpm, the travel speed is decreased even more, to 19.625 inches per second. Now that's more reasonable. Bear in mind that the actual travel speed, once the robot is all put together, may be lower than this. The heavier the robot, the larger the load on the motors, so the slower they will turn.

DIRECTION CONTROL

It's fairly easy to change the rotational direction of a dc motor. Simply switch the power lead connections to the battery, and the motor turns in reverse. The small robots discussed in Chapters 5, 6, and 7 dealt with using a double-pole double-throw (DPDT) switch to control the direction of drive motors. Two such switches were used, one for each of the drive motors. The wiring diagram is duplicated in Fig. 13-3 for your convenience. The DPDT

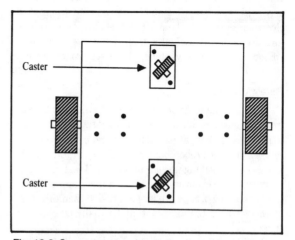

Fig. 13-2. Support casters located underneath a two-wheel drive robot platform.

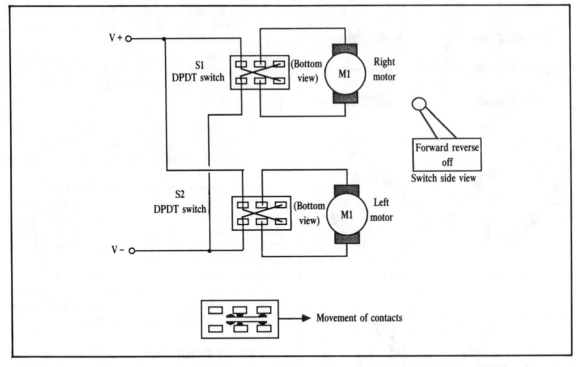

Fig. 13-3. The basic wiring diagram for controlling the robot drive motors. Note that the switches are DPDT, spring return to center off.

switches used here have a center-off position. When in the center position, the motors receive no power so the robot does not move.

You can use the direction control switch for experimenting, but you'll soon want to graduate to more automatic control of your robot. Fortunately, that's not hard, either. There are a number of ways to accomplish electronic or electrically-assisted direction control of motors. None are really better than the others; they have advantages and disadvantages. Let's see what they are.

Relay Control

Perhaps the most straightforward approach to controlling motors is to use relays. It may seem rather daft to install something as old-fashioned and cumbersome as relays in a hi-tech robot, but you should try the other techniques before making up your mind. You'll find that while relays may wear out in time (after a few hundred thousand switchings), they are less expensive than the other methods, easier to implement, and actually take up less space.

Basic on/off motor control can be accomplished with a single-pole relay. Rig up the relay so that current is bro-

ken when the relay is not activated. Turn on the relay, and the switch closes, thus completing the electrical circuit. The motor turns.

How you activate the relay isn't the important consideration here. You could control it by a pushbutton switch, but that's no better than the manual switch method above. Relays can easily be driven by digital signals. Figure 13-4 shows the complete driver circuit for a relay-controller motor. Logical 0 (LOW) turns the relay off; logical 1 (HIGH) turns it on. The relay can be operated from any digital gate, including a computer or microprocessor port. Chapters 32 and 33 deal more extensively in using a computer for external control. Use this wiring diagram with the information supplied in the forthcoming chapters.

Table 13-1. On-Off Relay Control Parts List.

RL1	SPDT relay, contacts rated 2 amps or more
Q1	2N2222 npn transistor
R1	1KΩ resistor
D1	1N4003 diode

Table 13-2. Direction Relay Control Parts List.

RL2	DPDT relay, contacts rated 2 amps or more
Q2	2N2222 npn transistor
R1	1KΩ resistor
D1	1N4003 diode

All resistors 5 or 10 percent tolerance, 1/4-watt.

Controlling the direction of the motor is only a little more difficult. This requires a DPDT relay, wired in series after the on/off relay described above (see Fig. 13-5). With the contacts in the relay in one position, the motor turns clockwise. Activate the relay and the contacts change positions, turning the motor counterclockwise. Again, you can easily control the direction relay with digital signals. Logical 0 makes the motor turn in one direction (let's say forward), and logical 1 makes the motor turn in the other direction. Both on/off and direction relay control is shown combined in Fig. 13-6.

You can quickly see how to control the operation and direction of a motor using just two data bits from a computer. Since most robot designs incorporate two drive motors, you can control the movement and direction of your robot with just four data bits.

When selecting relays, make sure the contacts are rated for the motors you are using. All relays carry contact ratings, and it will vary from a low of about .5 amp to over 10 amps, at 125 volts. Higher capacity relays are larger and may require bigger transistors to trigger them (the very small reed relays can often be triggered by digital control without the addition of the transistor). For most applications, you don't need a relay rated higher than

two or three amps. The parts lists for these controls are in Tables 13-1 and 13-2.

Transistor

Transistors provide true solid-state control of motors. For the purpose of motor control, you use the transistor as a simple switch.

There are two common ways to implement transistor control of motors. One way is shown in Fig. 13-7. Here, two transistors do all the work. The motor is connected so that when one transistor is switched on, the shaft turns clockwise. When the other transistor is turned on, the shaft turns counterclockwise. When both transistors are off, the motor stops turning. Notice that this setup requires a dual-polarity power supply. The schematic calls for a 6 volt motor, and a +6 volt and −6 volt power source.

As with relay control, you can switch the transistors on and off easily with digital control. The only caveat to remember here is that both transistors should never be on at the same time. This condition will surely cause damage to the transistors.

Note the resistor used to bias the base of each transistor. These are necessary to prevent the transistor from pulling excessive current from the gate controlling it (computer port, logic gate, whatever). Without the resistor, the gate would overheat and be destroyed.

The actual value of the resistor depends on the voltage and current draw of the motor, as well as the characteristics of the particular transistors used. For ballpark computations, the resistor is usually in the 1K to 3K ohm range. You can calculate the exact value of the resistor using Ohm's Law, taking into consideration the gain and

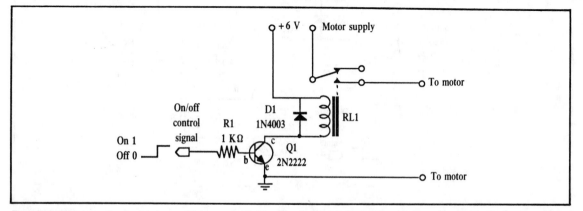

Fig. 13-4. Using a relay to turn a motor on and off. The input signal is TTL compatible, and can be directly controlled by a microprocessor or computer.

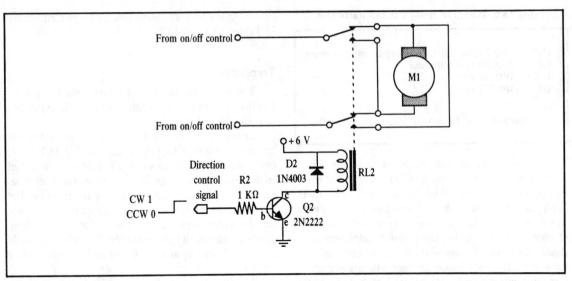

Fig. 13-5. Using a relay to control the direction of a motor. The input signal is TTL compatible, and can be directly controlled by a microprocessor or computer.

current output of the transistor, or you can experiment until you find a resistor value that works. Start high and work down, noting when the controlling electronics seem to get too hot. Don't go below 1K.

Despite the flagrant disregard for textbook electronic etiquette with the latter approach, I have had good luck

with both ways, and have yet to burn out a logic gate, transistor, or motor (but my fingers have gotten burned in the process!). Use whatever method makes you the most happy and comfortable.

Another transistor scheme is shown in Fig. 13-8. This is a familiar sight to electronic buffs. The "H" network

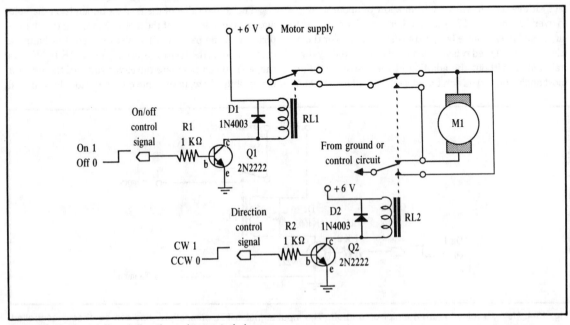

Fig. 13-6. Both on/off and direction relay controls in one.

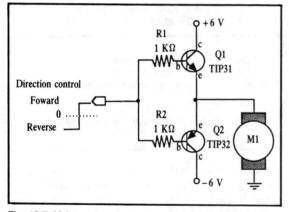

Fig. 13-7. Using a complementary pair of transistors to control the direction of a motor. Note the double-ended power supply.

ing resistors and avoid turning on all transistors at once. They will overheat and be damaged.

The choice of transistors really makes little overall difference, as long as they comply to some general guidelines. First, they must be capable of handling the current draw demanded by the motors. Most drive motors draw about one to three amps continuous, so the transistors you choose should be able to handle this. This immediately rules out the small signal transistors, which are rated for no more than 500 or 600 mA.

A good npn transistor for medium-duty applications is the TIP31, which comes in a TO-220 style case. It's pnp counterpart is the TIP32. Both of these transistors are universally available. Use them with suitable heat sinks. For high power jobs, the npn transistor that's almost universally used is the 2N3055 (get the version in the TO-3 case; it handles more power). Its close pnp counterpart is the MJ2955 (or 2N2955). Both transistors can handle up to 10 amps (115 watts), when used with a heat sink, such as the one in Fig. 13-9.

The driving transistors should be located off the main circuit board, preferably directly on a large heat sink or at least on a heavy board with clip-on or bolt-on heat sinks

is wired in such a way that only two transistors are on at a time. When transistor 1 and 4 are on, the motor turns in one direction. When transistor 2 and 3 are on, the motor spins the other way. When all transistors are off, the motor remains still. Again, note the addition of the bias-

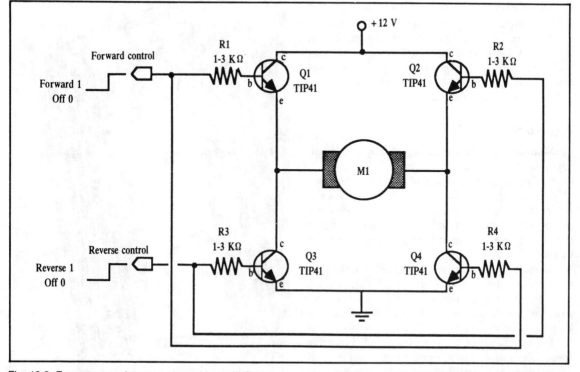

Fig. 13-8. Four npn transistors connected in an "H" pattern can be used to control the direction of a motor. The power supply is single-ended.

99

Table 13-3. Dual Transistor Motor Direction Control Parts List.	
Q1	TIP31 npn power transistor
Q2	TIP32 pnp power transistor
R1,R2	1K transistor
Misc.	Heat sinks for transistors
All resistors 5 or 10 percent tolerance, 1/4-watt.	

Table 13-4. Four-Transistor Motor Direction Control Parts List.	
Q1-Q4	TIP41 npn power transistor
R1-R4	1K transistor
Misc.	Heat sinks for transistors
All resistors 5 or 10 percent tolerance, 1/4-watt.	

attached to the transistors. Use the proper mounting hardware when attaching transistors to heat sinks.

Remember that with most power transistors, the case is the collector terminal. This is particularly important when there are more than one transistor on a common heat sink and they aren't supposed to have their collectors connected together. It's also important when that heat sink is connected to the grounded metal frame of the robot. You can avoid any extra hassle by using the insulating washer provided in most transistor mounting kits.

The power leads from the battery and to the motor should be 12 to 16 gauge wire. Use solder lugs or crimp-on connectors to attach the wire to the terminals of T0-3 style transistors. Don't tap off power from the electronics for the driver transistors; get it directly from the battery or main power distribution rail. See Chapter 9, "All About Batteries," for more details about robot power distribution systems. Parts for the two and four transistor direction controls are in Tables 13-3 and 13-4.

Power MOSFET

Wouldn't it be nice if you could use a transistor without bothering with biasing resistors? Well, you can, as long as you use a special brand of transistor, the power MOSFET. The MOSFET part stands for metal oxide semiconductor field effect transistor. The power part means you can use them for motor control without worrying about burning them, or the controlling circuitry, up in smoke.

MOSFET's look a lot like transistors, but there are a few important differences. First, like CMOS ICs, it is entirely possible to damage a MOSFET device by zapping it with static electricity. When handling it, always keep the protective foam around the terminals.

Further, the names of the terminals are different than transistors. Instead of base, emitter, and collector, MOSFET's have a gate, source, and drain. Broadly speaking, the gate is the same as the base of a transistor, and the

source and drain are the same as the emitter and collector, respectively. You can easily damage a MOSFET by connecting it in the circuit improperly. Always refer to the pin-out diagram before wiring the circuit, and double-check your work.

A commonly available power MOSFET is the IRF-511. It comes in a T0-220 style transistor case and can control several amps of current (when on a suitable heat sink). A practical circuit that uses MOSFET's is shown in Fig. 13-10. Note the similarity between this design and the transistor design on Fig. 13-8.

Fig. 13-9. Power transistors mounted on an aluminum heat sink.

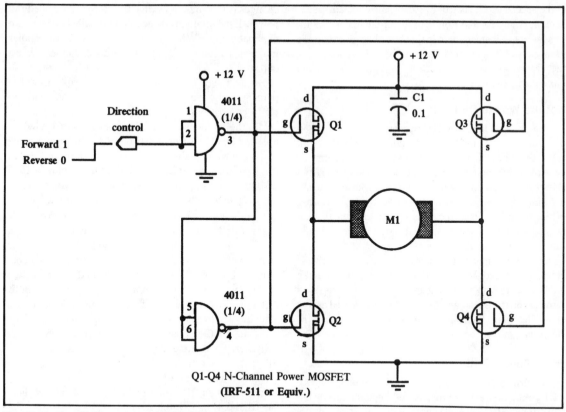

Fig. 13-10. Four N-channel power MOSFET transistors in an "H" pattern can be used to control the direction of a motor. In a circuit application such as this MOSFET devices do not require biasing resistors, as do standard transistors.

Here, a 4011 NAND CMOS gate has been added to provide positive-action control. When the control signal is LOW, the motor turns clockwise. When the control signal is HIGH, the motor turns counterclockwise. We'll elaborate on this design, the most flexible of the bunch, in the next section.

About the only real problem with MOSFET's is their price. You can purchase 2N3055 power transistors for less than 75 cents if you look long enough, but it's hard to find power MOSFET's for under $2.00. Seeing how you need four of them for each motor, it's obvious that costs can mount quickly. The MOSFET direction-control parts list is in Table 13-5.

MOTOR SPEED CONTROL

There will be plenty of times when you'll want the motors in your robot to go a little slower, or perhaps track a pre-defined speed. Speed control with continuous dc motors is a science in its own, and there are literally hundreds of ways to do it. If you are lucky enough to grab hold

Table 13-5. MOSFET Motor Direction Control Parts List.

Q1-Q4	IRF511 (or equiv) power MOSFET
U1	4011 CMOS Quad NAND Gate IC

Table 13-6. MOSFET Motor Speed/Direction Control Parts List.

Q1-Q4	IRF511 (or equiv) power MOSFET
U1	4011 CMOS Quad NAND Gate IC
R1	330 Ω resistor
R2	1 megohm resistor
R3	100K potentiometer
C1	0.033 ceramic capacitor
LED1	Light Emitting Diode

All resistors 5 or 10 percent tolerance, 1/4-watt; all capacitors 10 percent tolerance, rated 35 volts or higher.

of the special speed control ICs that some companies sell, such as Signetics and Sprauge, by all means use them. I chose not to include them in this book because they are extremely difficult to find. Rather, the following designs use readily available components and offer modest accuracy and capability.

Not the Way to Do It

Before exploring the right ways to control the speed of motors, let's examine the way not to do it. Nearly all robot experimenters, myself included, first attempt to vary the speed of a motor using a potentiometer. While this scheme certainly works, it wastes a lot of energy. Turning up the resistance of the pot decreases the speed of the motor, but it also causes excess current to flow through the pot. That current creates heat and abnormally draws precious battery power.

Another, similar approach is shown in Fig. 13-11. Here, a transistor is added to the basic circuit to make it a little more responsive, but again, excess current flows through the transistor, and the energy is dissipated as lost heat. There are, fortunately, far better ways of doing it. Read on.

Basic Speed Control

Figure 13-12 shows a schematic, based on the MOSFET circuit above, that provides rudimentary speed control. The 4011 NAND gate acts as an astable multivibrator, a pulse generator. By varying the value of R3, you increase or decrease the duration of the pulses emitted by the gates of 4011. The longer the duration of the pulses, the faster the motor, because it is getting full power for a longer period of time. The shorter the duration of the pulses, the slower the motor.

Notice that the power or voltage delivered to the motor does not change, as it does with the pot-only or pot-transistor scheme described earlier. The only thing that

changes is the amount of time the motor is provided full power. Incidentally, this technique is called pulse width modulation, or PWM, and is the basis for most popular motor speed control circuits. There are a number of ways of providing PWM; this is just one of dozens. Figure 13-13 shows a timing diagram of the PWM technique, from 100 percent duty cycle (100 percent on), to 0 percent duty cycle (0 percent on).

In the previous circuit, R3 is shown surrounded by a dotted box. You can substitute R3 with a fixed resistor, if you want to always use a certain speed, or you can use the circuit in Fig. 13-14. This circuit employs a 4066 CMOS analog switch IC. The 4066 allows you to select any of up to four speeds by computer or electronic control.

Resistors of various values are connected to one side of the switches; the other side of the switches are collectively connected to the 4011. To modify the speed of the motor, activate one of the switches by bringing its control input HIGH. The resistor connected to that switch is then brought into the circuit. You can omit the 3.3K pull-down resistors on the control inputs if your control circuitry is always activated and connected.

The 4066 is just one of several CMOS analog switches. There are other versions of this IC with different features and capabilities. The 4066 was chosen here because it adds very little resistance of its own when the switches are on. Note that the 4066 specifications sheet says that only one switch should be closed at a time. Table 13-6 list the parts needed.

Improved Speed Control

The main problem with the basic speed control technique described above is that the robot has no way of knowing exactly how fast it is going. In fact, the brains on board your robot really have no way of knowing that the motors are turning at all. Such is the problem of open-loop speed control systems. The term open-loop means that there is no mechanism for conveying the actual speed back to the controlling electronics.

Figure 13-15 shows one approach to a closed-loop motor speed control system. The term closed-loop means there is some type of mechanism that reports the speed of the motor back to the controlling electronics. Variations are corrected, so the motor operates closer to the desired speed.

Here is how it works: Attached to the motor shaft is an encoder disc, also called a shaft encoder. An example of one is shown in Fig. 13-16. This disc is composed of numerous holes or slots along the outside edge. An infrared LED is placed on one side of the disc, so that its

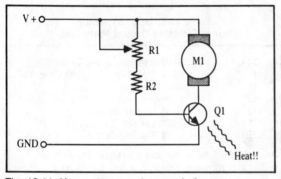

Fig. 13-11. How *not* to vary the speed of a motor.

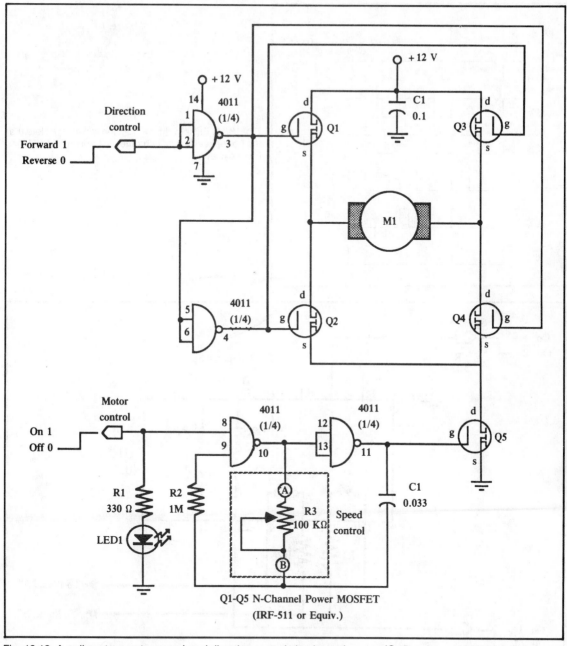

Fig. 13-12. A rudimentary motor speed and direction control circuit requires one IC, five power MOSFETs, and a small handful of resistors and capacitors. R1 and LED1 serve to indicate that the motor is on.

light shines through the holes. The number of holes or slots is not a consideration here, but for increased speed resolution, there should be as many around the outer edge of the disc as possible.

An infrared-sensitive phototransistor is positioned directly opposite the LED so that when the motor and disc turn, the holes pass the light intermittently. The result, as seen by the phototransistor, is a series of flash-

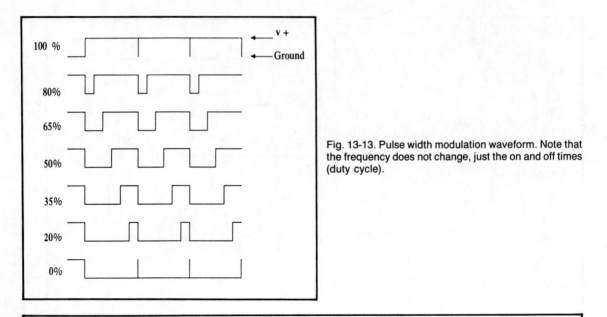

Fig. 13-13. Pulse width modulation waveform. Note that the frequency does not change, just the on and off times (duty cycle).

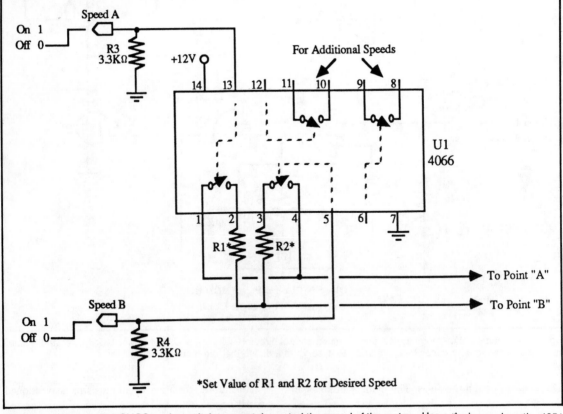

Fig. 13-14. Using a 4066 CMOS analog switch to remotely control the speed of the motor. Use a device such as the 4051 for even more speed choices.

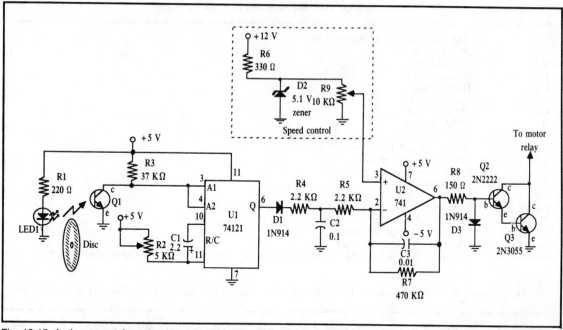

Fig. 13-15. An incremental motion controller based on an optical shaft encoder, a 74121 monostable multivibrator, and a 741 op amp. The output stage of the controller can be connected directly to the motor, or routed through an on/off and direction control circuit. The circuit can drive a motor with a current draw in excess of eight amps.

ing light. The Big Trak toy used a similar technique to control speed and monitor distance traveled.

The light pulses from the collector of the phototransistor are routed to a 74121 monostable one-shot. This special-purpose chip is a short-duration timer, and produces output pulses or a specified duration no matter how long the input pulse is. The duration of the output

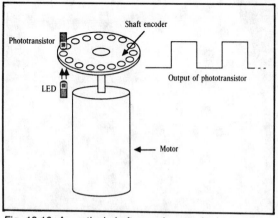

Fig. 13-16. An optical shaft encoder attached to a motor. Alternatively, you can place a series of reflective strips on a black disc, and bounce the LED light into the transistor.

pulse is determined by just two components: R2 and C1. You can vary the on-time of the pulse by turning potentiometer R2. In actual use, you'd adjust this pot to better match the characteristics of the motor you are using.

The output of the one-shot is applied to the inverting input of a 741 op-amp. The non-inverting input is provided a fairly well regulated reference voltage. Resistor R6 and diode D2 provide a regulated 5.1 volt supply. R9 is a precision ten-turn 10K pot. Adjusting this pot varies the reference voltage provided to the op-amp, and hence adjusts the speed, or the "setpoint," of the motor.

The output of the op-amp is a series of pulses directly proportional to the reference voltage (speed control) and the output of the 74121. When the motor begins to turn too fast, the 741 shortens its output pulses, which in turn slows the motor down. When the motor begins to turn too slow, the 741 lengthens its output pulses, which in turn speeds up the motor.

Finishing the circuit are driving transistors. The collectors of the transistors can be directly connected to the motor, or as in most cases, to the contacts of a control relay. Use the relay wiring scheme in Fig. 13-6 for providing on/off and direction control to your motor. Alternatively, you can use one of the transistor or MOSFET control circuits.

Table 13-7. Shaft Encoder Motor Speed Control Parts List.

U1	74121 Monostable Multivibrator IC
U2	LM741 Op Amp IC
R1	220Ω resistor
R2	5K potentiometer
R3	37KΩ resistor
R4,R5	2.2KΩ resistor
R6	330 Ω resistor
R7	470KΩ resistor
R8	150Ω resistor
R9	10K 10-turn precision potentiometer
C1	2.2 μF tantalum capacitor
C2	0.1 μF ceramic capacitor
D1,D3	1N914 signal diode
D2	5.1 volt 1 watt zener diode
Q1	Infrared sensitive phototransistor
Q2	2N2222 npn transistor
Q3	2N3055 npn power transistor
LED1	Infrared Light Emitting Diode
Misc.	Heat sink for 2N3055 transistor, infrared filter for phototransistor (if needed), shaft encoder

All resistors 5 or 10 percent tolerance, 1/4-watt; all capacitors 10 percent tolerance, rated 35 volts or higher.

The biggest disadvantage to this system is what engineers call *overshoot* and *undershoot*. Besides the *slewing* error that is always present as the servo overcompensates to maintain the desired setpoint speed, the servo system is susceptible to overshoot when the motor is allowed to run at appreciably faster speeds than what the speed control pot is set for. The circuit will tend to maintain the faster-than-normal speed until the motor is turned off, reverse, or slowed down by some mechanical means. Undershoot occurs when the motor is allowed to turn too slowly.

Note that the speed adjustment circuit can be replaced by a number of all-electronic controls. You can replace potentiometer R7 with a 4066 analog switch IC and a couple of resistors, as shown in Fig. 13-14, and control the IC from a computer. Or, you can replace the entire ensemble outlined in the dotted box with a digital-to-analog (D/A) converter. This single-chip converter exchanges digital data for voltage levels. Read more about digital-to-analog converters, as well as their close cousins analog-to-digital converters, in Chapter 32, "Build a Robot Interface Card."

Making the Shaft Encoder

By far, the hardest part about this circuit is making the shaft encoders. Sure, you can buy shaft encoders ready-made but cost is astronomical. For homebrew robotics projects, you're better off constructing your own.

The shaft encoder you make may not have the fine resolution of a commercially available disc, which often have 256 or 360 slots in them, but the home-made versions will be more than adequate. You may even be able to find already-machined parts that closely fit the bill. For example, Fig. 13-17 shows a sink strainer, if you can believe it, being used as a shaft encoder. The sink strainer

Fig. 13-17. An optical shaft encoder made from a sink strainer. The disc has a resolution of 18 degrees.

Table 13-8. Up/Down Shaft Encoder Reader Parts List.

U1	7414 Hex Schmitt Inverter IC
U2	7474 Dual "D" Flip-Flop IC
R1,R2	220 Ω resistor
R3,R4	22 KΩ resistor (experiment with value for best response)
Q1,Q2	Infrared sensitive phototransistor
Misc.	Infrared filter for phototransistor (if needed), shaft encoder

All resistors 5 or 10 percent tolerance, 1/4-watt.

comes with predrilled holes. To use it, the stem was cut off and a hole was drilled in the middle.

Lacking a sink strainer, you can make your own shaft encoder by taking a two- or three-inch disc of plastic or metal, and drilling holes in it. Remember that the disc material must be opaque to infrared light. Some things that may look opaque to you may actually pass infrared light. When in doubt, add a coat or two of flat black or dark blue paint. That should block stray infrared light from reaching the phototransistor.

Mark the disc for at least 20 holes. The more holes the better. Minimum hole size is about 1/16-inch. Use a compass to scribe an exact circle for drilling. The infrared light might not pass through holes not on this scribe line.

Mounting the Hardware

Secure the shaft encoder to the shaft of the drive motor or wheel. Using brackets, attach the LED so that it fits snugly on the back side of the disc. You can bend the lead of the LED a bit to line it up with the holes. Do the same for the phototransistor. You must mask the phototransistor so that it doesn't pick up stray light, nor reflected light from the LED, as shown in Fig. 13-18. You can increase the effectiveness of the phototransistor placing an infrared filter (a dark red filter will do in a pinch) between the lens of the phototransistor and the disc. You can also use the type of phototransistor that has its own built-in infrared filter.

If you find that the circuit isn't sensitive enough, check for stray light hitting the phototransistor. Baffle it with a piece of black construction paper if necessary. Also, try experimenting with the value of R3 (from the circuit in Fig. 13-15). It affects the overall sensitivity of the phototransistor.

Encoder Disc Number Two

Another approach is to use graphic arts film cemented

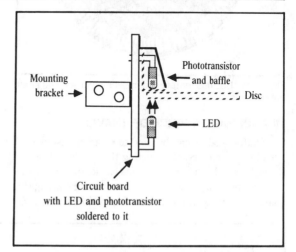

Fig. 13-18. How to mount an infrared LED and phototransistor on a circuit board for use with an optical shaft encoder.

to a clear plastic disc. Use the template in Fig. 13-19 as a master. Take it to any printer and have them make a photolith *positive* copy of it. The exact size of the copy depends on the size of the clear plastic disc you use. Make sure that the printer understands that the black background is to be as black as possible. A smokey gray background won't do.

Cut the photolith into a circle, punch a hole in the center, and mount it on the clear plastic disc. Use glue only on the black portions. Make sure the edges of the photolith film are flush against the surface of the plastic. Apply a small amount of glue to help assure a tight fit.

Some infrared light may peek through the dark background, but it won't be much. You can experiment with the circuit by moving the LED or phototransistor further or closer apart, by experimenting with the value of R3, and by baffling the phototransistor to block out ambient light. You can also try two photolith copies of the disc sandwiched together.

107

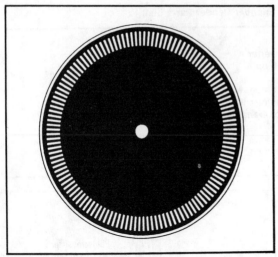

Fig. 13-19. Use this template to make an incremental shaft encoder disc.

MEASURING DISTANCE OF TRAVEL

Shaft encoders can be used to measure distance of travel of the motors. By counting the number of on/off flashes, the robot's control circuits can keep track of the revolutions of the drive wheels. Instead of mounting the shaft encoders on the motor shafts, mount them on the wheel shafts (if they are different). The number of slots in the disk determines the maximum accuracy of the travel circuit. The more slots, the better the accuracy.

Let's say the encoder disc has 50 slots around its circumference. That represents a minimum sensing angle of 7.2 degrees. As the wheel rotates, it provides a signal to the counting circuit every 7.2 degrees. Stated another way, if the robot is outfitted with a 7-inch wheel (circumference = 21.98 inches), the maximum travel resolution is approximately 0.44 linear inches. Not bad at all! The figure was calculated by taking the circumference of the wheel and dividing it by the number of slots in the shaft encoder.

You can use the left half of the circuit in Fig. 13-15 (the 74121 and timing components) to receive and condition the pulses. In fact, you can tap the pulses directly off the Q output of the 74121 chip and feed them to the robot's distance counter or computer.

Another approach is shown in Fig. 13-20. Two phototransistors are used to provide speed and direction of the encoder wheel. Two LEDs and phototransistors are used, as shown in Fig. 13-21. The two phototransistors are positioned so that when one senses the light from its LED, the light to the other is cut off.

The outputs of the phototransistors are conditioned by Schmitt triggers. This smooths out the waveshape of

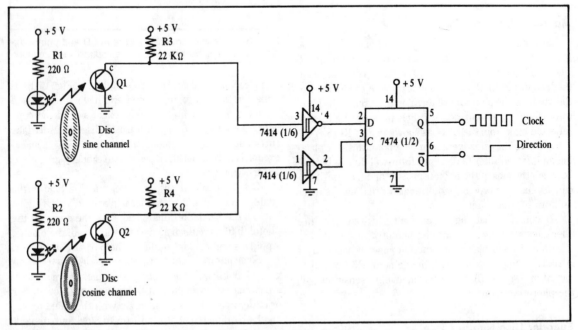

Fig. 13-20. A two-channel shaft encoder circuit. The outputs of the flip-flop indicate speed (one clock transition for every slot detected in the disc) and the direction of rotation.

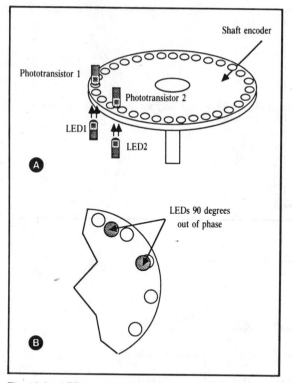

Fig. 13-21. LEDs and phototransistors mounted on a two-channel optical shaft encoder disc setup. A. The LEDs and phototransistors can be placed anywhere about the circumference of the disc; B. The two LEDs and phototransistors must be 90 degrees out of phase.

the light pulses. The output of the triggers are applied to the "D" and clock inputs of a 7474 flip-flop. The outputs of the flip-flops provide a pulsed, conditioned train representing the speed of the encoder disc, as well as an indication of the direction of the disc.

The direction sensing may come in handy so your robot can tell if it is drifting forward or backward, even though no power is applied. In fact, the shaft encoders can always be activated, even when the motors are turned off, to alert the robot that it is in motion.

The Distance Counter

The pulses from the flip-flops or the output of the 74121 do not in themselves carry distance measurement. The pulses must be counted, and the count converted to distance. Counting and conversion are the ideal tasks for a computer. Most single-chip computer/microprocessors, or their interface adapters, are equipped with counters. Connect the pulsed flip-flop or 74121 output to the counter of the microprocessor and have it keep track of distance for you.

If your robot lacks a computer, or the on-board microprocessor lacks a counter, you can use the basic circuit shown in Fig. 13-22 as a central counter. Here, the outputs of the flip-flop or 74121 is routed to a 4040 12-stage binary ripple counter. This CMOS chip has 12 binary weighted outputs and can count to 4,096. You'd probably use just the first eight outputs, to count to 256. To start counting, bring the Reset line (pin 11) HIGH,

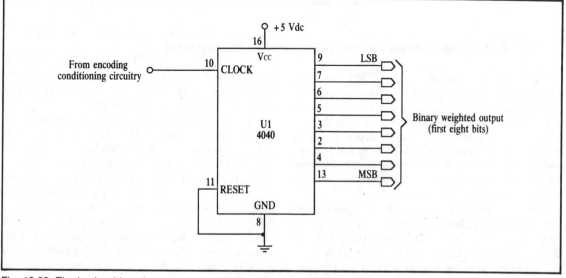

Fig. 13-22. The basic wiring diagram of the 4040 CMOS 12-stage ripple counter IC.

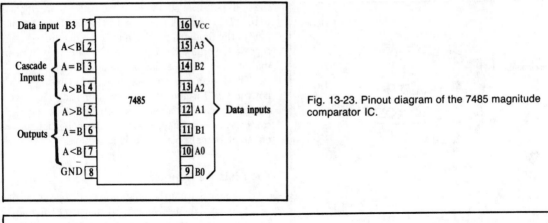

Fig. 13-23. Pinout diagram of the 7485 magnitude comparator IC.

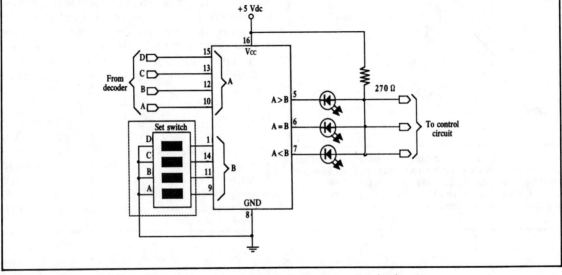

Fig. 13-24. The basic wiring diagram of a single-stage magnitude comparator circuit.

to clear the outputs, then back to LOW. The pulses from the disc assembly are routed to the Clock line (pin 10).

Using the 4040 isn't without its limitations. Computers can't easily connect to it without reading the outputs in two passes (high order byte, low order byte). You could also use a 4024 CMOS IC to count to 128. A more conventional counter, which can also be used to measure the travel speed of the robot, is presented in Chapter 29, "Navigating Through Space."

Any counter with a binary or BCD output can be used with a 7485 magnitude comparator. A pin-out of this versatile chip is shown in Fig. 13-23, and a basic hook-up diagram in Fig. 13-24. In operation, the chip will compare the binary weighted number at its "A" and "B" inputs. One of the three LEDs will then light up, depending

on the result of the difference between the two numbers. In a practical circuit, you'd replace the DIP switches (in the dotted box) with a computer port.

You can cascade comparators to count to just about any number. If counting in BCD, three packages can be used to count to 999, which should be enough for most distance recording purposes. Using a disc with 25 slots in it, and a seven-inch drive wheel, the travel resolution is 0.84 linear inches. Therefore, the counter system will stop the robot within 0.84 inches of the desired distance (give or take coasting and slip between the wheels and ground) up to a maximum working range of 69.93 feet. You can increase the distance by building a counter with more BCD stages or decreasing the number of slots in the encoder disc.

Chapter 14

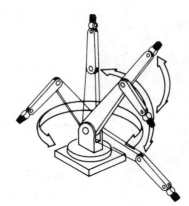

Robot Locomotion
With Stepper Motors

In past chapters, we've taken a look at powering robots using everyday continuous dc motors. Dc motors are cheap, deliver a lot of torque for their size, and are easily adaptable to a variety of robot designs.

By their nature, however, the common dc motor is rather imprecise. Without a servo feedback mechanism or tachometer, there's no telling how fast a dc motor is turning. Furthermore, it's difficult to command the motor to turn exactly a specific number of revolutions, let alone a fraction of a revolution. Robotics work, particularly arm designs, often requires this kind of accuracy.

Enter the stepper motor. Stepper motors are, in effect dc motors with a twist: Instead of being powered by a continuous flow of current—as with regular dc motors—they are driven by pulses of electricity. Each pulse drives the shaft of the motor a little bit. The more pulses that are fed to the motor, the more the shaft turns.

Stepper motors are inherently "digital" devices, which comes in handy when you want to control your robot by computer. By the way, there are ac stepper motors as well, but they aren't really suitable for robotics work, and won't be discussed here.

Stepper motors aren't as easy to use as standard dc motors, however, and they're harder to get and are more expensive. But for applications requiring them, stepper motors can solve a lot of problems with a minimum of fuss. Let's take a closer look at steppers and learn how to apply them to your robot designs.

INSIDE A STEPPER MOTOR

There are several designs of stepper motors. For the time being, we'll concentrate on the most popular variety, the four-phase stepper, like the one in Fig. 14-1.

A four-phase stepper motor is really two motors sandwiched together, as shown in Fig. 14-2. Each motor is composed of two windings. Wires connect to each of the four windings of the motor pair, so there are eight wires coming from the motor. The commons from the windings are often ganged together, which reduces the wire count to five or six instead of eight (see Fig. 14-3).

Wave Step Sequence

In operation, the common wires are attached to the positive (sometimes the negative) side of the power supply. Each winding is then energized in turn, by grounding it to the power supply, for a short period of time. The motor shaft turns a fraction of a revolution each time a winding is energized. For the shaft to turn properly, the windings must be energized in *wave step sequence*. For ex-

Fig. 14-1. A typical four-phase stepper motor.

ample, energize wires 1, 2, 3, and 4 in sequence, as shown in Fig. 14-4, and the motor turns clockwise. Reverse the sequence, and the motor turns the other way.

On-on/Off-off Sequence

The wave step sequence is the basic actuation technique of four-phase stepper motors. Another, and far better, approach actuates two windings at once in an on-on/off-off sequence, as shown in Fig. 14-5. The enhanced actuation sequence increases the driving power of the motor, and provides greater shaft rotation precision.

There are other varieties of stepper motors, as well, and they are actuated in different ways. One you may encounter is two-phase. It has four wires that are pulsed by reversing the polarity of the power supply for each of the four steps. The actuation technique for these motors is covered later in this chapter.

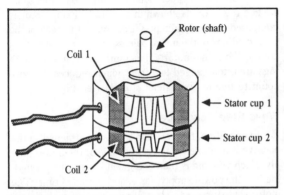

Fig. 14-2. Inside a four-phase stepper motor. Note the two sets of coils and stators. The four-phase stepper is really two motors in one.

DESIGN CONSIDERATIONS OF STEPPER MOTORS

Stepping motors differ in their design characteristics over continuous dc motors.

Stepper Phasing

A four-phase stepper requires a sequence of four pulses applied to its various windings for proper rotation. By their nature, all stepper motors are at least two-phase. The majority are four-phase; some are six phase. Usually, but not always, the more phases in a motor, the more accurate it is.

Step Angle

Stepper motors vary in the amount of rotation the shaft turns each time a winding is energized. The amount or rotation is called the *step angle* and can vary from as small as 0.9 degrees (1.8 degrees is more common) to 90 degrees.

The step angle determines the number of steps per revolution. A stepper with a 1.8 degree step angle, for example, must be pulsed 200 times for the shaft to turn one complete revolution. A stepper with a 7.5 degree step angle must be pulsed 48 times for one revolution, and so on.

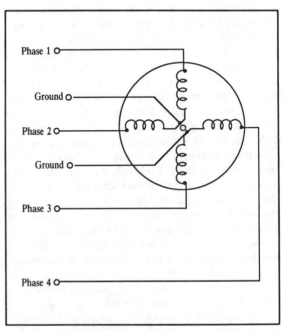

Fig. 14-3. The wiring diagram of the four-phase stepper. The ground connections may be separate or combined.

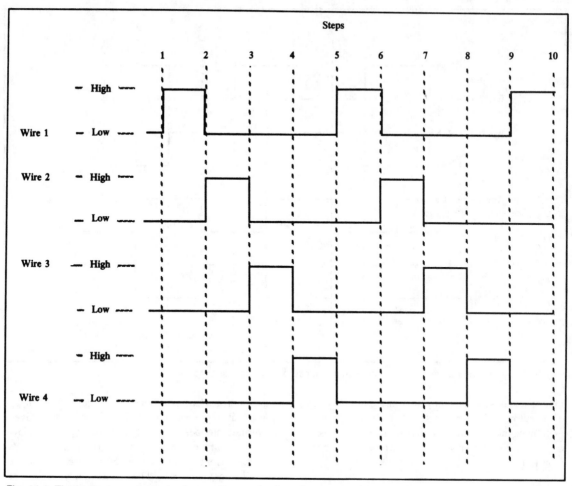

Fig. 14-4. The basic wave-step actuation sequence of a four-phase stepper motor.

Pulse Rate

Obviously, the smaller the step angle, the more accurate the motor is. But stepper motors have an upper limit to the number of pulses they can accept per second. Heavy-duty steppers usually have a maximum *pulse rate* (or step rate) of 200 or 300 steps per second, so they have an effective top speed of one to three revolutions per second (60 to 180 rpm). Some smaller steppers can accept a thousand or more pulses per second, but they don't provide very much torque and aren't suitable as driving or steering motors.

It should be noted that stepper motors can't be motivated to run at their top speeds immediately from a dead stop. Applying too many pulses right off the bat simply causes the motor to freeze up. To achieve top speeds, the motor must be gradually accelerated. The accelera-

tion can be quite swift in human terms. The speed can be 1/3 for the first few milliseconds, 2/3 for the next 50 or 75 milliseconds, then full blast after that.

Running Torque

Steppers can't deliver as much running torque as standard dc motors of the same size and weight. A typical 12 volt, medium size stepper motor may have a running torque of only 25 oz-in. The same 12 volt, medium size standard dc motor may have a running torque three or four times more.

Steppers are, however, at their best when turning slowly. With a stepper, the slower the motor revolves, the higher the torque. The reverse is usually true of continuous dc motors. Figure 14-6 shows a graph of the run-

113

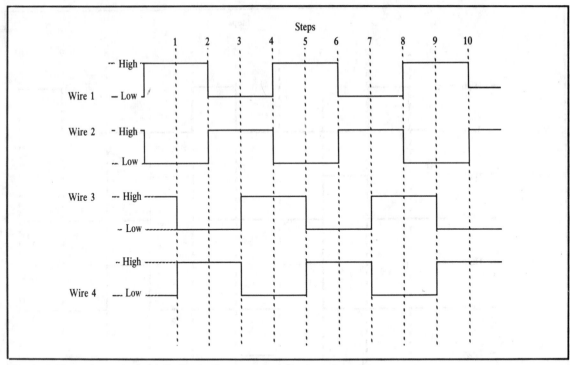

Fig. 14-5. The enhanced on-on/off-off actuation sequence of a four-phase stepper motor.

ning torque of a medium-duty, four-phase 12 volt stepper. This unit has a top running speed of 550 pulses per second. Since the motor has a step angle of 1.8 degrees, that ends up being a top speed of 2.75 revolutions per second (165 rpm).

Braking Effect

Actuation of one of the windings in a stepper motor advances the shaft. Continue to apply current to the winding and the motor won't turn any more. In fact, the shaft will be locked, as if you've applied brakes. As a result of this interesting locking effect, you never need to add a braking circuit to a stepper motor, because it has its own brakes built in.

The amount of braking power of a stepper motor is expressed as *holding torque*. Small stepper motors have a holding torque of a few oz-in. Larger, heavier duty models have holding torques exceeding 400 oz-in.

Voltage, Current Ratings

Like dc motors, stepper motors vary in their voltage and current ratings. Steppers for 5, 6, and 12 volt operation are not uncommon. But unlike dc motors, using a

higher voltage than specified doesn't result in faster operation, but more running and holding torque. Overpowering a stepper by more than 40 or 60 percent above the rated voltage can eventually burn up the motor.

The current rating of a stepper is expressed in amps (or milliamps) per phase. The power supply driving the motor needs to deliver at least as much as the per-phase

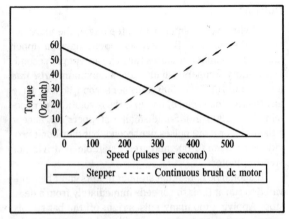

Fig. 14-6. With a stepper motor, torque increases as the speed of the motor is reduced.

specification, preferably more if the motor is driving a heavy load. The four step actuation sequence powers two phases at a time, requiring the supply to deliver at least twice as much current as the per-phase specification. If, for example, the current per phase is 0.25 amps, the power requirement at any one time is 0.50 amps.

CONTROLLING A STEPPER MOTOR

Steppers have been around for a long time. In the old days, stepper motors were actuated by a mechanical switch, a solenoid-driven device that pulsed each of the windings of the motor in the proper sequence. Now, stepper motors are invariably controlled by electronic means. Basic actuation can be accomplished via computer control by pulsing each of the four windings in turn. The computer can't directly power the motor, so transistors must be added to each winding, as shown in Fig. 14-7.

Using an LSI Controller Chip

Lacking direct computer control, the easiest approach to providing the proper sequence of actuation pulses is by using a custom LSI stepper motor chip, such as the Signetics SAA-1027 or Sprague UCN 4202. Both of these chips are designed expressly for use with the common four-phase stepper motor and provide a four step actuation sequence. But take note: The chips are expensive—$5 to $10 a piece. They are, however, designed to power a small- or medium-size stepper without adding lots of external components. All things considered, they're worth the price.

Typical schematics using these two chips are shown in Figs. 14-8 and 14-9. A 555, wired to send from 1 to about 500 pulses per second, serves as the pulse generator. Heavier duty motors (more than 350 mA per phase) can be driven by adding power transistors to the four outputs of the chips, as shown in the manufacturer's application notes.

Both chips have a rotation Direction pin. Pulling this pin high or low reverses the rotation of the motor. The SAA-1027 has a Set pin that, when pulled low, continuously supplies power to two of the windings, and brakes the motor.

Using Gates to Control Stepper Motors

If you can't locate the Signetics SAA-1027 or Sprague UCN 4202, the next best approach is to use discrete gates and clock ICs. You can assemble a stepper motor trans-

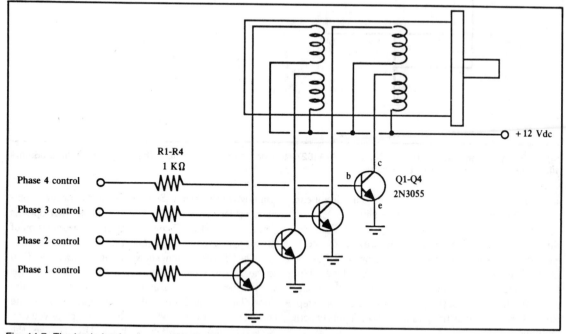

Fig. 14-7. The basic hookup connection to drive a stepper motor from a computer. The phasing sequence is provided by software output through a port in the following four-bit binary sequence: 1010, 0110, 0101, 1001 (reverse the sequence to reverse the motor).

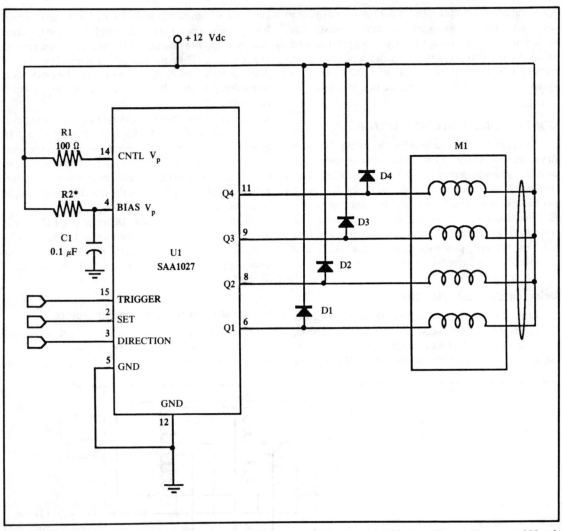

Fig. 14-8. The wiring diagram for the Signetics SAA1027 stepper motor translator IC. The IC can drive up to 350 mA without needing external driver transistors.

lator circuit using just two IC packages. The circuit can be constructed using TTL or CMOS chips.

The TTL version is shown in Fig. 14-10. Four Exclusive OR gates from a single 7486 IC provides the steering logic. You set the direction pulling pin 12 HIGH or LOW. The stepping actuation is controlled by a 7476, which contains two JK flip-flops. The Q and $\overline{Q}$ outputs of the flip-flops control the phasing of the motor. Stepping is accomplished by triggering the clock inputs of both flip-flops.

The 7476 can't directly power a stepper motor. You must use power transistors or MOSFET's to drive the windings of the motor. See the section below for a complete power driving schematic as well as other options you can add to this circuit. A pictorial representation of the stepping sequence is shown in Fig. 14-11.

The CMOS version, shown in Fig. 14-12, is identical to the TTL version, except that a 4070 chip is used for the Exclusive OR gates and a 4027 is used for the flip-flops. The pin-outs are slightly different, so follow the correct schematic depending on the type of chips you use. Note that another CMOS Exclusive OR package, the 4030, is also available. Don't use this chip; it behaves erratically in this, as well as other pulsed, circuits.

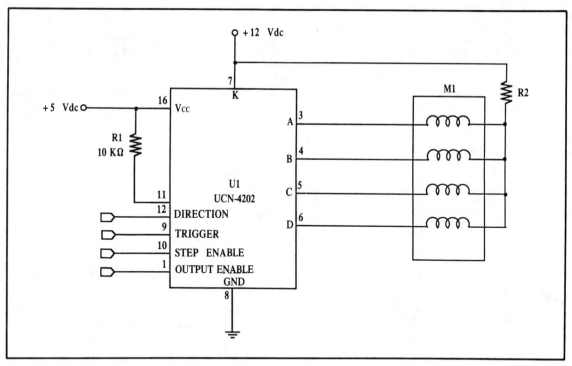

Fig. 14-9. The wiring diagram for the Sprague UCN-4202 stepper motor translator IC. The IC can drive up to 500 mA without needing external driver transistors.

In both the TTL and CMOS circuits, the stepper motor itself can be operated from a supply voltage wholly different than the voltage supplied to the ICs.

Translator Enhancements

Four npn power transistors, four resistors, and a handful of diodes are all that's needed to provide driving power for the translator circuits described in the last section (you can also use this scheme to increase the driving power of the SAA-1027 or UCN 4202 chips). The schematic for the circuit is shown in Fig. 14-13. Note that you can substitute the bi-polar transistors and resistors

with power MOSFETs. See Chapter 12, "Choosing the Right Motor for the Job," for more information on using power MOSFETs.

You can use just about any npn power transistor that will handle the motor. The TIP31 is a good choice for applications requiring up to one amp of current. Use the 2N3055 for heavier duty motors. Mount the drive transistors on a suitable heat sink.

You must insert a biasing resistor in series between the outputs of the translation circuit and the base of the transistors. Values between about 1KΩ and 3KΩ should work with most motors and most transistors. Experiment

Table 14-1. SAA1027 Stepper Motor Translator Parts List.

U1	Signetics SAA-1027 Stepper Motor Translator IC
R1	100KΩ resistor
R2	Bias Resistor 470 Ω 1/2 watt for small motors (under 200 mA current per phase); 150 Ω 1 watt for larger motors (200 to 350 mA current per phase)
C1	0.1 μF ceramic capacitor
D1-D4	1N4004 diode
M1	Four-phase stepper motor

Sprague UCN-4202 Stepper Motor Translator IC	Table 14-2. UCN-4202
10K resistor, 1/2 watt	**Stepper Motor Translator Parts List.**
Four-phase stepper motor	

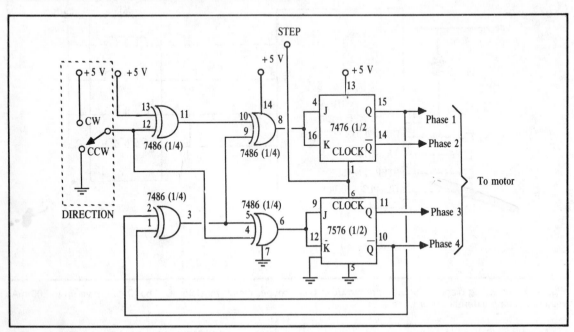

Fig. 14-10. Using a pair of commonly available TTL ICs to construct your own stepper motor translator circuit.

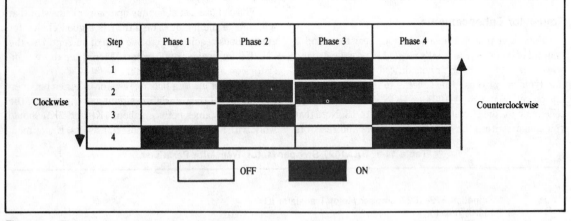

Fig. 14-11. The phasing sequence for a four-phase stepper.

until you find the value that works without causing the flip-flop chips to overheat. You can also apply Ohm's Law, figuring in the current draw of the motor and the gain of the transistor, to accurately find the correct value of the resistor. If this is new to you, see Appendix B, "Further Reading," for a list of books on electronic design and theory.

It is sometimes helpful to see a visual representation

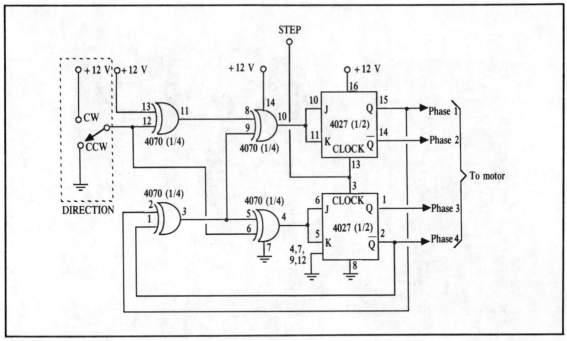

Fig. 14-12. Using a pair of commonly available CMOS ICs to construct your own stepper motor translator circuit.

of the stepping sequence. Adding an LED and current limiting resistor in parallel with the outputs provides such an indication. See Fig. 14-14 for a wiring diagram.

Figure 14-15 shows two stepper motor translator boards. The small board controls up to two stepper motors and is designed using TTL chips. The LED option is used to provide a visual reference of the step sequence. The large board uses CMOS chips and can accommodate

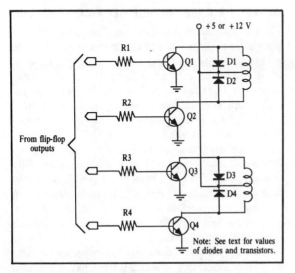

Fig. 14-13. Add four transistors and resistors to provide a power output stage for the TTL or CMOS stepper motor translator circuit.

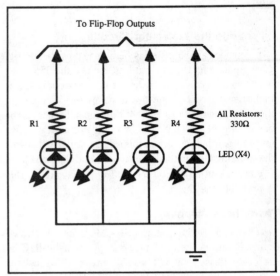

Fig. 14-14. Add four LEDs and resistors to provide a visual indication of the stepping action.

119

Fig. 14-15. Two finished stepper motor translator boards, complete with indicator LEDs.

up to four motors. The boards were wire wrapped; the driving transistors are placed on a separate board and heat sink.

Triggering the Translator Circuits

You need a squarewave generator to provide the triggering pulses for the motors. You can use the 555 timer wired as an astable multivibrator (see Appendix C), or make use of a control line in your computer. When using the 555, remember to add the 0.1 μF capacitor across the power pins of the chip. The 555 puts out a lot of noise into the power supply, and this noise regularly disturbs the counting logic in the Exclusive OR and flip-flop chips. If you are getting erratic results from your circuit, this is probably the cause.

Two-Phase Stepping

You can use the custom LSI, CMOS, or TTL translator circuits to operate 2-phase steppers, as well. Connect the "Phase 2" and "Phase 3" wires of the translator circuits to the control inputs in Fig. 14-16. The pictorial

truth table in Fig. 14-17 shows the actuation sequence.

USING THE CY500 INTELLIGENT CONTROLLER

A stepper motor translator must be operated by using a pulse generator, which admittedly doesn't provide a lot of flexibility over stepping speed, exact number of steps, and so forth. You can always use a computer to trigger the translator chips, but what if you are not using a computer or microprocessor in your current design?

One of the most sophisticated stepper motor controller ICs is the CY500, from Cybernetic Micro Systems. This single integrated circuit IC combines a stepper motor translator with the intelligence and programmability of a computer. You can control the chip with just a keyboard, typing in letter commands. If you happen to have a personal computer lying around (just about any will do), you can connect the CY500 to it via the RS232C serial or parallel printer port.

Among the commands you can program into this handy (but expensive) little chip is stepping rate, ramped (or sloped) stepping rate (to allow the motor to come up

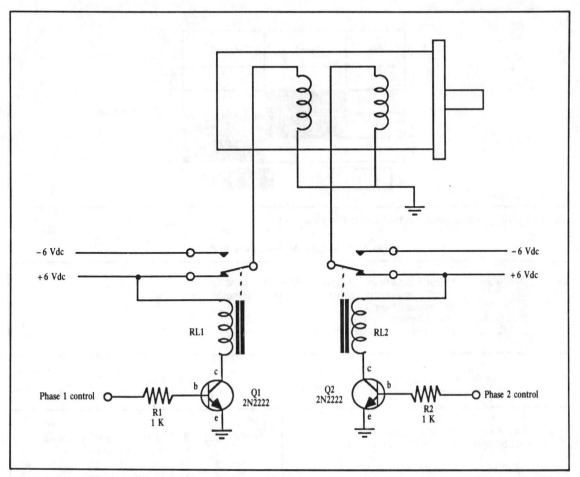

Fig. 14-16. The most straightforward approach to controlling a two-phase stepper motor (you can substitute the − 6 Vdc supply with ground with some motors).

to high speed or slow down), number of steps, and more. You can program a series of commands and the chip will perform them, one after the other. Table 14-8 presents a more complete command summary. Additional information on using this elaborate chip may be obtained from the manufacturer. Refer to Appendix A, "Sources," for the address. Tables 14-1 through Table 14-7 give the parts needed.

BUYING AND TESTING A STEPPER MOTOR

Spend some time with a stepper motor and you'll invariably come to admire them, thinking up all sorts of ways to make them work for you in your robot designs. But to use a stepper, you have to buy one. That in itself is not always easy. Then after you have purchased it and taken it home, there's the question of figuring out where all the wires go! Let's take each problem one at a time.

Table 14-3. TTL
Stepper Motor Translator Parts List.

U1	7486 Quad Exclusive OR Gate IC
U2	7476 Dual "JK" Flip-Flop IC

Table 14-4. CMOS
Stepper Motor Translator Parts List.

U1	4070 Quad Exclusive OR Gate IC
U2	4037 Dual "JK" Flip-Flop IC

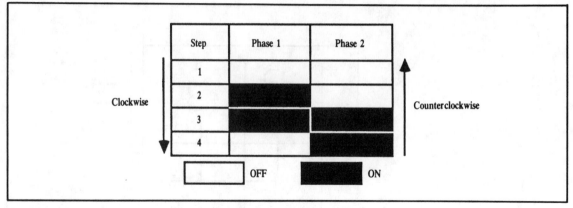

Step	Phase 1	Phase 2
1		
2	■	
3	■	■
4		■

Clockwise ← → Counterclockwise

☐ OFF   ■ ON

Fig. 14-17. The phasing sequence for a two-phase stepper.

Table 14-5. Stepper Motor Driver Parts List.

Q1-Q4	Under 1 amp draw per phase: TIP32 npn transistor
	1 to 3 amp draw per phase: TIP120 npn Darlington transistor
R1-R4	1K resistor, 1 watt
D1-D4	1N4004 diode
Misc.	Heat sinks for transistors

Table 14-6. Stepper Motor Phasing Indicator Parts List.

R1-R4	300 Ω resistor
LED1-4	Light Emitting Diode

All resistors 5 or 10 percent tolerance, 1/4-watt unless noted.

Table 14-7. Two-Phase Stepper Motor Controller Parts List.

RL1,RL2	SPDT Fast-Acting relay, 2 amp contacts (minimum)
R1,R2	1K resistor, 1/4-watt
Q1,Q2	2N2222 npn transistor

Sources for Stepper Motors

Despite their many advantages, stepper motors aren't nearly as popular as the trusty dc motor, so they are hard to find. And when you do find them, they're expensive. A new, light-duty model lists for about $30; heavier-duty jobs carry pricetags upwards of $100.

The surplus market is by far the best source for stepper motors for hobby robotics. Four of the best sources are Jerryco, H&R, Edmund Scientific, and BNF Enterprises. They carry most of the name brand steppers: Airpax (North American Philips), Molon, Haydon, and Superior Electric. The cost of surplus steppers is much more reasonable, usually a quarter or fifth of the original list price.

The disadvantage of buying surplus is that you don't

always get a hookup diagram or adequate specifications. Purchasing surplus stepper motors is largely a hit and miss affair, but most outlets let you return the goods if they aren't what you need. If you like the motor, yet it still lacks a hookup diagram, read the following section on how to decode the wiring.

The Signetics and Sprague stepper ICs covered earlier are, at this time, not available on the surplus market. These chips are special purpose, and it's doubtful you'll find them at your local Radio Shack, or most any other electronics specialty store for that matter. Write the manufacturers for a list of local distributors or dealers.

Wiring Diagram

The internal wiring diagram of both a two- and four-

Table 14-8. CY-500 Command Set.

ASCII CODE	NAME	DESCRIPTION
A	ATHOME	SET CURRENT LOCATION EQUAL ABSOLUTE ZERO
B	BITSET	TURN ON PROGRAMMABLE OUTPUT LINE
C	CLEARBIT	TURN OFF PROGRAMMABLE OUTPUT LINE
D	DOITNOW	BEGIN PROGRAM EXECUTION
E	ENTER	ENTER PROGRAM MODE
F	FACTOR	DECLARE RATE DEVISOR FACTOR
G	GOSTEP	BEGIN STEPPING OPERATION
H	HALFSTEP	SET HALF-STEP MODE OF OPERATION
I	INITIALIZE	INITIALIZE SYSTEM
J	JOG	SET EXT. START/STOP CONTROL MODE
L	LEFTRIGHT	SET EXT. DIRECTION CONTROL MODE
N	NUMBER	DECLARE RELATIVE NUMBER OF STEPS TO BE TAKEN
O	ONESTEP	TAKE ONE STEP IMMEDIATELY
P	POSITION	DECLARE ABSOLUTE TARGET POSITION
Q	QUIT	QUIT PROGRAMMING/ENTER COMMAND MODE
R	RATE	SET RATE PARAMETER
S	SLOPE	SET RAMP RATE FOR SLEW MODE OPERATION
T	LOOP TILL	LOOK TILL EXT. START/STOP LINE GOES LOW
U	UNTIL	PROGRAM WAITS UNTIL SIGNAL LINE GOES LOW
+	CW	SET CLOCKWISE DIRECTION
–	CCW	SET COUNTERCLOCKWISE DIRECTION
0	COMMAND MODE	EXIT PROGRAM MODE/ENTER COMMAND MODE

phase stepper motor is shown in Fig. 14-18. The wiring in a two-phase stepper is actually easy to decode. You use a volt-ohm meter to do the job right. You can be fairly sure the motor is two-phase if it has only four wires leading to it. You can identify the phases by connecting the leads of the meter to each wire and noting the resistance. You can readily identify mating phases when there is a small resistance through the wire pair.

Four-phase steppers behave the same, but with a

slight twist. Let's say, for argument sake, that the motor has eight wires leading to it. Each winding, then, has a pair of wires. Connect your meter to each wire in turn to identify the mating pairs. As illustrated in Fig. 14-19, no reading (infinite ohms) signifies that the wires do not lead to the same winding; a reading indicates a winding.

If the motor has six wires, then four of the leads go to one side of the windings; the other two are commons and connect to the other side of the windings (see Fig.

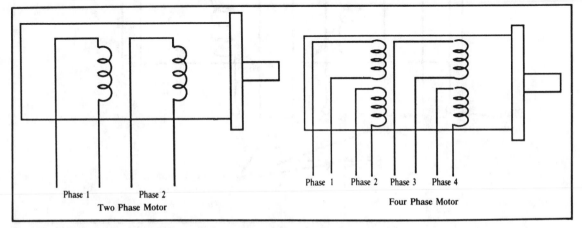

Fig. 14-18. Pictorial diagrams of the coils in a two- and four-phase stepper motor.

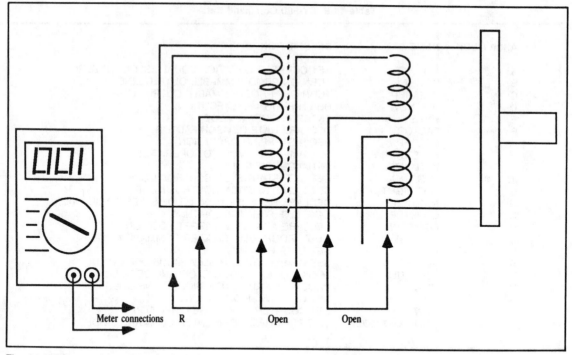

Fig. 14-19. Connection points and possible readings on an 8-wire, four-phase motor.

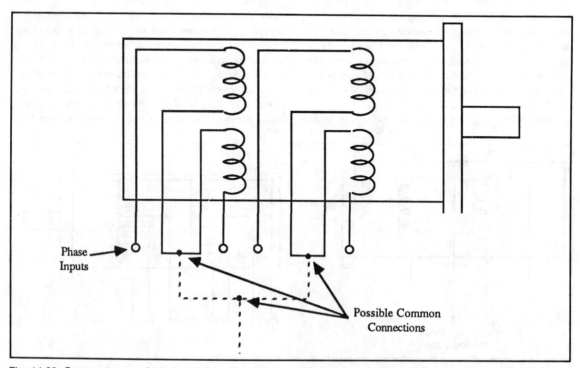

Fig. 14-20. Common connections may reduce the wire count of the stepper motor to five or six instead of eight.

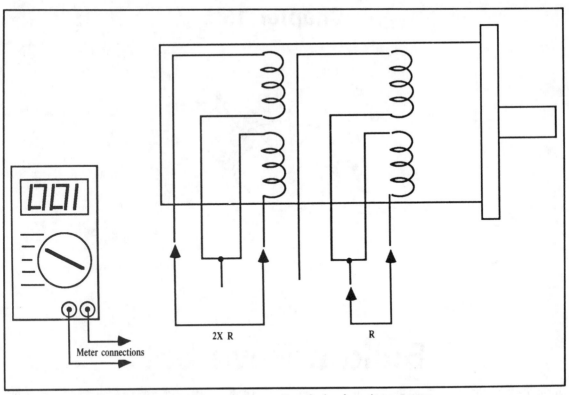

Fig. 14-21. Connection points and possible readings on a 5- or 6-wire, four-phase stepper.

14-20). Decoding this wiring scheme takes some patience, but it can be done. First, separate all those wires where you get an open reading. At the end of your test, there should be two three-wire sets that provide some reading among each of the leads. You should not get a reading when testing wires from each of the two sets.

Locate the common wire by following these steps. Take a measurement of each combination of the wires and note the results. You should end up with three measurements: wires 1 and 2, wires 2 and 3, and wires 1 and 3. The meter readings will be the same for two of the sets. For the third set, the resistance should be roughly doubled. These two wires are the main windings. The remaining wire is the common.

Decoding a five-wire motor is the most straightforward. Measure each wire combination, noting the results of each. When testing the leads to one winding, the result will be a specified resistance (let's call it "R"). When testing the leads to two of the windings, the resistance will be double the value of "R," as shown in Fig. 4-21. Isolate this common wire with further testing and you've successfully decoded the wiring.

Chapter 15

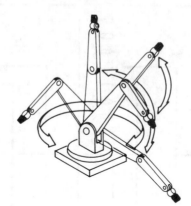

Build a Roverbot

Imagine a robot that can vacuum the floor for you, relieving you of that time-consuming household drudgery and freeing you to do other, more dignified tasks. Imagine a robot that patrols your house, inside or out, listening and watching for the slightest trouble, and sounding the alarm if anything goes amiss. Imagine a robot that knows how to look for fire, and when it finds one, puts it out. Impossible? A dream?

Think again. The compact and versatile Roverbot introduced in this chapter serves the foundation for building any of these more advanced robots. You can easily add a small dc operated vacuum cleaner to the robot, then set it free in your living room. Only the sophistication of the control circuit or computer running the robot limits its effectiveness in actually cleaning the rug.

Light and sound sensors can be attached to the robot, providing it with eyes that help detect potential problems. These sensors, as it turns out, can be the same kind used in household burglar alarm systems. Your only job is connecting them to the robot's other circuits. Similar sensors can be added so your Roverbot actively roams the house, barn, office, or other enclosed area looking for the heat, light, and smoke of fire. An electronically actuated fire extinguisher containing Halon is used to put out the fire.

The Roverbot described in the following pages of this chapter represent the base model only. As usual, other chapters in this book show you how to add onto the basic framework to create a more sophisticated automaton. The Roverbot borrows from techniques covered in Chapter 7, "Building a Metal Platform." If you haven't yet read that chapter, do so now. It will help you get more out of this one. Table 15-1 contains the roverbot parts list.

BUILDING THE BASE

The base of the Roverbot is constructed using shelving standards or extruded aluminum channel stock. The prototype Roverbot used aluminum shelving standards. Aluminum was chosen to keep the weight of the robot to a minimum. The size of the machine didn't seem to call for the heavier duty steel shelving standards.

The base measures 12 5/8-inches by 9 1/8-inches. The unusual dimensions were used to accommodate the galvanized nailing (mending) plates, discussed below. Cut two each 12 5/8-inch stock, with 45 degree miter edges on both sides, as shown in Fig. 15-1. Do the same with the 9 1/8-inch stock. Assemble the pieces using 1 1/4- by 3/8-inch flat corner irons and 8/32 by 1/2-inch nuts and bolts. Be sure the dimensions are as precise as possible,

Table 15-1. Roverbot Parts List.

Frame:

2	12 5/8-inch length aluminum or steel shelving standard
2	9 1/8-inch length aluminum or steel shelving standard
3	4 3/16- by 9-inch galvanized nailing (mending) plate
4	1 1/2- by 3/8-inch flat corner iron
18	1/2-inch by 8/32 stove bolts, nuts, tooth lockwashers

Riser:

4	15-inch length aluminum or steel shelving standard
2	7-inch length aluminum or steel shelving standard
2	10 1/2-inch length aluminum or steel shelving standard
4	1- by 3/8-inch corner angle iron
4	1/2-inch by 3/8 stove bolts, nuts, tooth lockwashers
4	1 1/2-inch by 8/32 stove bolts, nuts, tooth lockwashers

Motors:

2	Gear reduced output 12 volt dc motors
4	2 1/2- by 3/8-inch corner angle iron
12	1/2-inch by 8/32 stove bolts, nuts, tooth lockwashers
2	5- to 7-inch diameter rubber wheels
4	1/2-inch 20 nuts, fender washers, tooth lockwashers (for mounting wheels)

Caster:

2	1 1/4-inch swivel caster
4	1/2-inch by 3/8 stove bolt, nuts, tooth lockwashers washers (as spacers)

Power:

2	6 volt, 8 amp-hour gel-cell or lead-acid batteries
2	Battery clamps (see text)

and that the cuts are straight and even. Because of the mending plates used as a platform, a perfectly square frame is doubly important with this design. Don't bother to tighten the nuts and bolts at this point.

Attach one 4 3/16- by 9-inch mending plate to the left third of the base. Temporarily undo the nuts and bolts on the corners to accommodate the plate. Drill new holes for the bolts in the plate if necessary. Repeat the process for the center and left mending plate. When the three plates are in place, tighten all the hardware. Make sure the plates are secure on the frame by drilling additional holes near the inside corners (don't bother if the corner already has a bolt securing it to the frame). Use 8/32 by 1/2-inch bolts and nuts to attach the plates into place. The finished frame should look something like that depicted in Fig. 15-2. The underside should look like Fig. 15-3.

MOTORS

The Roverbot uses two drive motors for propulsion and steering. These motors, shown in Fig. 15-4, are attached in the center of the frame. The center of the robot was chosen to help distribute the weight evenly across the platform. Tipping over is less likely if the center of gravity is kept as close as possible to the center column of the robot.

The 12 volt motors used in the prototype are commonly available surplus finds (Edmund, Jerryco, H&R, and several other sources have carried them in the past). The motor comes with a built-in gear box that reduces the speed to about 38 rpm. The shafts are 1/4-inch. They were threaded using a 1/4-inch 20 die to secure the 6-inch diameter lawn mower wheels in place. You can skip the threading if the wheels you use have a set screw, or can

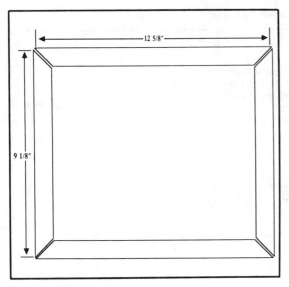

Fig. 15-1. Cutting diagram for the Roverbot.

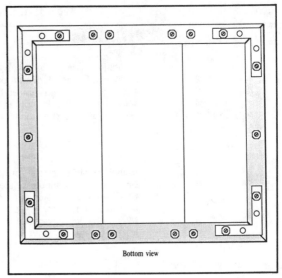

Fig. 15-3. Hardware detail for the frame of the Roverbot (bottom view).

be drilled to accept a set screw. Either way, make sure that the wheels aren't too thick for the shaft. The wheels used in the prototype were 1 1/2-inches wide, perfect for the 2-inch long motor shafts.

Mount the motors using two 2 1/2- by 3/8-inch corner irons, as illustrated in Fig. 15-5. Cut about one inch off one leg of the iron so it will fit against the frame of the motor. Secure the irons to the motor using 8/32 by 1/2-inch bolts. Finally, secure the motors in the center

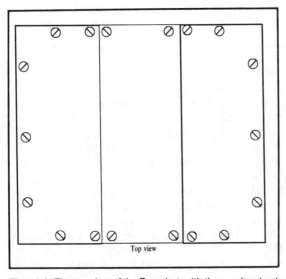

Fig. 15-2. The top view of the Roverbot, with three galvanized mending plates added (holes in the plate not shown!).

Fig. 15-4. One of the drive motors, with wheel, attached to the base of the Roverbot.

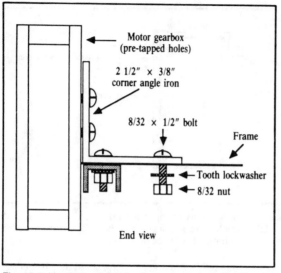

Fig. 15-5. Hardware detail for the motor mount. Cut the angle iron, if necessary, to accommodate the motor.

of the platform using 8/32 by 1/2-inch bolts and matching nuts. Be sure that the shafts of the motors are perpendicular to the side of the frame. If either motor is on crooked, the robot will crab to one side when rolling on the floor. There is generally enough play in the mounting holes on the frame to adjust the motors for proper alignment.

Now attach the wheels. Use reducing bushings if the hub of the wheel is too large for the shaft. If the shaft has been threaded, twist a 1/4-inch 20 nut onto it, all the way to the base. Install the wheel using the hardware shown in Fig. 15-6. Be sure to use the tooth lockwasher. The wheels may loosen and work themselves free otherwise. Repeat the process for the other motor.

SUPPORT CASTERS

The ends of the Roverbot must be supported by swivel casters. Use a two-inch diameter ball-bearing swivel caster, available at most hardware stores. Attach the caster by marking holes for drilling on the bottom of the left and right mending plate. You can use the baseplate of the caster as a drilling guide. Attach the casters using 8/32 by 1/2-inch bolts and 8/32 nuts (see Fig. 15-7). You may need to add a few washers between the caster baseplate and the mending plate to bring the caster level with the drive wheels (the prototype used a 5/16-inch spacer). Do the same for the opposite caster.

If you use different motors or drive wheels, you'll probably need to choose a different size caster to match.

Otherwise, the four wheels may not touch the ground all at once as they should. Before purchasing the casters, mount the motors and drive wheels, then measure the distance from the bottom of the mending plate to the ground. Buy casters to match. Again, add washers to increase the depth, if necessary.

BATTERIES

Each of the drive motors in the Roverbot consume one amp of continuous current with a moderate load. The batteries chosen for the robot, then, need to easily deliver two amps for a reasonable length of time, say one or two hours of continuous use of the motors. A set of hi-capacity Ni-Cad's would fit the bill. But the Roverbot is designed so that subsystems can be added to it. Those subsystems haven't been planned yet, so it's impossible to know how much current they will consume. The best approach to take is to overspecify the batteries, allowing for more current than is probably necessary.

Six and eight amp-hour lead-acid batteries are somewhat common on the surplus market. As it happens, six or eight amps is about the capacity that would handle intermittent use of the drive motors, plus supply continuous power to various electronic subsystems, such as an on-board computer and alarm sensors. These batteries are typically available in six-volt packs, so two are re-

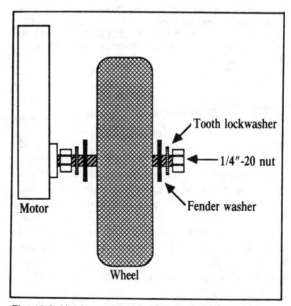

Fig. 15-6. Hardware detail for attaching the wheels to the motor shafts. The wheels can be secured by threading the shaft and using 1/4-inch 20 hardware, as shown, or secured to the shaft using a set screw.

129

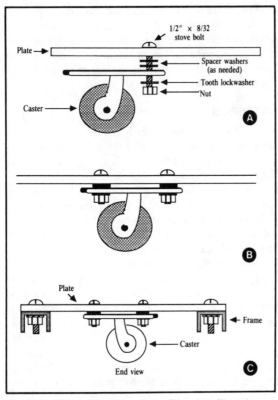

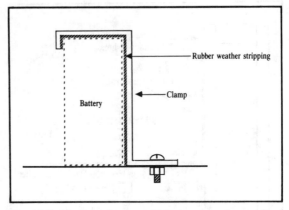

Fig. 15-8. A battery clamp made from a strip of galvanized plate, bent to the contours of the battery. Line the strip with weather stripping for a positive grip.

Fig. 15-7. Adding the casters to the Roverbot. There is one caster on each end, and both must match the depth of the drive wheels. A. Hardware detail; B. Caster mounted on mending plate; C. Mending plate shown secured to the frame of the robot.

or two four-cell "C" battery packs, should they be necessary.

RISER FRAME

The riser frame extends the height of the robot by approximately 15 inches. Attached to this frame will be the sundry circuit boards and support electronics, sensors, fire extinguisher, vacuum cleaner motor, or anything else you care to add. The dimensions are large enough

quired to supply the 12 volts needed by the motors.

Supplementary power, for some of the linear ICs, like op amps, can come from separate batteries, such as a Ni-Cad pack. A set of "C" Ni-Cad's don't take up much room, but it's a good idea to leave space for them now, instead of redesigning the robot later on to accommodate them.

The main batteries are rechargeable so they don't need to be immediately accessible to replace them. But you'll want to use a mounting system that allows you to remove the batteries should the need arise. The clamps shown in Fig. 15-8 allows such accessibility. The clamps are made from 1 1/4-inch wide galvanized mending plate, bent to match the contours of the battery. Rubber weather strip is used on the inside of the clamp to hold the battery firmly in place.

The batteries are positioned off to either side of the drive wheel axis, as shown in Fig. 15-9. This arrangement maintains the center of gravity to the inside center of the robot. The gap also allows for placement of one

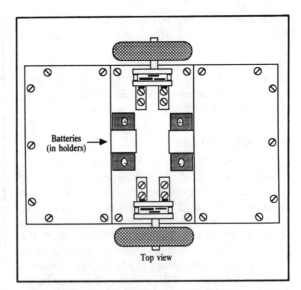

Fig. 15-9. Top view of the Roverbot, showing the mounted motors and batteries. Note the even distribution of weight across the drive axis. This promotes stability, and keeps the robot from tipping over. The wide wheel base doesn't hurt, either.

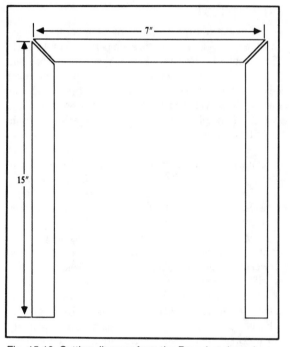

Fig. 15-10. Cutting diagram from the Roverbot riser pieces (two sets).

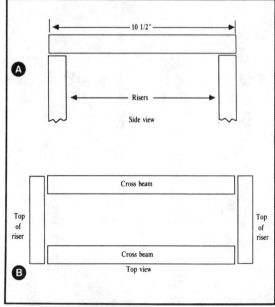

Fig. 15-11. Construction details for the top of the riser. A. Side view showing the cross piece joining the two riser sides; B. Top view showing the cross beams and the tops of the risers.

to assure easy placement of at least a couple of full-size circuit boards, a 2 1/2 pound fire extinguisher, and a Black & Decker DustBuster. You can alter the dimensions of the frame, if desired, to accommodate other add-ons.

Make the riser by cutting four 15 inch lengths of channel stock. One end of each length should be cut at 90 degrees, the other end at 45 degrees. Cut the mitered corners to make pairs, as shown in Fig. 15-10. Make the cross piece by cutting a length of channel stock to exactly seven inches. Miter the ends as shown in the figure.

Connect the two side pieces and cross piece using a 1 1/2-inch by 3/8-inch flat angle iron. Secure the angle iron by drilling matching holes in the channel stock. Attach the stock to the angle iron by using 8/32 by 1/2-inch bolts on the cross pieces, and 8/32 1 1/2-inch bolts on the riser pieces. Don't tighten the screws yet. Repeat the process for the other riser.

Construct two beams by cutting the angle stock to 10 1/2-inches, as illustrated in Fig. 15-11. Do not miter the ends. Secure the beams to the top corners of the risers by using 1- by 3/8-inch corner angle irons. Use 8/32 by 1/2-inch bolts to attach the iron to the beam. Connect the angle irons to the risers using the 8/32 by 1 1/2-inch bolts installed earlier. Add a spacer between the inside of the

channel stock and the angle iron if necessary, as shown in Fig. 15-12. Use 8/32 nuts to tighten everything in place.

Attach the riser to the baseplate of the robot using 1- by 3/8-inch corner angle irons. As usual, use 8/32 by 1/2-inch bolts and nuts to secure the riser into place. The finished Roverbot body and frame should look at least something like the one in Fig. 15-13.

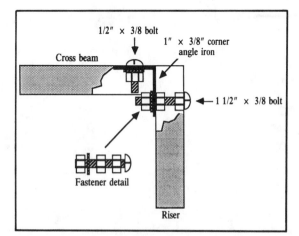

Fig. 15-12. Hardware detail for attaching the risers to the cross beams.

131

Fig. 15-13. The finished Roverbot (minus the batteries), ready for just about any enhancement you see fit.

STREET TEST

You can test the operation of the robot by connecting the motors and battery to a temporary control switch. See Chapter 5, "Robots of Plastic," for a wiring diagram.

With the components listed in the text, the robot should travel at a speed of about one foot per second. The actual speed will probably be under that because of the weight of the robot. Fully loaded, the Roverbot will probably travel at a moderate speed of about eight or nine inches per second. That's just right for a robot that vacuums the floor, roams the house for fires, and protects against burglaries. If you need your Roverbot to go a bit faster, the easiest (and cheapest) way is to use larger wheels. Using eight-inch wheels will make the robot travel at a top speed of 15 inches per second.

One problem with using larger wheels, however, is that it raises the center of gravity of the robot. Right now, the center of gravity is kept rather low, thanks to the low position of the two heaviest objects, the batteries and motors. Jacking up the robot using larger wheels puts the center of gravity higher, so there is a slightly greater chance of tipping over. You can minimize any instability by making sure that subsystems are added to the robot from the bottom of the riser, and that the heaviest parts are positioned closest to the base. You can also mount the motor on the bottom of the frame instead of on top.

Chapter 16

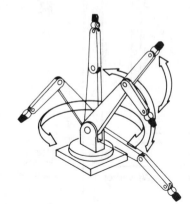

Build A Six-Legged Walking Robot

Let's be honest with each other. Do you like challenges? Do you like being faced with problems that demand decisive action on your part? Do you like spending many long hours tinkering in the garage or workshop? Do you like the idea of building the ultimate robot, one that will amaze you and your friends? If the answer is yes to all the questions, then maybe you're ready to build the Walkerbot, described in-depth in this chapter.

This strange and unique contraption walks on six legs, turning corners with an ease and grace that belies its rather simple design. The basic Walkerbot frame and running gear can be used to make other types of robots as well. In Chapters 17, "Advanced Locomotion Systems," you'll see how to convert the Walkerbot to tracked or wheeled drive. The conversion is simple and straightforward. In fact, you can switch back and forth between drive systems.

The Walkerbot designs described in this chapter are for the basic frame, motor, battery system, running gear, and legs. You can embellish the robot with additional components, such as arms, a head, computer control, you name it. The frame is oversized (in fact, it's too large to fit through some inside doors!), and there's plenty of room to add new subsystems. The only requirement is that the

weight doesn't exceed the driving capacity of the motors and batteries, and that the legs and axles don't bend. The prototype Walkerbot, as shown on the cover of this book, weighs about 50 pounds. It moves along swiftly and no structural problems have yet occurred. Another 10 or 15 pounds could be added without worry.

FRAME

The completed Walkerbot frame measures 18 inches wide by 24 inches long by 12 inches deep. Construction is all aluminum, using a combination of 41/64- by 1/2- by 1/16-inch channel stock and 1- by 1- by 1/16-inch angle stock.

Build the bottom of the frame by cutting two 18-inch lengths of channel stock and two 24-inch lengths of channel stock, as shown in Fig. 16-1. Miter the ends. Attach the four pieces using 1 1/2- by 3/8-inch flat angle irons and secure with 3/8 by 1/2-inch bolts and nuts. For added strength, use four bolts on each corner.

In the prototype Walkerbot, I replaced many of the nuts and bolts with aluminum pop rivets. Until the entire frame is assembled, however, use the bolts as temporary fasteners. Then, when the frame is assembled, square it up and replace the bolts and nuts with rivets

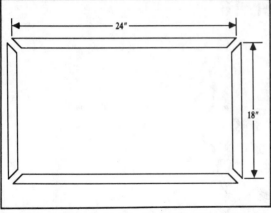

Fig. 16-1. Cutting diagram for the frame of the Walkerbot (two sets).

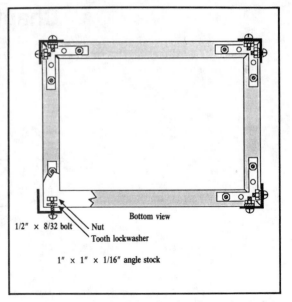

Fig. 16-2. Hardware detail for securing the angle stock to the top and bottom frame pieces.

one at a time. Construct the top of the frame in the same manner.

Connect the two halves with four 12-inch lengths of angle stock, as shown in Fig. 16-2. Secure the angle stock to the frame pieces by drilling holes at the corners. Use 8/32 by 1/2-inch bolts and nuts initially; exchange for pop rivets after the frame is complete. The finished frame should look like the one diagrammed in Fig. 16-3.

Complete the basic frame by adding the running gear mounting rails. Cut four 23 7/8-inch lengths of 1- by 1- by 1/16-inch angle stock and two 17 5/8-inch lengths of the same angle stock. Drill 1/4-inch holes in four long pieces as shown in Fig. 16-4. The spacing between the sets of holes is important. If the spacing is incorrect, the U-bolts won't fit properly.

Refer to Fig. 16-5. When the holes are drilled, mount two of the long lengths of angle stock as shown. The holes should point up, with the side of the angle stock flush against the frame of the robot. Mount the two short lengths on the ends. Tuck the short lengths immediately under the two long pieces of angle stock you just secured. Use 8/32 by 1/2-inch bolts and nuts to secure the pieces together. Dimensions, drilling, and placement are criti-

cal with these components. Put the remaining two long lengths of drilled angle stock aside for the time being. The parts needed for the frame are in Table 16-1.

LEGS

You're now ready to construct and attach the legs. This is probably the hardest part of the project, so take your time and measure everything twice to assure accuracy. Cut six 14-inch lengths of 57/64- by 9/16- by 1/16-inch aluminum channel stock. Do not miter the ends. Drill a hole with a #19 bit 1/2-inch from one end (the top); drill a 1/4-inch hole 4 3/4-inches from the top (see Fig. 16-6). Make sure the holes are in the center of the channel stock.

With a 1/4-inch bit, drill out the center of six 4 5/8-inch diameter circular electric receptacle plate covers. The plate cover should have a notched hole near the outside,

Table 16-1. Walkerbot Frame Parts List.

4	24-inch lengths 41/64- by 1/2- by 1/16-inch aluminum channel stock
4	18-inch lengths 41/64- by 1/2- by 1/16-inch aluminum channel stock
4	12-inch lengths 1- by 1- by 1/16-inch aluminum angle stock
8	1 1/2- by 3/8-inch flat angle iron
48	1/2-inch by 8/32 stove bolts, nuts, tooth lockwashers
4	23 7/8-inch lengths 1- by 1- by 1/16-inch aluminum angle stock
2	17 5/8-inch lengths 1- by 1- by 1/16-inch aluminum angle stock

134

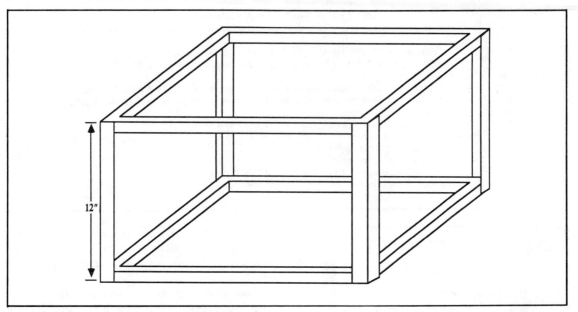

Fig. 16-3. How the Walkerbot frame should look so far.

used to secure it to the receptacle box. If the cover doesn't have the hole, drill one with a 1/4-inch bit 3/8-inch from the outside edge. The finished plate cover becomes a cam for operating the up and down movement of the legs.

Assemble four legs as follows: Attach the 14-inch long leg piece to the cam using a 1/2-inch length of 1/2 Schedule 40 PVC pipe and hardware, as shown in Fig. 16-7. Be sure the ends of the pipe are filed clean and that the cut is as square as possible. The bolt should be tightened against the cam but should freely rotate within the leg hole.

Assemble the remaining two legs in a similar fash-

ion, except use a 2-inch length of PVC pipe and a 3-inch stove bolt. These two legs will be placed in the center of the robot, and will stick out from the others. This allows the legs to cross one another without interfering with the gait of the robot. The bearings used in the prototype were 1/2-inch diameter closet door rollers.

Now refer to Fig. 16-8. Thread a 5-inch by 1/4-inch 20 carriage bolt through the center of the cam, using the hardware shown. Install the wheel bearings to the shafts at this time, 1-inch from the cam. The 1 1/4-inch diameter bearings are the kind commonly used in lawn mowers and are readily available. The bearings used in the prototype

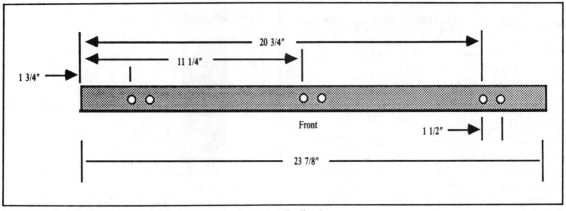

Fig. 16-4. Cutting and drilling guide for the motor mount rails (four).

135

Table 16-2. Walkerbot Legs Parts List.

6	14-inch lengths 57/64- by 9/16- by 1/16-inch aluminum channel stock
6	6-inch lengths 41/64- by 1/2- by 1/16-inch aluminum channel stock
6	Roller bearings
4	1 1/4-inch by 10/24 stove bolts, nuts, tooth lockwashers, locking nuts, flat washers
4	1 3/4-inch by 10/24 stove bolts, nuts, tooth lockwashers, locking nuts, flat washers
4	3-inch 10/25 stove bolts
12	1/2-inch by 8/32 stove bolts, nuts, tooth lockwashers
6	Steel electrical covers (4 5/8-inch diameter)
6	5-inch hex-head carriage bolt
6	2- by 3/8-inch flat mending iron
6	1 1/4-inch 45 degree "ELL" Schedule 40 PCV pipe fitting
Misc.	1/2-inch Schedule 40 PVC cut to length (see text)

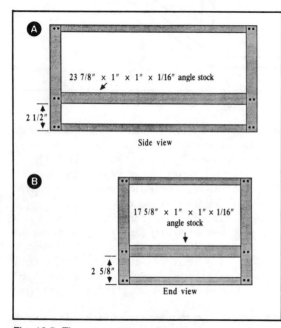

Fig. 16-5. The motor mount rails secured to the robot. A. The long rail mounts 2 1/2 inches from the bottom of the frame (the holes drilled earlier point up); B. The short end cross piece rail mounts 2 5/8 inches from the bottom of the frame.

had 1/2-inch hubs. A 1/2-inch-to-1/4-inch reducing bushing was used to make the bearings compatible with the diameter of the shaft.

Install 3 1/2-inch diameter 30 tooth #25 chain sprocket (another size will also do, as long as all the leg mechanism sprockets in the robot are the same size). Like the bearings, a reducing bushing was used to make the 1/2-inch I.D. hubs of the sprockets fit on the shaft. The exact positioning of the sprockets on the shaft is not im-

portant at this time, but follow the spacing diagram shown in Fig. 16-9 as a guide. You'll have to fine-tune the sprockets on the shaft as a final alignment procedure anyway.

Once all the legs are complete, install them on the robot using U-bolts. The 1 1/2-inch wide by 2 1/2-inch long by 1/4-inch 20 thread U-bolts fit over the bearings perfectly. Secure the U-bolts using the 1/4-inch 20 nuts supplied.

Refer to Fig. 16-10 for the next step. Cut six 6-inch lengths of 41/64- by 1/2- by 1/16-inch aluminum channel stock. With a #19 bit, drill holes 3/8-inch from the top and bottom of the rail. With a nibbler tool, cut a 3 1/2-inch

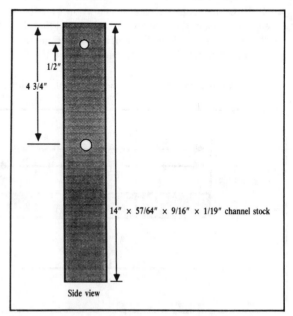

Fig. 16-6. Cutting and drilling guide for the six legs.

136

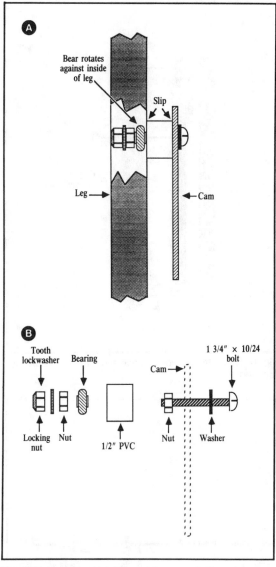

Fig. 16-7. Hardware detail for the leg cam. A. Complete cam and leg; B. Exploded view. Note that two of the legs use a 2-inch piece of PVC and a 3-inch bolt.

do it for you (it'll save you an hour or two of blister-producing nibbling!). An alternative method, which requires no slot cutting, is shown in Fig. 16-11. Be sure to mount the double rails parallel to them.

Mount the rails using 8/32 by 2-inch bolts and 8/32 nuts. Make sure the rails are directly above the shaft of each leg, or the legs may not operate properly. You'll have to drill through both walls of the channel in the top of the frame.

The rails serve to keep the legs aligned in the up and down piston-like stroke of the legs. Attach the legs to the rails using 3/8-inch by 1 1/2-inch bolts. Use nuts and locking nut fasteners as shown in Fig. 16-12. This finished leg mechanism should look like the one depicted in Fig. 16-13. Use grease or light oil to lubricate the slot. Be sure that there is sufficient play between the slot and the bolt stem. The play cannot be excessive, however, or the leg may bind as the bolt moves up and down inside the slot. Adjust the sliding bolt on all six legs for proper clearance.

Drill small pilot holes in the side of six 45 degree 1 1/4-inch PVC pipe elbows. These serve as the feet of

slot in the center of each rail. The slot should start 1/2-inch from one end.

Alternatively, you can use a router, motorized rasp, or other tool to cut the slot. In any case, make sure the slot is perfectly straight. Once cut, polish the edges with a piece of 300 grit wet-dry emory paper, used wet. Use your fingers to find any rough edges. There can be none. This is a difficult task to do properly, and you may want to take this portion to a sheet metal shop and have them

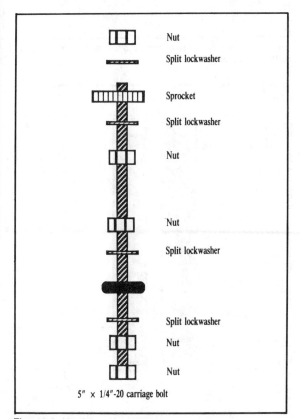

Fig. 16-8. Hardware detail of the leg shafts.

137

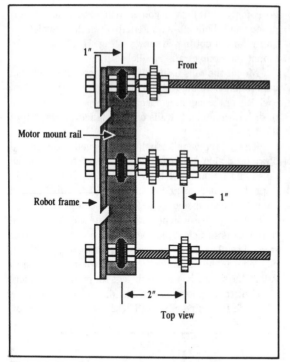

Fig. 16-9. The leg shafts attached to the motor mount rails (left side shown only).

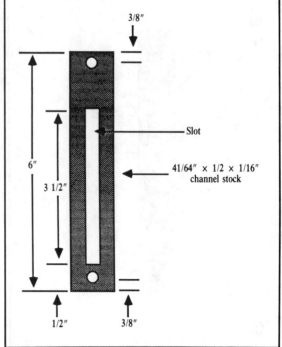

Fig. 16-10. Cutting and drilling guide for the cam sliders (six required).

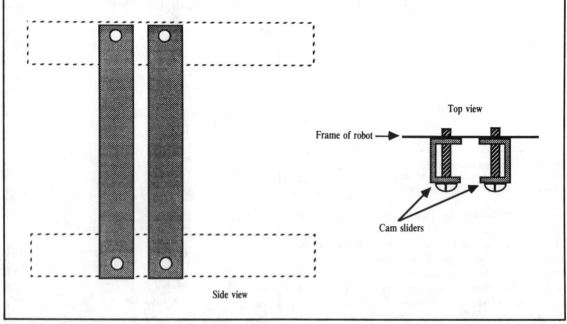

Fig. 16-11. An alternate approach to the slotted cam sliders.

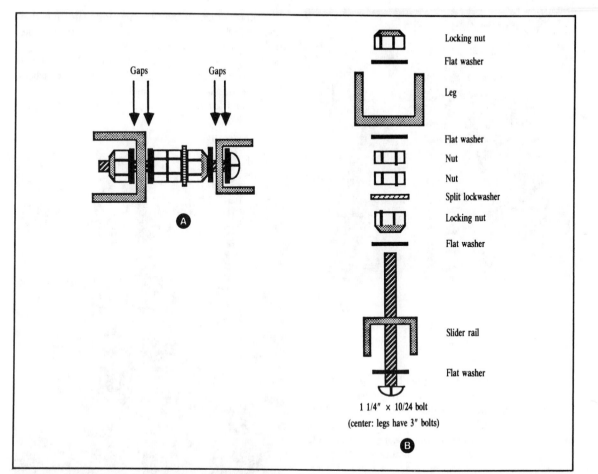

Fig. 16-12. Cam slider hardware detail. A. Complete assembly; B. Exploded view. Note that the center legs have 3-inch bolts.

Fig. 16-13. The slider cam and hardware. The slot must be smooth and free of burrs, or the leg will snag.

legs. Paint the feet if you wish at the point. Using #10 wood screws, attach a 2- by 3/8-inch flat mending iron to each of the elbow feet. Drill 1/4-inch holes 1 1/4-inches from the bottom of the leg. Secure the feet onto the legs using 1/2-inch by 1/4-inch 20 machine bolts, nuts, and lockwashers. Apply a 3-inch length of rubber weather strip to the bottom of each foot for better traction. The leg should look like the one in Fig. 16-14. The legs should look like the one in Fig. 16-15. A close-up of the cam mechanism is shown in Fig. 16-16.

MOTORS

The motors used in the prototype Walkerbot were surplus finds originally intended as the driving motors in a child's motorized bike or go-cart. The motors have a fairly high torque at 12 volts dc and a speed of about 600 rpm. A one-step reduction gear was added to bring

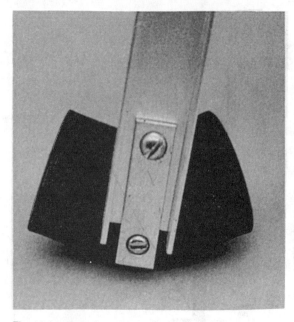

Fig. 16-14. PVC plumbing fittings used as feet. Secure the feet using a flat mending iron.

Before attaching the two mending plates together, thread a 28 1/2-inch length of #25 chain over the sprocket (the exact length can be one or two links off; you can correct for any variance later on). Assemble the two plates using 8/32 by 3-inch bolts and 8/32 nuts and lockwashers. Separate the plates using 2-inch spacers.

Attach the two remaining 27 7/8-inch lengths of an-

Fig. 16-15. One of six legs, completed (shown already attached to the robot).

the speed down to about 230 rpm. The output speed is further reduced to about 138 rpm with the use of a drive sprocket. For a walking machine, that's about right, although it could stand to be a bit slower. Electronic speed reduction can be used to slow the motor output down to about 100 rpm. You can use other motors and other driving techniques as long as the motors have a (pre-reduced) torque of at least 6 lb-ft. and a speed that can be reduced to 140 rpm or so.

The motors are mounted inside two 6 1/2- by 1 1/2-inch mending plate T's. Drill a large hole, if necessary, for the shaft of the motor to stick out, as shown in Fig. 16-17. The motors used in the prototype came with a 12 pitch 12 tooth nylon gear. The gear was not removed for assembly, so the hole had to be large enough for it to pass through. The 30 tooth 12 pitch metal gear and 18 tooth 1/4-inch chain sprocket were also sandwiched between the mending plates.

The 1/4-inch shaft of the driven gear and sprocket is free-running; you can install a bearing on each plate, if you wish, or have the shaft freely rotate in oversize holes. The sprocket and gear have 1/2-inch I.D. hubs, so reducing bushings were used. The sprocket and gear are held in place with compression. Don't forget the split washers. They provide the necessary compression to keep things from working loose.

Fig. 16-16. A closeup detail of the leg cam.

as possible for the idler; otherwise, you may need to shorten or lengthen the chain). Thread the chain around the sprocket, and find a position along the rail until the chain is taut (but not overly tight). Make a mark using the center of the sprocket as a guide and drill a 1/4-inch hole in the rail. Attach the sprocket to the robot. Figures

gle bracket on the robot, as shown in Fig. 16-18. The stock mounts directly under the two end pieces. Use 1/2-inch by 8/32 bolts and nuts to secure the cross pieces in place. Secure the leg shafts using 1 1/4-inch bearings and U-bolts.

Mount the motor to the newly added inner mounting rails using 3- by 1/2-inch mending plate T's. Fasten the plates onto the motor mount, as shown in Fig. 16-19, with 8/32 by 1/2-inch bolts and nuts. Position the shaft of the motor approximately 7 inches from the back of the robot (you can make any end the back, it doesn't matter). Thread the chain over the center sprocket and the end sprocket. Position the motor until the chain is taut. Mark holes and drill. Secure the motor and mount to the frame using 8/32 by 1/2-inch bolts and nuts. Repeat the process for the opposite motor. The final assembly should look like Fig. 16-20.

Thread a 28 1/2-inch length of #25 chain around sprockets of the center and front legs. Attach an idler sprocket 7 1/2 inches from the front of the robot in line with the leg mounts (use a diameter as close to 2-inches

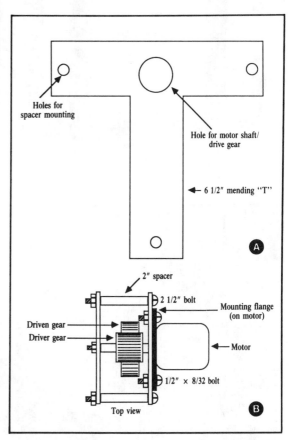

Fig. 16-17. Motor mount detail. A. Drilling guide for the Mending "T"; B. The motor and drive gear/sprocket mounted with two mending "T"s.

Table 16-3. Walkerbot Motor Mount/Drive System Parts List.

4	6 1/2-inch galvanized mending plate "T"
4	3-inch galvanized mending plate "T"
2	Heavy-duty gear-reduction dc motors
12	3 1/2-inch diameter 30-tooth #15 chain sprocket
4	28 1/2-inch length #25 chain
12	2 1/2- by 1 1/2- by 1/4-inch 20 "U" bolts, with nuts and tooth lockwashers
12	1 1/2-inch O.D. 1/4- to 1/2-inch ID bearing
Misc.	Reducing bushings

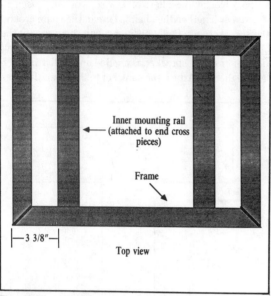

Fig. 16-18. Mounting location of the inner rails.

16-21 through 16-23 show the motor mount, idler sprocket, and chain locations.

BATTERIES

The Walkerbot is not a lightweight robot, and its walking design requires about 30 percent more power than a wheeled robot. The batteries for the Walkerbot are not trivial. You have a number of alternatives. One workable approach is to use two 6 volt motorcycle batteries, each rated at about 30 Ah. The two batteries together equal in size and weight a slimmed-down version of a car battery.

You can also use a 12 volt motorcycle or dune buggy battery, rated at more than 20 Ah. The prototype Walkerbot used 12 Ah six volt gel cell batteries. The amp-hour capacity is a bit on the low side, considering the two-amp draw from each motor, and the planned heavy use of electronics and support circuits. In tests, the 12 Ah batteries provided about two hours of use before requiring a recharge.

Fig. 16-19. One of the drive motors mounted on the robot using smaller galvanized mending "T"'s.

Fig. 16-20. Drive motor attached to the Walkerbot, with drive chain joining the motor to the leg shafts.

There is plenty of room to mount the batteries. A good spot is slightly behind the center legs. By off-setting the batteries a bit in relation to the drive motors, you re- store the center of gravity to the center of the robot. Of course, other components you add to the robot can throw the center of gravity off. Add one or two articulated arms

Fig. 16-21. Mounting locations for idler sprocket.

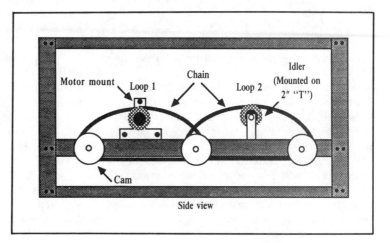

TESTING AND ALIGNMENT

You can test the operation of the Walkerbot by temporarily installing a wired control box. The box consists of two DPDT switches wired to control the forward and backward motion of the two legs. See Chapter 5, "Robots of Plastic," for more details and a wiring diagram.

Before you test the Walkerbot, you need to align its legs. The legs on each side should be positioned so that either the center leg touches the ground, or the front and back leg touches the ground. When the two sets of legs are working in tandem, the walking gait should look like Fig. 16-25. This gait is the same as an insect's, and provides a great deal of stability. To turn, one set of legs stop (or reverse) while the other set continues. During this time, the tripod arrangement of the gait will be lost, but the robot will still be supported by at least three legs.

An easy way to align the legs is to loosen the chain sprockets (so you can move the legs independently) and position the middle leg all the way forward and the front and back legs all the way back. Retighten the sprockets, being on the lookout for misalignment of the chain and sprockets. If a chain bends to mesh with a sprocket, it is likely to pop off when the robot is in motion.

During testing, be on the lookout for things that rub, squeak, and work loose. Keep your wrench handy and adjust gaps and tighten bolts as necessary. Add a dab of silicone spray oil to those parts that seem to be binding. You may find that a sprocket or gear doesn't stay tightened on a shaft. Look for ways to better secure the component to the shaft, such as a set screw or another split lockwasher. It may take several hours of tuning up to get the robot working to top efficiency.

Once the robot is aligned, run it through its paces walking over level ground, stepping over small rocks and ditches, and navigating tight corners. Keep an eye on your watch to see how long the batteries provide power. You may need to upgrade the batteries if they cannot provide more than an hour of fun and games.

The Walkerbot is ideally suited for expansion. Figure 16-26 shows an arm attached to the front side of the robot. You can add a second arm on the other side for more complete dexterity. Attach a dome on the top of the robot, and you've added a head on which you can attach a video camera, ultrasonic ears and eyes, and lots more. The enhanced version of the Walkerbot is shown in the front cover of this book. The black plastic top is for looks. Additional panels can be added to the front and back ends. Attach the panels using Velcro strips. That way, you can easily remove the panels should you need quick access to the inside of the robot.

Fig. 16-22. A view of the mounted motor, with chain drive.

to the robot, and the weight suddenly shifts toward the front. For flexibility, why not mount the batteries on a sliding rail, thus allowing you to shift their position forward or back depending on the other weight you add to the Walkerbot.

The complete Walkerbot, minus the batteries, is shown in Fig. 16-24. Some additional hardware and holes may be apparent on this version. Pay no attention to it. These were either my mistakes (! !), or were for components removed for the picture.

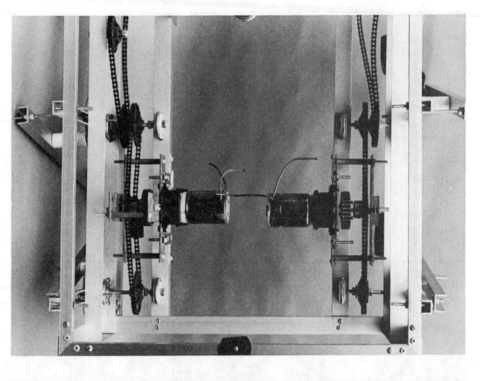

Fig. 16-23. Left and right motors attached to the robot.

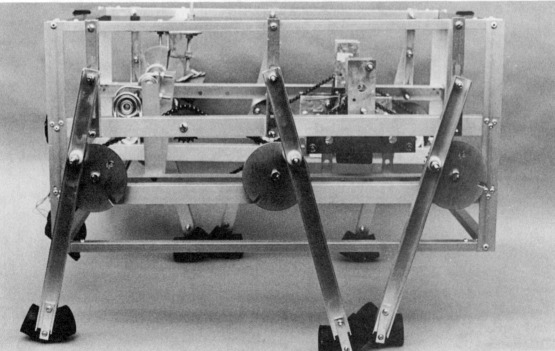

Fig. 16-24. The completed Walkerbot.

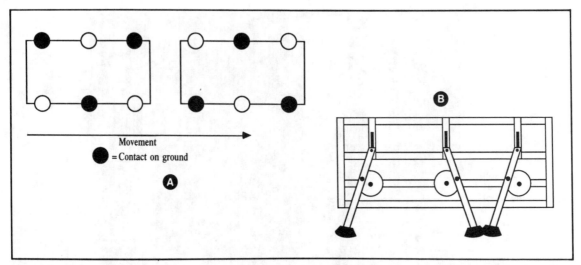

Fig. 16-25. The walking gait. A. The alternating tripod walking style of the Walkerbot, shared with thousands of crawling insects; B. The positioning of the legs for proper walking (front and back legs in synchronism, middle leg 180 degrees out of sync). The middle leg doesn't hit either the front or back leg because it is further from the body of the robot.

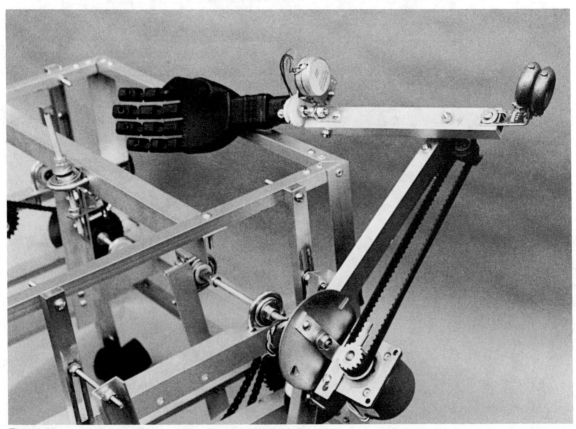

Fig. 16-26. An arm attached to the front side of the Walkerbot.

Chapter 17

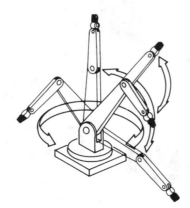

Advanced Locomotion Systems

Two drive wheels aren't the only way to move a robot across the living room or workshop floor. If you read Chapter 16, you learned about how to build a six-legged walking robot. Here, in this chapter, you'll learn the basics of how to apply some other unique drive systems to propel your robot designs, including a stair-climbing robot, an outdoor tracked robot, even a six-wheeled buggybot.

TRACKS O' MY ROBOT

There is something exciting about seeing a tank climb embankments, bounding over huge boulders as if they were tiny dirt clods. A robot with tracked drive is a perfect contender for an automaton that's designed for outdoor use. Where a wheeled or legged robot can't go, the tracked robot can roll in with relative ease. Experimental tracked robots, using metal tracks just like tanks, have been designed for the U.S. Navy and Army, and are even used by many police and fire departments. Everyone has liked what they've seen so far; development of tracked autonomous vehicles continues still.

Using a heavy metal track for your personal robot is decidedly a bad idea. The track is too heavy and much too hard to fabricate. For a homebrew robot, a rubber track is more than adequate. A large timing belt, even an automotive fan belt, can be used for the track. Another alternative I've used with some success is rubber wetsuit material. Most diving shops have long strips of the rubber lying around that they'll sell or give to you. The rubber can be mended using a special water-proof adhesive. You can glue the strip together to make a band, then glue on small rubber cleats to the band. Figure 17-1 shows the basic idea.

The drive train for a tracked robot must be specifically engineered or modified for the task. The Walkerbot described in Chapter 16 makes a good base for a tracked robot. Remove the legs and install three small drive pulleys, as diagrammed in Fig. 17-2. The track fits inside the groove of the pulleys, so the track won't easily slip out. Rollers must be added to the bottom of the carriage against which the track passes. Unless the track is thick, there can be no groove in the rollers. Otherwise, the track would ride inside the rollers, instead of the outside. Wide rubber tires make good rollers.

With this design, and under certain circumstances, the track may pop off the rollers and drive wheels. To help minimize throwing the track every few minutes, add a guide roller to the bottom of the carriage, as di-

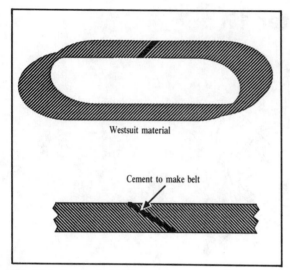

Fig. 17-1. A wetsuit drive belt.

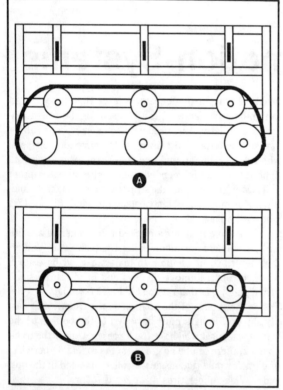

Fig. 17-2. Two ways to add track drive to the Walkerbot presented in Chapter 16. A. Track roller arrangement for good traction and stability, but relatively poor turning radius; B. Track roller arrangement for good turning radius, but hindered traction and stability.

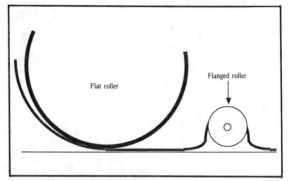

Fig. 17-3. A close-up view of the flanged roller used to prevent the track from popping off the drive.

agrammed in Fig. 17-3. The track rides inside a groove (or flange) in the small guide roller, thus preventing the track from popping out of place.

To propel the robot, you activate both motors so that the tracks move in the same direction and at the same speed. To steer, you simply stop or reverse one side. For example, to turn left, stop the left track. To make a hard left turn, reverse the left track.

The Walkerbot has six driven wheels. The three wheels on each side are linked together, so they all provide power to the track. But you don't need a three-wheel drive system. In fact, you can usually get by with just one driver wheel on each side of the robot.

STEERING WHEEL SYSTEMS

Using dual motors to affect propulsion and steering is just one method of getting your robot around. Another approach is to use a pivoting wheel to steer the robot. The same wheel can provide power, or it can come from two wheels in the rear (the latter is much more common). The arrangement is not unlike golf carts, where the two rear wheels provide power and a single wheel in the front provides steering. See Fig. 17-4 for a diagram of a typical steering wheel robot. Figure 17-5 shows a detail of the steering mechanism.

The advantage to a steering wheel robot is that you need only one powerful drive motor. The motor can power both rear wheels at once. The steering wheel motor needn't be as powerful, since all it has to do is swivel the wheel back and forth a few degrees.

The biggest disadvantage to steering wheel systems is the steering! You must build stops into the steering mechanisms (either mechanical or electronic) to prevent the wheel from turning more than 50 or 60 degrees to either side. Angles greater than about 60 degrees causes

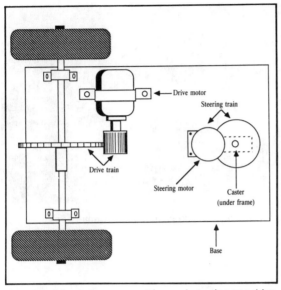

Fig. 17-4. A basic arrangement for a robot using one drive motor and a steering wheel.

the robot to suddenly steer in the other direction. It may even cause the robot to lurch to a sudden stop because the front wheel is at a right angle to the rear wheels.

The servo mechanism controlling the steering wheel must know when the wheel is pointing forward. The wheel must return to this exact spot when the robot is

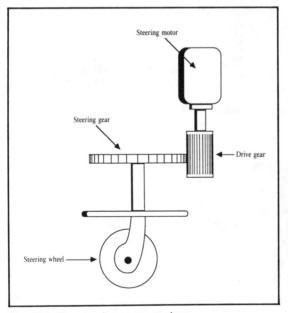

Fig. 17-5. The steering gear, up close.

commanded to forge straight ahead. Most servo mechanisms aren't this accurate. The motor may stop one or more degrees off the center point, and the robot may never actually travel in a straight line. A good steering motor, and a more sophisticated servo mechanism, can reduce this limitation.

A number of robot designs with steering wheel mechanisms have been described in other robot books. Check out Appendix B, "Further Reading," for more information.

SIX-WHEELED ROBOT CART

The Walkerbot described in Chapter 16 can also be modified into a six-wheeled rugged terrain cart, or buggybot. Simply remove the legs and attach wheels, as diagrammed in Fig. 17-6. The larger the wheels the better, as long as they aren't over about nine inches (the centerline diameter between each drive shaft).

Pneumatic wheels are the best choice, because they provide more bounce and handle rough ground better than hard rubber tires. Most hardware stores carry a full assortment of pneumatic tires. Most are designed for things like wheelbarrows and hand dollies. Cost can be high, so you may want to check out the surplus or used industrial supply houses. Just be sure the tire doesn't have a flat!

Steering is accomplished as with a two wheeled robot. The series of three wheels on each side acts as a kind of track tread, so the vehicle behaves much like a tracked vehicle. Maneuverability isn't as good as with a two-wheeled robot, but you can still turn the robot in a radius a little longer than its length. Sharp turns require you to reverse one set of wheels while applying forward motion to the other.

TRI-STAR WHEELS

In the science fiction film "Damnation Alley," starring Jan-Michael Vincent and George Peppard, the earth

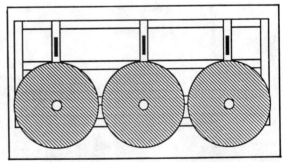

Fig. 17-6. Converting the Walkerbot (from Chapter 16) into a six-wheeled all-terrain buggybot.

Fig. 17-7. The "Landmaster," from the motion picture "Damnation Alley." Photo courtesy 20th Century Fox.

Fig. 17-8. A close-up of the tri-star wheels used in the Landmaster.

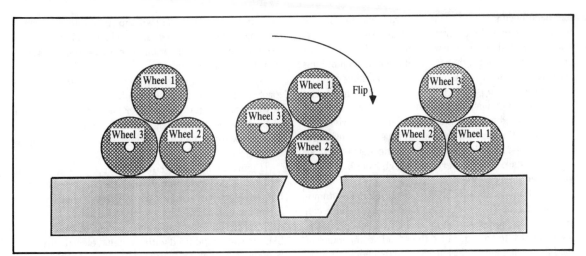

Fig. 17-9. How the wheel gang "flips" when it encounters a hole or obstacle. The same basic motion can be used to climb stairs.

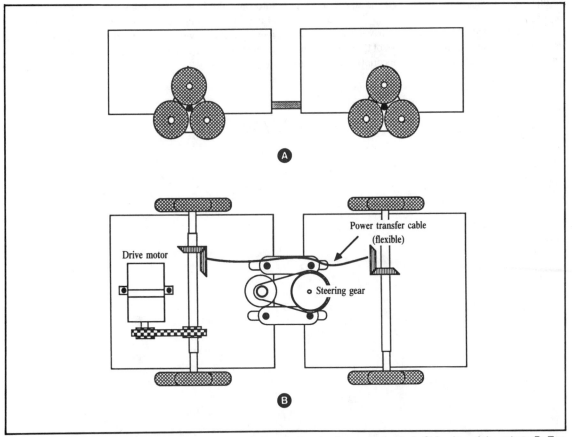

Fig. 17-10. A preliminary design for a robot based on the tri-star wheel arrangement. A. Side view of the robot; B. Top view of the robot showing the one drive motor and the central steering motor.

has been decimated by an atomic war, and the heroes must trek across the country, amid radioactive storms, marauders, and other post-war denizens. To help them get there in one piece, they use an incredible 35-foot long steel vehicle called the Landmaster. This thing crashes through walls, hops over large ditches as if they were nothing but potholes, even swims across the water (see the movie production still in Fig. 17-7). The drive system used by the Landmaster is an unusual tri-star arrangement that's perfectly adept for robotic tasks such as climbing stairs, maneuvering through rough terrain, yes—even going through water.

The Landmaster was built by Dean Jeffries Automotive Styling, of Hollywood, CA. The full-size vehicle can still be seen parked in the lot of his workshop off Cahuenga Blvd. Jeffries modeled the wheel arrangement of the Landmaster after a design patented (but never actively used) by Lockheed Aircraft for an all-terrain vehicle. Each of the four wheels on the vehicle is actually a set of three smaller wheels, clustered in a triangle, as shown in Fig. 17-8.

All three tires continually rotate, driven by a central shaft, but in normal operation, only two of them are touching the ground. When the vehicle encounters an obstacle or hole, the wheel gang will flip and rotates the wheels into a new position, as diagrammed in Fig. 17-9. You have to either see the movie or build your own robot based on this design to believe it.

All four wheel gangs are powered by a central motor. Steering is accomplished by bending the midsection of the robot. In the Landmaster vehicle, this is accomplished with the use of hydraulic rams, but for a homebrew robot, you'd probably use a servo or stepping motor.

Figure 17-10 shows how such a robot might be constructed. One large motor, which rests in one of the robot halves, drives the four wheel gangs. Since the robot is made in two parts, and the midsection bends, the wheels in the other half of the robot are driven via a flexible cable, the same kind used with handheld electric drills. To balance the robot, the batteries are placed in the second half. A wiring harness connects the two halves together.

In his book *Android Design,* Marty Weinstein goes into some detail in constructing a robot based on the tri-star wheel arrangement. Consult his book for further ideas. Marty provides some interesting design equations to ensure that the robot will indeed climb stairs at a 45 degree angle.

Chapter 18

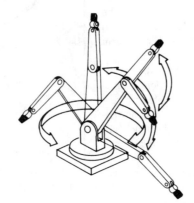

An Overview of Arm Systems

Robots without arms are limited to rolling or walking about, perhaps noting things than occur around them, but little else. The robot can't, as the slogan goes, "reach out and touch someone," and it certainly can't manipulate its world.

The more sophisticated robots in science, industry, and research/development have at least one arm for the purpose of grasping, reorienting, or moving objects. Arms extend the reach of robots and make them more human-like. For all the extra capabilities they provide a robot, it's interesting that arms aren't at all difficult to build. Your arm designs can be used as factory-style stationary "pick and place" robots, or they can be attached to a mobile robot as an appendage.

This chapter deals with the concept and design theory of robotic arms. Specific arm projects are given in Chapters 19 and 20. Incidentally, when we speak of arms, the discussion usually means just the arm mechanism, minus the hand (also called the gripper). Chapter 21, "Experimenting with Gripper Designs," talks about how to construct robotic hands, and how they can be added to arms to make a complete, functioning appendage.

THE HUMAN ARM

Take a close at your own arms for a moment or two.

You'll quickly notice a number of important points. First, your arms are amazingly adept mechanisms, no doubt about it. They are capable of being maneuvered into just about any position you want. Your arm has two major joints: the shoulder and the elbow (the wrist, as far as robotics is concerned, is considered part of the gripper mechanism). Your shoulder can move in two planes, both up and down and back and forth. Move your shoulder muscles up, and your entire arm is raised away from your body. Move your shoulder muscles forward, and your entire arm moves forward. The elbow joint is capable of moving in two planes, as well: back and forth, and up and down.

The joints in your arm, and your ability to move them, is called *degrees of freedom*. Your shoulder provides two degrees of freedom in itself: shoulder rotation and shoulder flexion. The elbow joint adds a third and fourth degree of freedom: elbow flexion and elbow rotation.

Robotic arms also have degrees of freedom. But instead of muscles, tendons, ball and socket joints, and bones, robot arms are made from metal, plastic, wood, motors, solenoids, gears, pulleys, and a variety of other mechanical components. Some robot arms provide but one degree of freedom; others provide three, four, even five separate degrees of freedom.

ARM TYPES

Robot arms are classified by the shape of the area that the end of the arm (where the gripper is) can reach. This accessible area is called the *work envelope*. For the sake of simplicity, the work envelope does not take into consideration motion by the robot's body, just the arm mechanics.

The human arm has a nearly spherical work envelope. We can reach just about anything, as long as it is within arm's length, within the inside of about three-quarters of a sphere. Imagine being inside a hollowed-out orange. You stand by one edge. When you reach out, you can touch the inside walls of about three quarters of the orange peel.

In a robot, such a robot arm would be said to have *revolute coordinates*. The three other main robot arm designs are *polar coordinate, cylindrical coordinate*, and *cartesian coordinate*. You'll note that there are three degrees of freedom in all four basic types of arm designs. Let's take a closer look at each one.

Revolute Coordinate

Revolute coordinate arms, such as the one depicted in Fig. 18-1, are modeled after the human arm, so they have many of the same capabilities. The typical design is somewhat different, however, because of the complexity of the human shoulder joint.

The shoulder joint of the robotic arm is really two different mechanisms. Shoulder rotation is accomplished by spinning the arm at its base, almost as if the arm were mounted on a record player turntable. Shoulder flexion

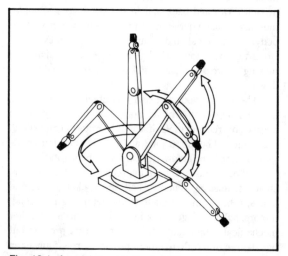

Fig. 18-1. A revolute coordinate arm.

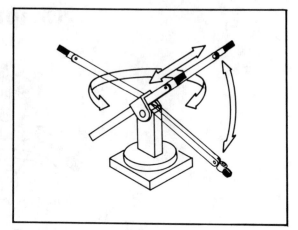

Fig. 18-2. A polar coordinate arm.

is accomplished by tilting the upper arm member backward and forward. Elbow flexion works just as it does in the human arm. It moves the forearm up and down.

Revolute coordinate arms are a favorite design choice for hobby robots. They provide a great deal of flexibility, and besides, they actually look like arms. For details on how to construct a revolute coordinate arm, see Chapter 19.

Polar Coordinate

The work envelope of the polar coordinate arm is half-sphere shaped. Next to the revolute coordinate design, polar coordinate arms are the most flexible in terms of being able to grasp a variety of objects scattered about the robot. Figure 18-2 shows a polar coordinate arm and its various degrees of freedom.

A turntable rotates the entire arm, just as it does with a revolute coordinate arm. This function is akin to shoulder rotation. The polar coordinate arm lacks a means for flexing or bending its shoulder, however. The second degree of freedom is the elbow joint, which moves the forearm up and down. The third degree of freedom is accomplished by varying the reach of the forearm. An inner forearm extends or retracts to bring the gripper closer to or farther away from the robot. Without the inner forearm, the arm would be able to grasp objects laid out in a finite two-dimensional circle in front of it, instead of a sphere. Not very helpful.

The polar coordinate arm is often used in factory robots, and finds its greatest application as a stationary device. It can, however, be mounted to a mobile robot for increased flexibility. Chapter 20 shows how to build a rather useful stationary polar coordinate arm.

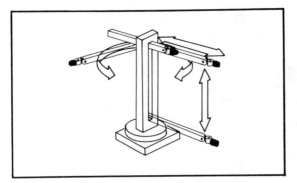

Fig. 18-3. A cylindrical coordinate arm.

Cylindrical Coordinate

The cylindrical coordinate arm looks a little like a robotic forklift. Its work envelope resembles a thick cylinder, hence its name. Shoulder rotation is accomplished by a revolving base, as in revolute and polar coordinate arms. The forearm is attached to an elevator-like lift mechanism, as depicted in Fig. 18-3. The forearm moves up and down this column to grasp objects of various heights. To allow the arm to reach objects in three dimensional space, the forearm is outfitted with an extension mechanism, similar to the one found in a polar coordinate arm.

Cartesian Coordinate

The work envelope of a cartesian coordinate arm (Fig. 18-4) resembles a box. It is the most unlike the human arm and least resembles the other three arm types. It has no rotating parts. The base consists of a conveyer belt-like track. The track moves the elevator column (like the one in a cylindrical coordinate arm) back and forth. The forearm moves up and down the column, and has an inner arm that extends the reach closer to or further away from the robot.

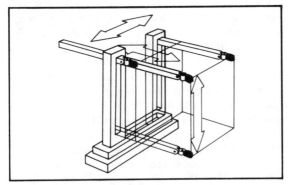

Fig. 18-4. A cartesian coordinate arm.

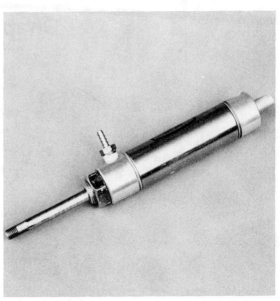

Fig. 18-5. One of hundreds of available sizes and styles of pneumatic cylinders. This one has a bore of about 1/2-inch and a stroke of three inches.

ACTIVATION TECHNIQUES

There are three general ways of moving the joints in a robot arm:

- Electrical
- Hydraulic
- Pneumatic

Electrical actuation is with the use of motors, solenoids, and other electro-mechanical devices. It is the most common and easiest to implement. The motors for elbow flexion, as well as the motors for the gripper mechanism, can be placed in or near the base. Cables, chains, or belts connect the motors to the joints they serve.

Hydraulic actuation using oil-reservoir pressure cylinders, similar to the kind used in earth moving equipment and automobile brake systems. Pneumatic actuation is similar to hydraulic, except that pressurized air is used instead of oil or fluid. Both hydraulic and pneumatic systems provide greater power than electrical actuation, but they are difficult to use. In addition to the actuation cylinders themselves, such as the one shown in Fig. 18-5, a pump is required to pressurize the air or oil, a holding tank is needed to stabilize the pressurization, and values are used to control the retraction or extension of the cylinders.

Chapter 19

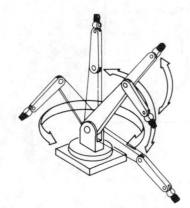

Build A Revolute Coordinate Arm

The revolute coordinate arm design provides a great deal of flexibility, yet with few components. The arm described in this chapter enjoys only two degrees of freedom. You'll find, however, that even with two degrees of freedom, the arm can do many things. It can be used by itself as a stationary pick-and-place robot, or it can be attached to a mobile platform. The construction details given are for a left hand; to build a right hand, simply make it in mirror image.

You'll note that the arm lacks a hand—a gripper. Just about any type of gripper can be used. In fact, you can design the forearm so it accepts many different grippers interchangeably. See Chapter 21, "Experimenting with Gripper Designs," for more information on robot hands.

DESIGN OVERVIEW

The design of the revolute coordinate arm is modeled somewhat after the human arm. A shaft-mounted shoulder joint provides shoulder rotation (degree of freedom #1). A simple swing-arm rotating joint provides the elbow flexion (degree of freedom #2).

You could add a third degree of freedom, shoulder flexion, by providing another joint immediately after the shoulder. In tests, however, I found that this basic two

degree of freedom arm to be quite sufficient for most tasks. It is best used, however, on a mobile platform where the robot can move closer to or further away from the object it's grasping. That's cheating, in a way, but it's a lot simpler than adding another joint.

SHOULDER JOINT AND UPPER ARM

The shoulder joint is a shaft that connects to a bearing mounted on the arm base or in the robot. Attached to the shaft is the drive motor for moving the shoulder up and down. The motor is connected by a single-stage gear system, as shown in Fig. 19-1. In the prototype arm, the output of the motor was approximately 22 rpm, or roughly one third of a revolution per second.

For a shoulder joint, that's a little on the fast side. A gear ratio of 3:1 was therefore chosen to decrease the speed by a factor of three (and increase the torque of the motor roughly by a factor of three). With the gear system used, the shoulder joint moves at about one revolution every eight seconds. That may seem slow, but remember that the shoulder joint swings in an arc a little less than 50 degrees, or roughly one seventh of a complete circle. Thus, the shoulder will go from one extreme to the other in one or two seconds.

Table 19-1. Revolute Arm Parts List.

1	10-inch length 57/64- by 9/16- by 1/16-inch aluminum channel stock
1	10-inch length 41/64- by 1/2- by 1/16-inch aluminum channel stock
1	8-inch length 57/64- by 9/16- by 1/16-inch aluminum channel stock
1	8-inch length 41/64- by 1/2- by 1/16-inch aluminum channel stock
1	7-inch length 1/4-inch 20 all-thread rod
2	1 3/4-inch by 10/24 stove bolt
2	1 1/2- by 3/8-inch flat corner iron
1	3- by 3/4-inch mending plate "T" (for motor mounting)
2	1/2-inch aluminum spacer
1	1/4-inch aluminum spacer
2	3/4-inch diameter 5 lugs-per-inch timing belt sprocket
1	20 1/2-inch length timing belt (5 lpi)
2	Stepper motors
1	3:1 gear reduction system (such as one 20 tooth 24 pitch spur gear and one 60 tooth 24 pitch spur gear)
Misc.	8/32, 10/24, and 1/4-inch 20 nuts, washers, tooth lockwashers, fishing tackle weights

Refer to Fig. 19-2. The upper arm is constructed from a 10-inch length of 57/64- by 9/16- by 1/16-inch aluminum channel stock, and a matching 10-inch length of 41/64- by 1/2- by 1/16-inch aluminum channel stock. The two stocks are sandwiched together to make a bar. Drill a 1/4-inch hole 1/2-inch from the end of the channel stock pieces. Cut a piece of 1/4-inch 20 all-thread rod to a length of seven inches (this measurement depends largely on the shoulder motor arrangement, but seven inches provides room for making changes). Thread a 1/4-inch 20 nut, flat washer, and locking washer onto one end of the rod. Leave a little extra—about 1/8-inch to 1/4-inch on the outside of the nut. You'll need the room.

Drill a 1/4-inch hole in the center of a 3 3/4-inch diameter metal electrical receptacle cover plate. Insert the

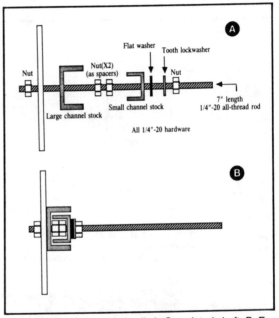

Fig. 19-2. Shoulder shaft detail. A. Completed shaft; B. Exploded view.

rod through it, and the hole of the larger channel aluminum. Thread two 1/4-inch #20 nuts onto the rod to act as spacers, then attach the smaller channel aluminum. Lock the pieces together using a flat washer, tooth washer, and 1/4-inch 20 nut. The shoulder is now complete.

ELBOW AND FOREARM

The forearm attaches to the end of the upper arm.

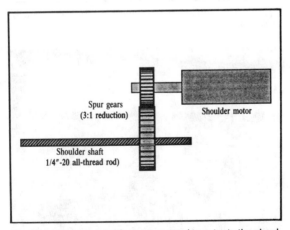

Fig. 19-1. The gear transfer system used to actuate the shoulder of the revolute arm. You can also use a motor with built-in gear reduction if the output of the motor is slow enough for the arm.

The joint there serves as the elbow. The forearm is constructed in a similar manner as the upper arm: cut the small and large pieces of channel aluminum to eight inches instead of ten inches. Construct the elbow joint as shown in Fig. 19-3 and 19-4, using two 1 1/2- by 3/8-inch flat corner angles, 1/2-inch spacers, and 8/32 hardware. The 3/4-inch timing belt sprocket (5 lugs per inch) is used to convey power from the elbow motor, which is mounted at the shoulder. The completed joint is shown in Fig. 19-5.

You can actually use just about any size of timing belt or sprocket. When using the size of sprockets specified in the text, the timing belt is 20 1/2-inches. If you use another size sprocket for the elbow or the motor, you may need to choose another length. You can adjust for some slack by mounting the elbow joint closer or further to the end of the upper arm.

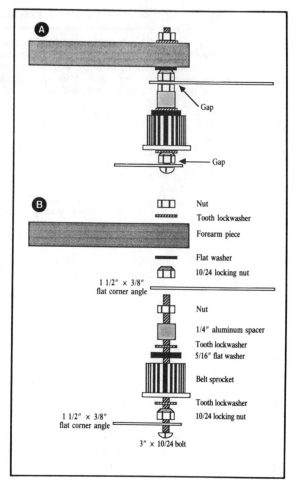

Fig. 19-4. Forearm elbow joint detail. A. Complete joint; B. Exploded View.

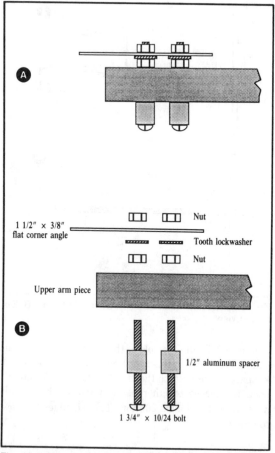

Fig. 19-3. Upper arm elbow joint detail. A. Complete joint; B. Exploded view.

Fig. 19-5. A close-up view of the elbow joint.

158

You may also use #25 chain to power the elbow. Use a sprocket on the elbow and a sprocket on the motor shaft. Connect the two with a #25 chain. You'll need to experiment based on the size of sprockets you use, to come up with the exact length for the chain.

When the elbow and forearm are complete, mount the motor on the shoulder, directly on the plate cover. The motor chosen for the prototype revolute coordinate arm was a one amp medium-duty stepper motor. Pre-drilled holes on the face of the motor facilitated mounting to the arm. A 3- by 3/4-inch mending plate T was used to secure the motor to the plate, as illustrated in Fig. 19-6. New holes were drilled in the plate to match the holes in the motor (1 7/8-inch spacing), and the T was bent at the cross.

Unscrew the nut holding the cover plate and upper arm to the shaft, place the T on it, and retighten. Make sure the motor is perpendicular to the arm. Then, using the other hole in the T as a guide, drill a hole through the cover plate. Secure the T in place with an 8/32 by 1/2-inch bolt and nut. The finished arm, with a gripper attached, is shown in Fig. 19-7.

Fig. 19-6. The motor mounted on the shoulder.

REFINEMENTS

As it is, the arm is unbalanced, and the shoulder motor must work extra to position the arm. You can help to rebalance the arm by relocating the shoulder rotation shaft and by adding counterweights. Before you do anything hasty, however, you may want to attach a gripper to the end of the forearm. Any attempts at balancing the arm now will be severely thwarted when you add the gripper.

The center of gravity for the whole arm, with the elbow drive motor included, is approximately midway along the length of the upper arm (at least, this is true of the prototype arm; your arm may be different). Remove the long shaft from the present shoulder joint, and replace it with a short 1 1/2-inch or 2-inch long 1/4-inch 20 bolt. Drill a new 1/4-inch hole through the upper arm at the approximate center of gravity, and thread the shoulder shaft through it. Attach it as before, using 1/4-inch 20 nuts, flat washers, and toothed lockwashers.

The forearm is also out of balance, and it can be corrected in a similar manner, by attaching the shoulder joint nearer to the center of the arm. This has an unfortunate side effect, however, of shortening the reach of the forearm. One solution is to make the arm longer, to compensate. In effect, you'll be keeping the elbow joint where it is, just adding extra length behind it.

This may interfere with the operation of the arm or robot, so you may want to opt for counterweights attached to the end of the arm. I successfully used two four-ounce fishing tackle weights attached to the arm with a 2- by 3/4-inch corner angle bracket (see Fig. 19-8).

POSITION CONTROL

The stepper motors used for the shoulder and elbow joints of the prototype provide a natural control over the position of the arm. Under electronic control, the motors can be commanded to rotate a specific number of steps, which in turn moves the upper arm and forearm a specified amount.

You should supplement the open-loop servo system with limit switches. These switches provide an indication when the arm joints have moved to their extreme positions. The most common limit switches are small leaf switches. You can also construct optical switches using photo-interrupters. A small patch of plastic or metal interrupts the flow of light between a LED and phototransistor, thus signaling the limit of movement.

The interrupters can be built by mounting an infrared LED and phototransistor on a small perforated board, or

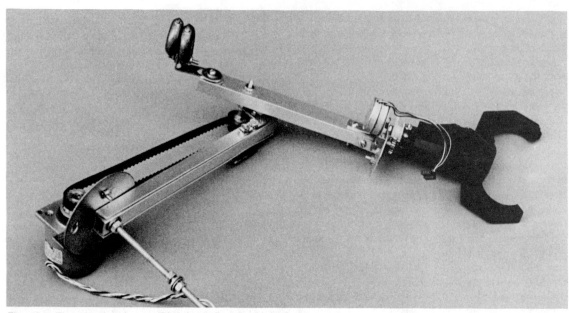

Fig. 19-7. The completed arm, with gripper (hand) attached.

purchased ready-made modules (common surplus finds).

When using continuous dc motors, you need to provide some type of feedback to report the position of the arm. Otherwise, the control electronics (almost always a computer) will never know where the arm is, or how far it has moved. There are several ways you can provide this feedback. The most popular methods are potentiometer, incremental shaft encoder, and absolute shaft encoder.

Potentiometer

Attach the shaft of a potentiometer to the shoulder or elbow joint or motor (see Fig. 19-9), and the varying resistance of the pot serves as an indication of the position of the arm. Just about any pot will do, as long as it has a travel rotation the same as or greater than the travel rotation of the joints in the arm. Otherwise, the arm will go past the internal stops of the potentiometer. Travel rotation is usually not a problem in arm systems, where joints seldom move more than 40 or 50 degrees. If your arm design moves greater than about 270 degrees, use a multi-turn pot. A three-turn pot should suffice.

Another method is to use a slider-pot. You operate a slider-pot by moving the wiper up and down, rather than turn a shaft. Slider-pots are ideal when you want to measure linear distance, like the amount of travel (distance) of a chain or belt. Figure 19-10 shows a slider-pot mounted

Fig. 19-8. Counter-balance weights attached to the end of the forearm help redistribute the weight.

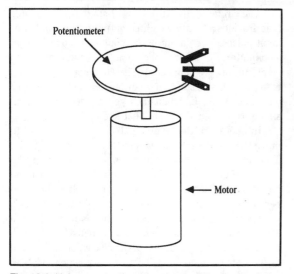

Fig. 19-9. Using a potentiometer as a position feedback device. Mount the potentiometer on a drive motor or a joint of the arm.

to a cleat in the timing belt used to operate the elbow joint.

The value of the pot is a function of the control electronics you have hooked up to it, but 10K to 100K potentiometers usually work well with most any circuit. The potentiometer may provide a relative measurement of the position of the arm, but the information is in analog form, as a resistance or voltage, both of which cannot be directly interpreted by a computer.

By connecting the pot as shown in Fig. 19-11, the output is a voltage between 0 and the positive supply voltage (usually 5 or 12 volts). The wiper of the pot can be connected to the input of an analog-to-digital converter (A/D converter), which translates voltage levels into bytes.

Now, before you go off screaming about the complexity of A/D converters, you should really try one first. The latest chips are relatively inexpensive (under $5) and require a very minimum of external components to oper-

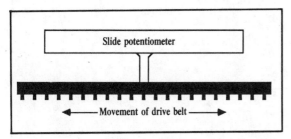

Fig. 19-10. Using a slide potentiometer to register position feedback. The wiper of the pot can be linked to any mechanical device, like a chain or belt, that moves laterally.

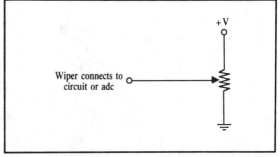

Fig. 19-11. The basic electrical hookup to provide a varying voltage from a potentiometer.

ate. The best part about A/D chips is that most have provisions for connection to eight or more analog signals. You select which signal input you want to convert into digital data. That means one $5 A/D chip can be used for all of the joints in a two-arm robot system.

To be useful, the A/D converter should be connected to a microprocessor or computer. You can also use your personal computer as the controlling electronics for your robot. For more information about A/D converters and computer control, see Chapters 31 through 33. You should also collect (and read!) the data sheets for the ADC0808 and ADC0809, commonly available A/D converters.

Incremental Shaft Encoder

The incremental shaft encoder was first introduced in this book in Chapter 13, "Robot Locomotion Using Dc Motors." The shaft encoder is a disc with many small holes or slots near the outside circumference. The disc is attached to a motor shaft or the shoulder or elbow joint. As shown in Fig. 19-12, an infrared LED-phototransistor pair is used as an optical switch. The light flashes on and off each time the slots pass by as the disc rotates.

A circuit connected to the phototransistor (the latter of which is baffled to block off ambient light) counts the number of on-off flashes, and then converts that into distance traveled. For example, one on-off flash may equal a two degree movement of the joint. Two flashes may equal a four degree movement, and so forth.

The advantage of the incremental shaft encoder is that its output is inherently digital. A computer, or even a simple counter circuit, can be used to simply count the number of on-off flashes. The result, when the movement ends, is the new position of the arm.

Figure 19-13 shows a two-channel up/down position sensing system. Two LED-photodetectors are mounted by the disc; they are positioned 90-degrees apart so that

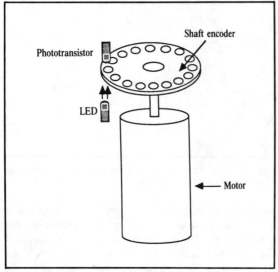

Fig. 19-12. Using an incremental shaft encoder as a position feedback device. Mount the potentiometer on a drive motor or a joint of the arm.

The circuit in Fig. 19-16 is a 12-stage ripple counter that provides a binary weighted output of the number of input pulses. The output of the counter is connected to a computer. For most applications, only the first five to eight outputs are connected. The 4040 counter goes up to 4096 with all 12 outputs, but few computers have 12 data lines. With eight data lines, you can decode 256 possible positions of the arm. That should be enough.

In operation, the computer is given a command to move the shoulder or elbow joint X number of degrees. The software running on the computer multiplies the number of degrees of movement desired (X) by the number of on-off flashes per degree. The result is the number of on-off flashes, or counts, for the given movement.

It's easiest when the encoder disc is marked off in 360 degree increments, but this is usually impractical. In most cases, the resolution of the disc will be about 150 to 200 slots per revolution. That's the equivalent of about one on-off flash of light for every two degrees. Obviously, the smallest movement obtaining with such a system is two degrees. For greater accuracy, invest in a commercially-made, high resolution shaft encoder.

The 4040 should be reset before each movement of the arm (the schematic shows the Reset line tied LOW; to reset the counter, momentarily bring the Reset line HIGH). The computer continually scans the output of the counter, looking for the binary number that equals the desired movement. When that number is reached, the motor is turned off.

In a practical application, you'll probably want to write the software so that the computer looks ahead a few degrees, so that it can stop the motor before the po-

when one photodetector sees light through a hole, the other is blocked off. The output of one Exclusive OR gate from the 7486 IC is a train of clock pulses that can be used to count the rotation of the motor or shaft. The output of another gate provides up/down information, and is decoded by determining whether the signal is HIGH or LOW at each clock transition. Figure 19-14 shows the relevant waveforms of the gate outputs. Another up/down counting circuit is shown in Fig. 19-15. This one uses the "JK" flip-flops in a 7473 IC package.

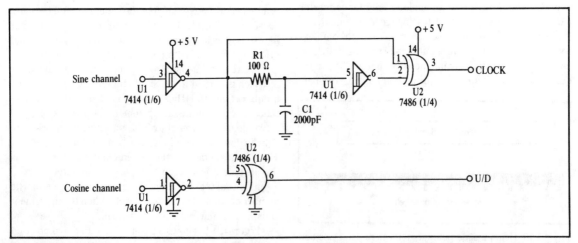

Fig. 19-13. A simple two-channel circuit for counting the number of revolutions of the encoder disc as well as determining the direction of rotation.

Table 19-2. Two-Channel Shaft Encoder Parts List.

U1	7414 Quad Schmitt Inverter IC
U2	7473 Dual "JK" Flip-Flop IC
R1,R3	270 Ω resistor
R2,R4	10 KΩ resistor
Q1,Q2	Infrared sensitive phototransistor
Misc.	Infrared filter for phototransistor (if needed), incremental shaft encoder

sition is reached. You will have to experiment with your arm designs to find the optimum cut-off point. If the motors tend to coast a great deal, you'll need to increase the look-ahead time. If the motors stop relatively quickly, the look-ahead time can be kept short.

Absolute Shaft Encoder

The absolute shaft encoder is similar in concept to the incremental shaft encoder, but instead of counting the number of on-off flashes to report position, the absolute encoder disc provides the information directly. The process is simple: contact switches, or more commonly LED-phototransistor pairs, detect a binary pattern en-coded on the disc. As the disc rotates, the pattern changes—the binary number increases or decreases.

Absolute shaft encoders are a little more difficult to engineer than incremental shaft encoders, but they do not require an intermediate counter circuit. The output of the disc can be directly routed to a computer or control circuit.

The optical absolute encoder disc is the most relia-ble, and that's the one we'll investigate here. The reso-lution of the encoder depends on the number of discrete binary digits encoded on its surface. For a homebrew ro-bot, a disc with 32 to 128 binary values is about the ex-tent of normal garage tinkering. You can purchase a ready-made encoder disc with a resolution of 256 (or more), but it can be expensive. Values of 128 should be

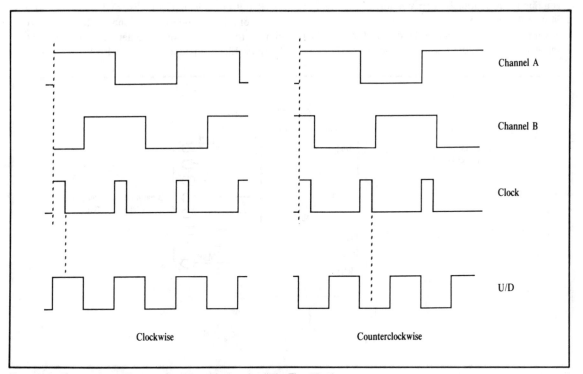

Fig. 19-14. A waveform diagram for use with the circuit in Fig. 19-13.

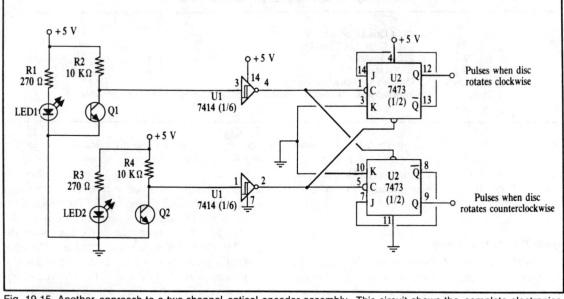

Fig. 19-15. Another approach to a two-channel optical encoder assembly. This circuit shows the complete electronics, including the infrared LEDs and phototransistors.

more than enough for positioning the arm in our robot. If you find the resolution isn't enough for your liking, use a 64 or 128 digit code.

Alternatively, you can use the disc in Fig. 19-17 as a template for a medium-resolution disc. Take the tem-

plate to a printer and have a photolith positive made of it. Glue it on a clear plastic disc and drill a hole in the center of the disc to mount the shaft. This procedure is similar to one used for incremental shaft encoders in Chapter 13, "Robot Locomotion with Dc Motors."

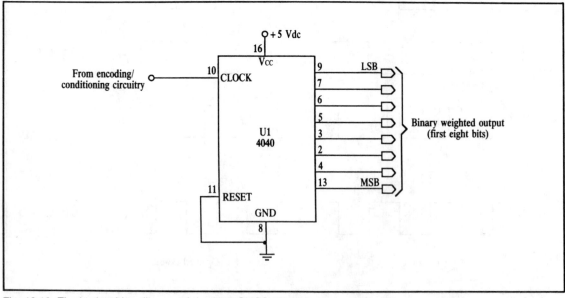

Fig. 19-16. The basic wiring diagram of the 4040 CMOS 12-stage ripple counter IC.

Table 19-3. Alternate Shaft Encoder Parts List.

U1	7414 Quad Schmitt Inverter IC
U2	7486 Quad Exclusive OR Gate IC
R1	100 Ω resistor
C1	2000 pF ceramic capacitor

All resistors 5 or 10 percent tolerance, 1/4-watt; all capacitors 10 percent tolerance, rated 35 volts or higher.

Fig. 19-17. Template art to make your own absolute shaft encoder disc.

The maximum value obtained from the disc determines the number of bits for each number. Five bits are required to count to 32. Five LEDs are placed behind the disc, and lined up with each row of holes. Matching phototransistors are mounted on the other side. You may have to tweak the positions of the LEDs and transistors to make sure everything is aligned. Be sure to baffle the phototransistors to protect them against ambient light or the light from the neighboring LEDs.

The on-off flashes received by each of the five phototransistors must be conditions with a Schmitt trigger inverter. As it turns out, one 7414 TTL contains six of these inverters, so the whole conditioning process takes just one chip. The output of the inverters connect directly to the data lines of the computer.

Chapter 20

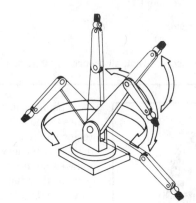

Build a Polar Coordinate Arm

Polar coordinate arms are ideal for use in a stand-alone robotic manipulator. They are fairly easy to build, and are adaptable to a number of useful applications, especially robotic training. The design described in this chapter is a three-degree-of-freedom polar coordinate arm that's mounted on a stationary base. You can, if you wish, attach the arm to a mobile base, or for an even more outrageous project, add wheels or track to the base itself, making a giant rolling arm.

The arm design presented below is devoid of a gripper, or hand, mechanism. You can attach any number of different grippers to the end of the arm. Choose the gripper depending on the application you have in mind. Read more about robotic grippers in Chapter 21, "Experimenting with Gripper Designs."

CONSTRUCTING THE BASE

The base measures 10 inches by 12 inches by 4 inches. It is made from aluminum shelving standards. You can also use 41/64- by 1/2- by 1/16-inch aluminum channel stock. Construct the base by cutting four each 10 inch and 12 inch lengths. Cut each end at a 45 degree angle. Cut four 2 1/2-inch riser pieces. Do not miter the ends of these lengths. Assemble the top and bottom frames

using 1 1/2- by 3/8-inch flat corner angles. Secure the stock to the corner angles with 8/32 by 1/2-inch bolts and 8/32 nuts.

Refer to Fig. 20-1. Attach a 1- by 1/2-inch corner angle bracket using 8/32 by 1/2-inch bolts and nuts to each one of the short riser pieces. Attach 2- by 1/2-inch flat mending plates to the top of the riser pieces. Connect the top and bottom frames with the risers spaced 2 3/4 inches from the corners. Use 8/32 by 1/2-inch bolts and 8/32 nuts.

SHOULDER ROTATION MECHANISM

The shoulder rotation mechanism consists of a motor, a turntable, and a chain gear system. Start by adding a cross brace to the top of the base. Cut a 10 5/8-inch length of 57/64- by 9/16- by 1/16-inch aluminum channel stock. Mount it lengthwise in the center base using two 2 1/2- by 1/2-inch flat mending iron T's. Use 8/32 by 1/2-inch bolts and 8/32 nuts to secure the T's and cross brace into place.

Drill a 3/8-inch hole in the center of the cross brace. Position one 3-inch diameter ball-bearing turntable (Lazy Susan) over the hole. Using the mounting holes on the baseplate of the turntable as a guide, mark corresponding mounting holes in the cross brace. Drill for 6/32 bolts

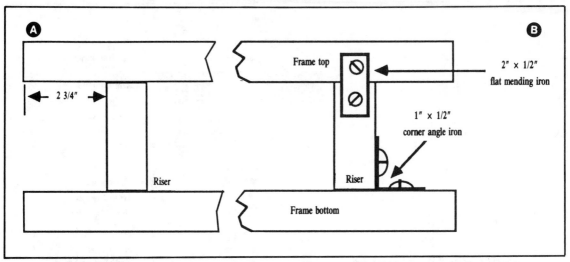

Fig. 20-1. Cutting and assembly for polar arm base risers. A. Riser placement; B. Hardware assembly detail.

(#28 bit) and attach the turntable using two 6/32 by 1/2-inch bolts and 6/32 nuts (see Fig. 20-2).

Now refer to Fig. 20-3. Construct the center shaft of the arm with a 3-inch by 10/24 pan-head stove bolt. Place a 1/2-inch diameter bearing on either side of the channel stock. Be sure the center (rotating part) of the bearings rest over the hole, or they won't turn properly, and that the head of the bolt is positioned over the inner wheel of the bearing. Add a 1/4-inch spacer and lock the assembly into place with a 10/24 nut.

Now for the drive mechanics. The drive sprocket (35 teeth, 3″ diameter, #25 chain) is sandwiched between two

Fig. 20-2. Cutting and assembly detail for cross piece and turntable.

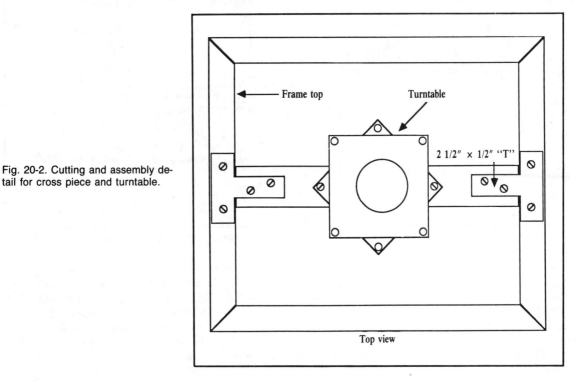

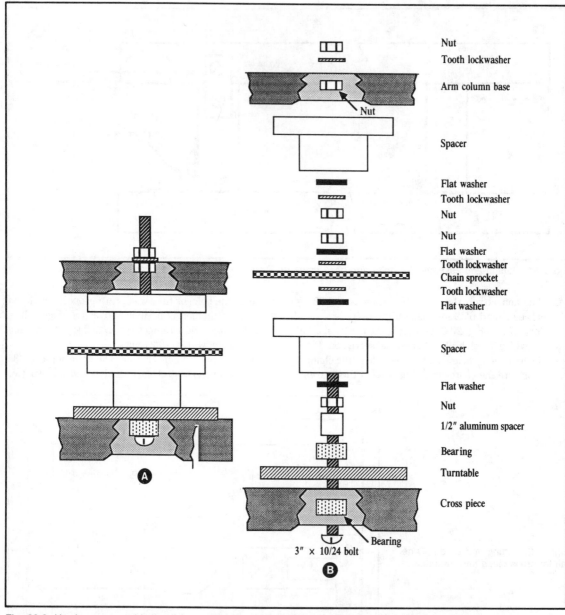

Nut
Tooth lockwasher
Arm column base
Nut
Spacer
Flat washer
Tooth lockwasher
Nut
Nut
Flat washer
Tooth lockwasher
Chain sprocket
Tooth lockwasher
Flat washer
Spacer
Flat washer
Nut
1/2″ aluminum spacer
Bearing
Turntable
Cross piece
Bearing
3″ × 10/24 bolt

A

B

Fig. 20-3. Hardware assembly detail for central shoulder shaft. A. Assembled shaft; B. Exploded view.

plastic spacers, as shown. These spacers are actually closet pole holders. They already have holes drilled in the center you can just plop them onto the shaft. The drive sprocket should be approximately one inch above the cross brace. Use a 10/24 bolt and tooth lockwasher, and flat washer to clamp the drive mechanism into place.

Attach a 20 tooth 1 3/4-inch diameter #25 chain

sprocket to the shaft of the motor. The prototype arm used a medium-duty stepper motor with a 1/4-inch shaft. The 1/2-inch I.D. hub of the sprocket was reduced to 1/4-inch with reducing bushings. If you use a similar motor (they are common on the new and surplus market), and the same size sprockets, the chain length should be a nominal 17 inches. You can use a slightly longer or

Table 20-1. Polar Coordinate Arm Parts List.

Frame/Base:

4	12-inch lengths 41/64- by 1/2- by 1/16-inch aluminum channel stock
4	10-inch lengths 41/64- by 1/2- by 1/16-inch aluminum channel stock
4	2 1/2-inch lengths 41/64- by 1/2- by 1/16-inch aluminum channel stock
4	2- by 1/2-inch flat mending iron
4	1- by 1/2-inch corner angle iron
8	1 1/2- by 3/8-inch flat corner angle iron
32	1/2-inch by 8/32 stove bolts, nuts, tooth lockwashers

Shoulder Base:

1	10 5/8-inch length 57/64- by 9/16- by 1/16-inch aluminum channel stock
1	9-inch length 57/64- by 9/16- by 1/16-inch aluminum channel stock
2	7-inch length 57/64- by 9/16- by 1/16-inch aluminum channel stock
2	2 1/2-inch mending plate "T"
2	1 1/4- by 5/8-inch corner angle iron
1	3-inch diameter ball-bearing turntable
1	3-inch by 10/24 stove bolt, nuts, flat washers, tooth lockwashers
12	1/2-inch by 8/32 stove bolts, nuts, tooth lockwashers
2	Plastic closet pole holders
2	1/2-inch bearings
1	1/2-inch aluminum spacer
1	3-inch diameter 35 tooth chain sprocket (#25 chain)
1	Stepper motor
1	1 3/4-inch diameter 20 tooth chain sprocket (#25 chain)
1	17-inch long (nominal) #25 chain

Elbow:

2	6-inch lengths 57/64- by 9/16- by 1/16-inch aluminum channel stock
2	3 1/2-inch lengths 1- by 1- 1/16-inch aluminum angle stock
1	2 1/2-inch length 1- by 1- by 1/16-inch aluminum angle stock
1	7 1/2-inch length 1/4-inch 20 all-thread rod, nuts, locking nuts, flat washers, tooth lockwashers
2	1 3/4-inch diameter 20 tooth chain sprocket (#25 chain)
1	17-inch long #25 chain
6	1/2-inch by 8/32 stove bolts, nuts, tooth lockwashers
1	Stepper motor

Forearm:

1	16-inch long (nominal) drawer rail
2	1-inch to 1 1/2-inch diameter spur gears, with set screw recessed in hub
1	3 1/2-inch length 1- by 1- 1/16-inch aluminum angle stock
1	18-inch length (approx) 1/16-inch diameter steel aircraft cable
2	14-16 gauge wire lug
4	1/2-inch by 8/32 stove bolts, nuts, tooth lockwashers
2	1/2-inch by 6/32 stove bolts, nuts
1	Stepper motor

shorter length of chain, because you can position the motor anywhere along the length of the frame to compensate.

Choose a mounting location with the chain in place. Move the motor along the edge of the frame until the chain is tight, and mark the mounting location. At the mark, attach the motor to the frame using two 1 1/2- by 3/8-inch flat corner braces. Elevate the braces using 1/2-inch spacers as shown in Fig. 20-4. Use 6/32 by

Fig. 20-4. The mounted shoulder motor, with chain sprocket and chain.

for the elbow rotation shaft. Cut a 7 1/2-inch length of 1/4-inch 20 all-thread rod and attach a locking nut to one end. Drill a 1/4-inch hole 1/2-inch down from the top of each 7-inch arm column piece. Thread the elbow rotation shaft through the pieces, using the hardware noted, as shown in Fig. 20-7. Secure a 20 tooth, 1 1/2-inch diameter #25 chain sprocket to the end.

Mount a matching 20 tooth #25 chain sprocket on the shaft of the elbow stepper motor (it is the same type as the one used for shoulder rotation). Attach a 17-inch length of #25 chain between the two sprockets, and mount the motor to the end of the 9-inch shoulder cross brace using a 2 1/2-inch length of 1- by 1- by 1/16-inch aluminum angle stock. Use 6/32 by 1/2-inch bolts and nuts to secure the motor to the angle stock. The angle stock is

1/2-inch bolts and nuts to secure the motor to the braces; 8/32 by 1/2-inch bolts and nuts to secure the braces to the frame.

Construct the arm column using two 7-inch lengths of 57/64- by 9/16- by 1/16-inch aluminum channel stock and one 9-inch length of the same. Mount a 7-inch length flush to one end of the 9-inch member. Use 1 1/4- by 5/8-inch corner angle brackets and 8/32 by 1/2-inch bolts and nuts to secure the pieces in place. Mount the other 7-inch length three inches from the opposite end of the 9-inch member (see Fig. 20-5 for details). Likewise, secure it using an angle bracket. Drill a 1/4-inch hole in the center of the 9-inch piece and mount the assembly on the shoulder rotation shaft (refer back to Fig. 20-3 for assembly detail).

BUILDING THE ELBOW MECHANISM

The elbow mechanism consists of a platform driven by a stepper motor. For weight distribution, the motor is mounted on the 9-inch shoulder member. Refer to Fig. 20-6. Construct the elbow platform by cutting two 6-inch lengths of 57/64- by 9/16- by 1/16-inch aluminum channel stock. Couple them together with two 3 1/2-inch lengths of 1- by 1- by 1/16-inch aluminum angle stock. Connect the pieces using 1/2-inch by 8/32 and 8/32 nuts.

Drill a 1/4-inch hole in the center of each angle stock

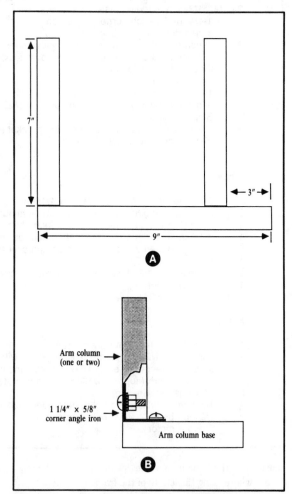

Fig. 20-5. Cutting and assembly for the arm column. A. Dimensions of pieces; B. Hardware assembly detail.

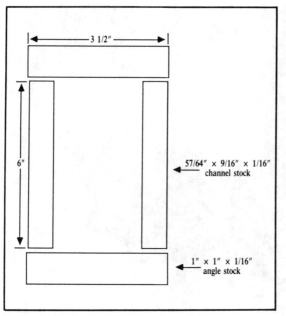

Fig. 20-6. Cutting detail for the elbow shaft.

riveted to the cross brace (the head of a machine bolt is too thick). The motor, as attached, should look like the one in Fig. 20-8.

BUILDING THE FOREARM

The retractable forearm is a rather simple mechanism, but it requires some patience when constructing it. The forearm uses commonly available parts and, to make your job easier, you may want to stick with the parts specified. No sense in both of us sweating this one out.

The retractable forearm is constructed out of a metal drawer rail. The rail is composed of two pieces: an 11-inch base and a 16 3/4-inch long retracting rail. The rail rides within the base on a set of ball bearings. With a little bit of grease added, the rail slides smoothly along the length of the rail without trouble. The drawer rail used in the prototype required no modification, but some rails have stops and locks that you may want to defeat. Usually, this consists of nothing more than filing down a piece of metal or drilling out the offending stop.

Drill mounting holes in the rail to match the bolts already in place on the elbow platform (you may need to remove the inner rail to get to some portions of the base). Unfasten the bolts on the side opposite the sprocket, as depicted in Fig. 20-9, and attach the rail. Retighten the bolts.

Mount the rail motor directly in the center of the el-

bow platform. Cut another 3 1/2-inch length of 1- by 1- by 1/16-inch aluminum angle stock and attach it to the platform using 8/32 by 1/2 inch bolts and nuts. Secure the motor using the mounting technique best suited for it. The stepper motor used in the prototype arm already had threaded mounting holes on the shaft end. These were used to secure the motor in place. Attach two 1 1/2-inch diameter gears to the motor shaft. Position the gears so

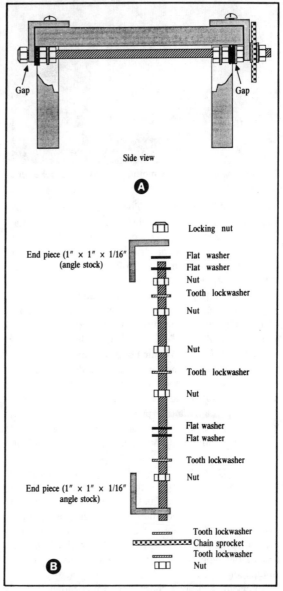

Fig. 20-7. Assembly detail for the elbow shaft. A. Assembled shaft; B. Exploded view.

171

Fig. 20-8. Mounted elbow motor, with chain sprocket and chain.

that the hubs face each other. The idea is to create a spool-like shaft for the forearm cable (see Fig. 20-10). Alternatively, you can use sprockets or fashion a real spool out of metal or wood. The main design consideration is that the inside of the spool must be flush. Set screws that bulge out will tangle with the cable.

Cut a length of 1/16-inch round steel aircraft cable to approximately 18 inches long. On both ends, clamp a 14-16 gauge wire lug using a pair of pliers or clamping tool (see Fig. 20-11). Secure one lug to the back end of the rail using 6/32 by 1/2-inch bolts and nuts (there may already be a hole for the hardware; if not drill your own).

Loop the cable once around the spool shaft and pull it tight to the other end. Remove as much slack as possible and make a mounting mark using the wire lug as a guide. Drill the hole and secure the lug using 6/32 by 1/2-inch bolts and 6/32 nuts. The assembly should look similar to Fig. 20-12.

You may find that, when using metal or plastic gears or sprockets for the spool, the cable slops around and doesn't have much traction. One solution that seems to work is to line the spool shaft with a couple of layers of masking tape. This approach has proved satisfactory for the prototype arm, even after a year's worth of use. Alternatively, you can rough up the shaft using coarse sandpaper.

The finished polar coordinate arm is shown in Fig. 20-13. Note that the shoulder is able to rotate continu-

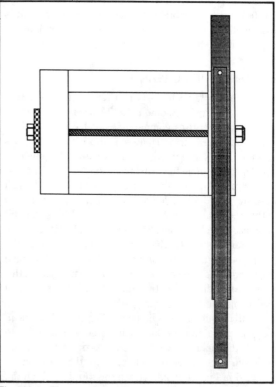

Fig. 20-9. The recommended mounting location for the drawer rail.

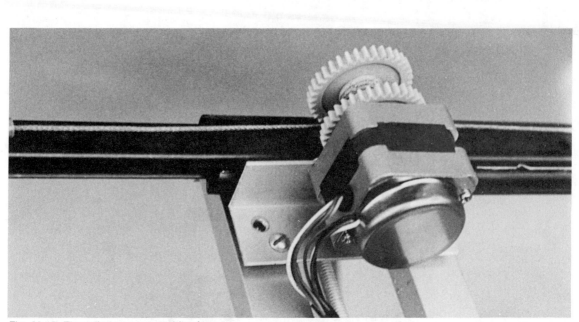

Fig. 20-10. The rail motor mounted in place.

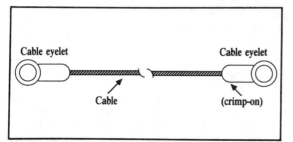

Fig. 20-11. A length of aircraft cable terminated with crimp-on electrical lugs. The 14-16 gauge size seems to work with 1/16-inch diameter steel cable.

ously in a 360 circle, and will keep on rotating indefinitely like a wheel. In actual use, the wires to the motors will prevent the shoulder from rotating more than one complete turn.

With the forearm completely extended, the reach is not quite to the ground (but it is to the arm's own base). This shouldn't be too short because most gripper designs will add at least five or six inches to the length. The robot should be able to pick up small objects placed a half foot or more from the base. You can increase the reach by making the base thinner, or by using a longer drawer rail.

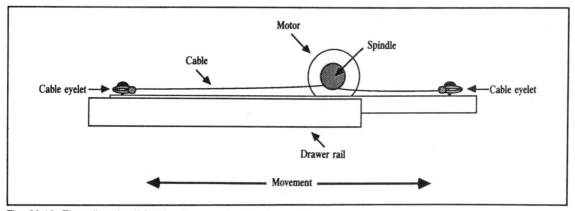

Fig. 20-12. Threading detail for the drawer rail cable assembly.

Fig. 20-13. The finished polar coordinate arm.

HELPFUL IMPROVEMENTS

There is room for improvement in this basic design of a polar coordinate arm. One improvement you can make quite easily is to add cross pieces to support the turntable used for shoulder rotation. As it is, there is a great deal of side-to-side slop and additional braces would largely eliminate it.

POSITION CONTROL

Arm systems need a great deal of position control if they are to manipulate objects without the direct intervention of you, its human master. See Chapter 19, "Build a Revolute Coordinate Arm," for more complete details on adding position control to the joints of arms.

Chapter 21

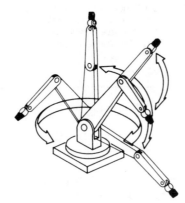

Experimenting With Gripper Designs

The arm systems detailed in Chapters 19 and 20 aren't much good without hands. In the robotics world, hands are usually called grippers, because the word more closely describes their function. Few robotic hands can manipulate objects with the fine motor control of a human hand; they simply grasp or grip the object, hence the name gripper. Never to be a stickler for semantics, we'll use the terms hands and grippers interchangeably.

Gripper designs are numerous, and none is ideal for all applications. Each gripper technique has unique advantages over the others and you must fit the gripper to the application. Here are a number of useful gripper designs you can use for your various robots. Most are fairly easy to build; some even make use of inexpensive plastic toys. The gripper designs encompass just the finger or grasping mechanisms. A section at the end of this chapter details how to add wrist rotation to any of the gripper designs.

THE CLAPPER

The "Clapper" gripper is a popular design, favored because of its easy construction and simple mechanics. The clapper can be built using metal, plastic or wood,

or a combination of all three. The details given below are for a metal and plastic clapper.

The clapper consists of a wrist joint (which, for the time being, we'll assume is permanently attached to the forearm of the robot). Connected to the wrist are two plastic plates. The bottom plate is secured to the wrist; the top plate is hinged. A small spring-loaded solenoid is positioned inside, between the two plates. When the solenoid is not activated, the spring pushes the two flaps out, and the gripper is open. When the solenoid is activated, the plunger pulls in, and the gripper closes. The amount of movement at the end of the gripper is minimal—about 1/2-inch with most solenoids. However, that is enough for general gripping tasks.

Cut two 1/16-inch thick acrylic plastic pieces to 1 1/2-inches by 2 1/3-inches. Attach the lower flap to two 1- by 3/8-inch corner angle brackets. Place the brackets approximately 1/8-inch from either side of the flap. Secure the pieces using 6/32 by 1/2-inch bolts and 6/32 nuts. Cut a 1 1/2-inch length of 1 1/2- by 1/8-inch aluminum bar stock. Mount the two brackets to the bottom of the stock as shown in the figure.

Attach the top flap to a 1 1/2- by 1-inch (approximate) brass or aluminum miniature hinge. Drill out the holes

Table 21-1. Clapper Gripper Parts List.

2	1 1/2- by 2 1/2-inch 1/16-inch thick acrylic plastic sheet
2	1- by 3/8-inch corner angle bracket
1	1 1/2- by 1-inch brass or aluminum hinge
1	Small 6 or 12 Vdc spring-loaded solenoid
8	1/2-inch by 6/32 stove bolts, nuts.

in the hinge with a #28 drill to accept 6/32 bolts. Secure using 6/32 bolts and nuts.

The choice of solenoid is important, because it must be small enough to fit within the two flaps and it must have a flat bottom to facilitate mounting. It must also operate with the voltage used in your robot, usually 6 or 12 volts. Some solenoids have mounting flanges opposite the plunger. If yours does, use the flange to secure the solenoid to the bottom flap. Otherwise, mount the solenoid in the center of the bottom flap approximately 1/2-inch from the back end (nearest the brackets) with a large gloop of household cement. Let stand to dry.

Align the top flap over the solenoid. Make a mark at the point where the plunger contacts the plastic. Drill a hole just large enough for the plunger. You want a tight fit. Insert the plunger through the hole and push down so that the plunger starts to peek through. Align the top and bottom flaps so that they are parallel to one another.

Using the mounting holes in the hinges as a guide, mark corresponding holes in the aluminum bar. Drill holes and mount using 1/2-inch by 6/32 bolts and nuts. The finished clapper should look like Fig. 21-1.

Test the operation of the clapper by activating the solenoid. If the plunger works loose, apply some household cement to keep it in place. You may want to add a short piece of rubber weather stripping to the inside ends of the clappers, to facilitate grasping objects. You can also use stick-on rubber feet squares, available at most hardware and electronics stores.

TWO-PINCHER GRIPPER

The two-pincher gripper consists of two movable fingers, somewhat like the claw of a lobster.

Basic Model

For ease in construction, the basic two-pincher gripper is made from extra Erector Set parts (components from a similar construction kit toy may also be used).

Cut two metal girders to 4 1/2 inches (this is a stan-

dard Erector Set size, you may not have to do any cutting). Cut a length of angle girder to 3 1/2 inches, as shown in Fig. 21-2. Use 6/32 by 1/2-inch bolts and nuts to make two pivoting joints. Cut two 3-inch lengths and mount them (see Fig. 21-3). Nibble the corner off both pieces to prevent the two from touching one another. Nibble or cut through two or three holes on one end to make a slot. As illustrated in Fig. 21-4, use 6/32 by 1/2-inch bolts and nuts to make pivoting joints in the fingers.

The basic gripper is finished. You can actuate it in a number of ways. One way is to mount a small eyelet between the two pivot joints on the angle girder. Thread two small cables or wire through the eyelet and attach the cables. Connect the other end of the cables to a solenoid or a motor shaft. Use a light compression spring to force the fingers apart when the solenoid or motor is not actuated.

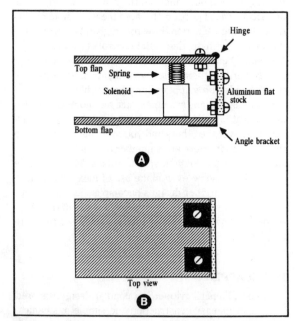

Fig. 21-1. The clapper gripper. A. Assembly detail; B. Top view.

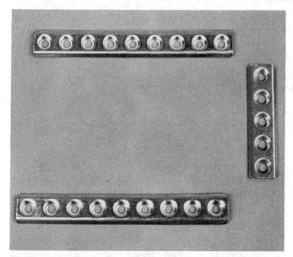

Fig. 21-2. An assortment of girders from an Erector Set toy construction kit.

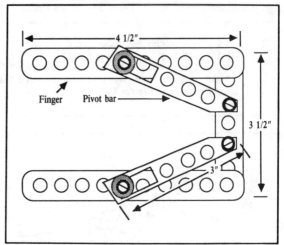

Fig. 21-3. Construction detail of the basic two-pincher gripper, made with Erector Set parts.

You can add pads to the fingers by using the corner braces included in most Erector Set kits, and attaching weather stripping or rubber feet to the brace. The finished gripper should look like the one depicted in Fig. 21-5.

Advanced Model #1

You can use a readily-available plastic toy and convert it to a useful two-pincher gripper for your robot arm. The toy is a plastic extension arm with the pincher claw on one end and a hand gripper on the other (see Fig. 21-6). To close the pincher, you pull on the hand gripper. The contraption is inexpensive— usually under $10—and it is available at many toy stores.

Chop off the gripper three inches below the wrist. You'll cut through an aluminum cable. Now cut off another 1 1/2 inches of tubing—just the arm, but not the cable. File off the arm tube until it's straight, then fashion a 1 1/2-inch length of 3/4-inch diameter dowel to fit into the rectangular arm. Drill a hole for the cable to go through. The cable is off-centered as it attaches to the pull-mechanism in the gripper, so allow for this in the hole. Place the cable through the hole, push the dowel at least 1/2-inch into the arm, then drill two small mounting holes to keep the dowel in place (see Fig. 21-7). Use 6/32 by 3/4-inch bolts and nuts to secure the pieces.

The dowel can now be used to mount the gripper on an arm assembly. You can use a small 3/4-inch U-bolt or

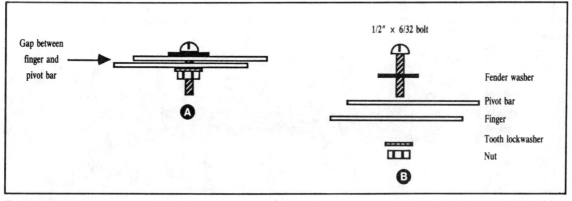

Fig. 21-4. Hardware assembly detail of the pivot bar and fingers of the two-pincher gripper.   A. Assembled sliding joint; B. Exploded view.

Table 21-2. Two-Finger Erector Set Gripper Parts List.

2	4 1/2-inch Erector Set girder
1	3 1/2-inch length Erector Set girder
4	1/2-inch by 6/32 stove bolts, fender washer, tooth lockwasher, nuts
Misc.	14-16 gauge insulated wire ring lugs, aircraft cable, rubber tabs, 1/2- by 1/2-inch corner angle brackets (galvanized or from Erector Set)

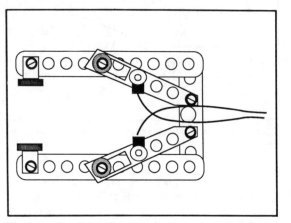

Fig. 21-5. The finished two-pincher gripper, with fingertip pads and actuating cables.

Advanced Model #2

This gripper design uses a novel worm gear approach, without using a hard-to-find (and expensive) worm gear. The worm is a length of 1/4-inch 20 bolt; the gears are standard 1-inch diameter 64 pitch aluminum spur gears (hobby stores have these for about $1 apiece). Turning the bolt opens and closes the two fingers of the gripper.

Construct the gripper by cutting two 3-inch lengths of 41/64- by 1/2- by 1/16-inch aluminum channel stock. Using a 3-inch flat mending T plate as a base, attach the fingers and gears to the T as shown in Fig. 21-9. The distance of the holes is critical and depends entirely on the diameter of gears you have. You may have to experiment with a different spacing if you use another gear diameter. Be sure the fingers rotate freely on the base, but that the play is not excessive. Too much play will cause the gear mechanism to bind or skip.

Secure the shaft using a 1 1/2- by 1/2-inch corner angle bracket. Mount it to the stem of the T using an 8/32 by 1-inch bolt and nut. Add a #10 flat washer between the T and the bracket to increase the height of the bolt shaft. Mount a 3 1/2-inch long 1/4-inch 20 machine bolt through the bracket. Use double nuts or locking nuts to form a free-spinning shaft. Reduce the play as much as possible without locking the bolt to the bracket. Align the finger gears to the bolt so that they open and close at the same angle.

To actuate the fingers, attach a motor to the base of

flatten one end of the dowel and attach it directly to the arm. The gripper opens and closes with only a 7/16-inch pull. Attach the end of the cable to a heavy-duty solenoid that has a stroke of at least 7/16-inch. You can also attach the gripper cable to a 1/8-inch round aircraft cable. Use a crimp-on connector designed for 14-16 gauge electrical wire to connect them end to end, as shown in Fig. 21-8. Attach the aircraft cable to a motor shaft and activate the motor to pull the gripper closed. The spring built into the toy arm opens the gripper when power is removed from the solenoid or motor.

Fig. 21-6. A commercially available plastic two-pincher robotic arm and claw. The gripper can be salvaged for use in your own designs.

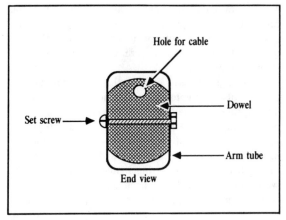

Fig. 21-7. Assembly detail for the claw gripper and wooden dowel. Drill a hole for the actuating cable to pass through.

the bolt shaft. The prototype gripper used a 1/2-inch diameter 48 pitch spur gear, and matching 1-inch 48 pitch spur gear on the drive motor. Operate the motor in one direction and the fingers close. Operate the motor in the other direction and the fingers open. Apply small rubber feet pads to the inside ends of the grippers to facilitate grasping objects. The finished gripper is shown in Fig. 21-10.

Figures 21-11 through 21-15 show other approaches to two-pincher grippers. By adding a second rail to the fingers, and allowing a pivot for both, the fingertips remain parallel to one another as the fingers open and close. There are several actuation techniques you can employ with such a gripper. Figure 21-16 shows the gripping mechanism of the Radio Shack/Tomy Armatron. Note that it uses double rails to affect parallel closure of the fingers. You can model your own gripper using the de-

sign of the Armatron, or amputate an Armatron and use its gripper for your own robot.

FLEXIBLE FINGER GRIPPERS

Clapper and two-pincher grippers are unlike human fingers. One thing they lack is a *compliant* grip: a means to contour the grasp to match the object. The digits in our fingers can wrap around just about any oddly shaped object, which is one of the reasons we are able to use tools successfully.

You can approximate the compliant grip by making articulated fingers for your robot. At least one toy is available that uses this technique; you can use it as a design base. The plastic toy arm described above is available with a hand-like gripper instead of a claw gripper. Pulling on the hand grip causes the four fingers to close around an object, as shown in Fig. 21-17. The opposing thumb is not articulated, but a thumb can be made to move in a compliant gripper of your own design.

The fingers are made from hollow tube stock cut at the knuckles. The mitered cuts allow the fingers to fold inward. The fingers are hinged by the remaining plastic on the topside of the tube. Inside the tube fingers is semi-flexible plastic, attached to the fingertips. Pulling on the hand grip exerts inward force on the fingertips. The result: the fingers collapse at the cut joints.

You can use the ready-made plastic hand for your projects. Mount it as detailed in the section above for the two-pincher claw arm.

You can make your own fingers from a variety of materials. One approach is to use the plastic pieces from some of the toy construction kits, such as Fastech. Cut notches into the plastic to make the joints. Attach a length of 20 or 22 gauge stove wire to the fingertip and keep

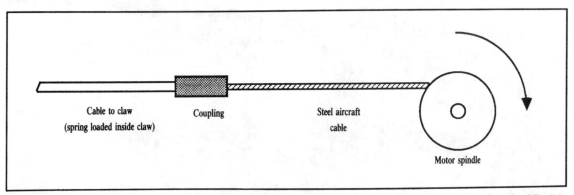

Cable to claw
(spring loaded inside claw)

Coupling

Steel aircraft
cable

Motor spindle

Fig. 21-8. One method for actuating the gripper: attach the solid aluminum cable from the claw to a length of flexible steel aircraft cable. Anchor the cable to a motor. Actuate the motor and the gripper closes. The spring in the gripper opens the claw when power to the motor is removed.

179

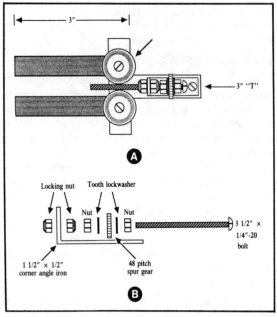

Fig. 21-9. A two-pincher gripper based on a home-made worm drive system. A. Assembled gripper; B. Worm shaft assembly detail.

it pressed against the finger using nylon wire ties. Do not make the ties too tight, or the wire won't be able to move. An experimental plastic finger is shown in Fig 21-18.

You can mount three of four such fingers on a plas-

tic or metal palm, and connect all the cables from the fingers to a central pull-rod. The pull-rod is activated by a solenoid or motor. Note that it takes a considerable pull to close the fingers, so the actuating solenoid or motor should be fairly powerful.

The finger opens again when the wire is pushed back out, and also by the natural spring action of the plastic. This springiness may not last forever, and may differ when using other materials. One way to guarantee opening the fingers is to attach an expansion spring, or a strip of flexible spring metal, to the tip and base of the finger, on the back side. The spring should give under the inward force of the solenoid or motor, but adequately return the finger to open position when power is cut.

WRIST ROTATION

The human wrist has three degrees of freedom: it can twist on the forearm, it can rock up and down, and it can rack from side to side. You can add some or all of these degrees of freedom to a robotic hand. A basic schematic of a three-degree of-freedom wrist is shown in Fig. 21-19.

With most arm designs, you'll just want to rotate the gripper at the wrist. Wrist rotation is usually performed by a motor, attached at the end of the arm, or at the base. When connected at the base (for weight considerations), a cable or chain joins the motor shaft to the wrist. The gripper and motor shaft are outfitted with mating spur gears. You can also use chains (miniature or #25) or tim-

Fig. 21-10. The finished two-pincher worm drive gripper.

180

Table 21-3. Worm Drive Gripper Parts List.

2	3-inch lengths 41/64- by 1/2- by 1/16-inch aluminum channel
2	1-inch diameter 64 pitch plastic or aluminum spur gear
1	2-inch flat mending "T"
1	1 1/2- by 1/2-inch corner angle iron
1	3 1/2-inch by 1/4-inch 20 stove bolt
2	1/4-inch 20 locking nuts, nuts, washers, tooth lockwashers
2	1/2-inch by 8/32 stove bolts, nuts, washers
1	1-inch diameter 48 pitch spur gear (to mate with gear on driving motor shaft)

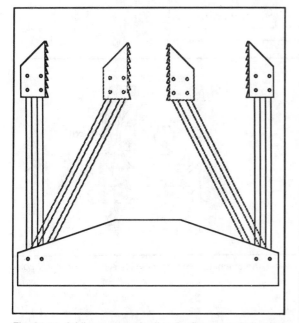

Fig. 21-11. Adding a second rail to the fingers and allowing the joints to freely pivot causes the fingertips to remain parallel to one another.

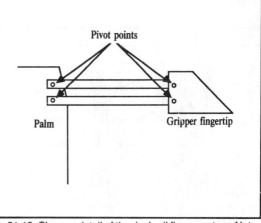

Fig. 21-12. Closeup detail of the dual-rail finger system. Note the pivot points.

ing belts to link the gripper to the drive motor. Figure 21-20 shows the wrist rotation scheme used to add a gripper to the revolute coordinate arm described in Chapter 19.

You can also use a worm gear on the motor shaft. Remember that worm gears introduce a great deal of gear reduction, so take this into account when planning your robot. The wrist should not turn too quickly or too slowly.

Another approach is to use a rotary solenoid. These special-purpose solenoids have a plate that turns 30 to 50 degrees in one direction when power is applied. The plate is spring loaded, so it returns to normal position when the power is removed. Mount the solenoid on the arm and attach the plate to the wrist of the gripper.

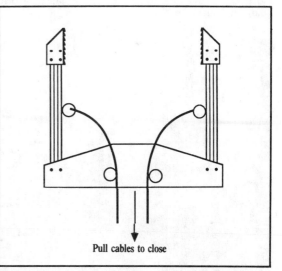

Fig. 21-13. A way to actuate the gripper: Attach cables to the fingers and pull the cables with a motor or solenoid. Fit a torsion spring along the fingers and palm to open the fingers when power is removed from the motor/solenoid.

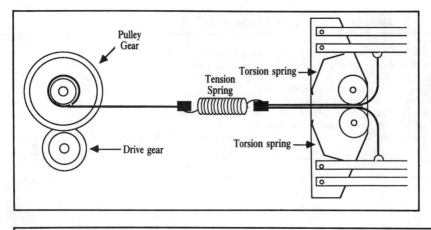

Fig. 21-14. Actuation detail of a basic two-pincher gripper using a motor. The tension spring prevents undue pressure on the object being grasped. Note the torsion springs in the palm of the gripper.

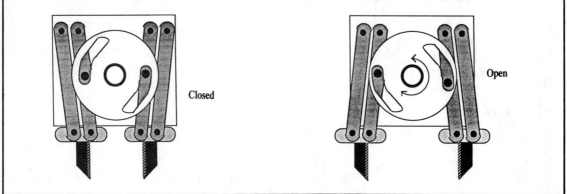

Fig. 21-15. A novel rotary gripper design. Rotating the disc causes the pinchers to open or close. A motor or rotary solenoid can actuate the disc.

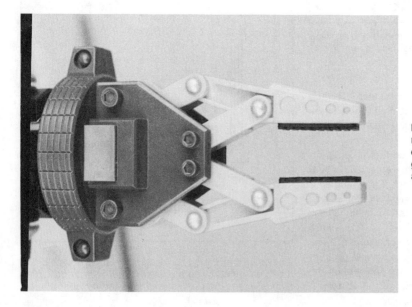

Fig. 21-16. A close-up view of the Armatron toy gripper. Note the use of the dual-rail finger system to keep the fingertips parallel. This gripper is moderately adaptable to your own designs.

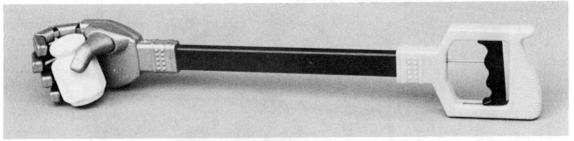

Fig. 21-17. A commercially available plastic robotic arm and hand. The gripper can be salvaged for use in your own designs. The opposing thumb is not articulated, but the fingers themselves have a semi-compliant grip.

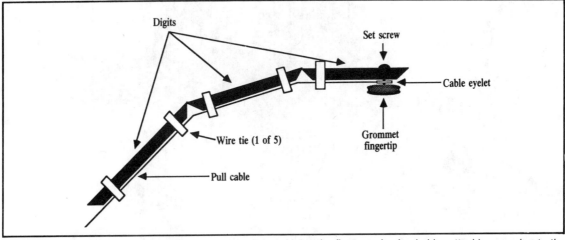

Fig. 21-18. A design for an experimental compliant finger. Make the finger spring-loaded by attaching a spring to the back of the finger (a strip of lightweight spring metal also works).

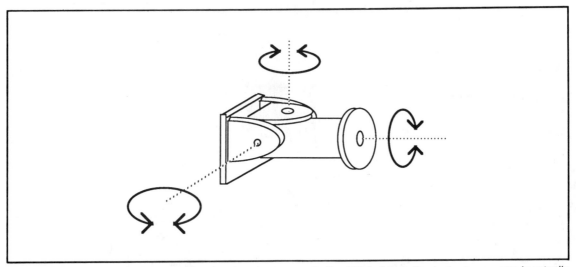

Fig. 21-19. The three basic degrees of freedom in a human or robotic wrist (wrist rotation in the human arm is actually accomplished by rotating the bones in the forearm).

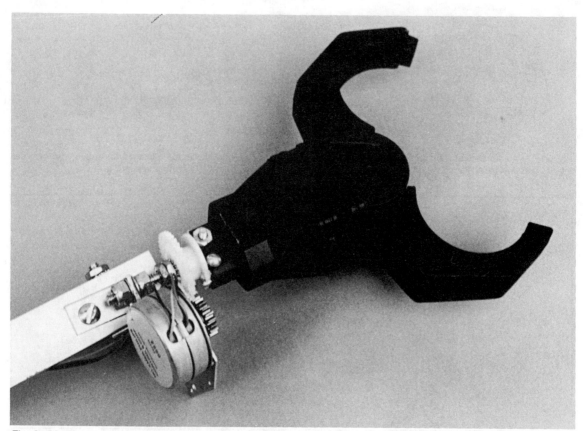

Fig. 21-20. A two-pincher gripper (from the plastic toy robotic arm), attached to the revolute arm described in Chapter 19. A stepper motor and gear system provide wrist rotation.

Chapter 22

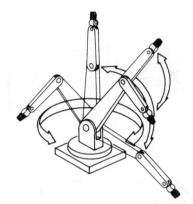

Adding the Sense of Touch

Like the human hand, robotic grippers often need a sense of touch to determine if and when it has something in its grasp. Knowing when to close the gripper to take hold of an object is only part of the story, however. The amount of pressure exerted on the object is also important. Too little pressure and the object may slip out of grasp; too much pressure and the object may be damaged.

The human hand—indeed, nearly the entire body—has an immense network of complex nerve endings that serve to sense touch and pressure. Touch sensors in a robot gripper are much more crude, but for most hobby applications, the sensors serve their purpose: to provide nominal feedback of the presence of an object and the pressure exerted on the object.

This chapter deals with the fundamental design approach of several touch sensing systems. Modify these systems as necessary to match the specific gripper design you are using, and the control electronics used to monitor the sense of touch.

OPTICAL SENSORS

Optical sensors use a narrow beam of light to detect when an object is within the grasping area of a gripper. Optical sensors provide the most rudimentary form of touch sensitivity, and are often used with other touch sensors.

Building an optical sensor into a gripper is easy. Mount an infrared LED in one finger or pincher; mount an infrared-sensitive phototransistor in another finger or pincher (see Fig. 22-1). Where you place the LED and transistor along the length of the finger or pincher determines the grasping area.

Mounting the infrared pair on the tips of the fingers or pinchers provides for little grasping area, because the robot is told that an object is within range when only a small portion of it can be grasped. In most gripper designs, two or more LEDs and phototransistors are placed along the length of the grippers or fingers to provide more positive control.

Alternatively, you may wish to detect when an object is closest to the palm of the gripper. You'd mount the LED and phototransistor accordingly.

Figure 22-2 shows the schematic diagram on a single LED-transistor pair. Adjust the value of R2 to increase or decrease the sensitivity of the phototransistor. You may need to place an infrared filter over the phototransistor to prevent it from triggering from ambient light sources (some phototransistors have the filter built into them already). Use a LED-transistor pair equipped with a lens

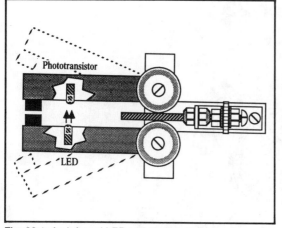

Fig. 22-1. An infrared LED and phototransistor pair can be added to the fingers of a gripper to provide go/no-go grasp information.

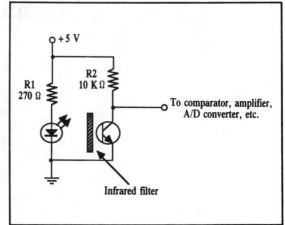

Fig. 22-2. The basic electronic circuit for an infrared touch system. Note the infrared filter. It prevents the phototransistor from being activated by ambient light.

to provide additional rejection of ambient light and to increase sensitivity.

During normal operation, the transistor is on, because it is receiving light from the LED. When an object breaks the light path, the transistor switches off. A control circuit connected to the conditioned transistor output detects the change, and closes the gripper. In a practical application, using a computer as a controller, you'd write a short software program to control the actuation of the gripper.

PRESSURE SENSORS

An optical sensor is a go/no-go device that can detect only the presence of an object, not the amount of pressure on it. A pressure sensor detects the force exerted by the gripper on the object. The sensor is connected to a converter circuit, or in some cases a servo circuit, to control the amount of pressure applied to the object.

Pressure sensors are best used on grippers where you have incremental control over the position of the fingers or pinchers. A pressure sensor would be of little value when used with a gripper that's actuated by a solenoid. The solenoid is either pulled in or it isn't; there are no in-between states. Grippers actuated by motors are the best choices when you must regulate the amount of pressure exerted on the object.

Conductive Foam

You can make your own pressure sensor (or transducer) out of a piece of discarded conductive foam, the stuff used to package CMOS ICs. The foam is like a resistor. Attach two pieces of wire to either end of a 1-inch square hunk and you get a resistance reading on your volt-ohm meter. Press down on the foam and the resistance lowers.

The foam comes in many thicknesses and densities.

Table 22-1. Sound Amplifier Parts List.

U1	LM741 Op Amp
R1	1 megohm potentiometer
R2,R3	6.8KΩ resistor
R4	1KΩ resistor
C1,C2	0.47 μF ceramic capacitor
Q1	2N2222 npn transistor
MIC1	Electret condenser microphone

All resistors 5 or 10 percent tolerance, 1/4-watt; all capacitors 10 percent tolerance.

U1	ADC0808 8-bit Analog-to-Digital Converter IC
R1	10KΩ resistor (experiment with value for best response of sensor)
Misc.	Conductive foam pressure sensor (see text)

I've had the best luck with the semi-stiff foam that bounces back to shape quickly after squeezing it. Very dense foam is not useful, because it doesn't quickly spring back to shape. Save the foam from the various ICs you buy and test various types until you find the right stuff for you.

Here's how to make a pressure sensor. Cut a piece of foam 1/4-inch wide by 1-inch long. Attach leads to it using 30 gauge wire-wrapping wire. Wrap the wire through the foam several places to ensure a good connection, then apply a dab of solder to keep it in place. Use flexible household adhesive to cement the transducer on the tips of the gripper fingers.

A better way is to make the sensor by sandwiching several pieces of material together, as depicted in Fig. 22-3. The conductive foam is placed between two thin sheets of copper or aluminum foil. A short piece of 30 AWG wire-wrapping wire is lightly soldered onto the foil

(when using aluminum foil, the wire is wound around one end). Mylar plastic, like the kind used to make heavy-duty garbage bags, is glued on the outside of the sensor to provide electrical insulation.

If the sensor is small, and the sense of touch does not need to be too great, you can encase the foam and foil in heat-shrink tubing. There are many sizes and thick-

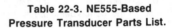

**Table 22-3. NE555-Based
Pressure Transducer Parts List.**

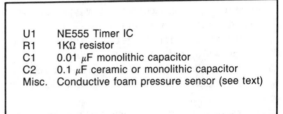

U1	NE555 Timer IC
R1	1KΩ resistor
C1	0.01 μF monolithic capacitor
C2	0.1 μF ceramic or monolithic capacitor
Misc.	Conductive foam pressure sensor (see text)

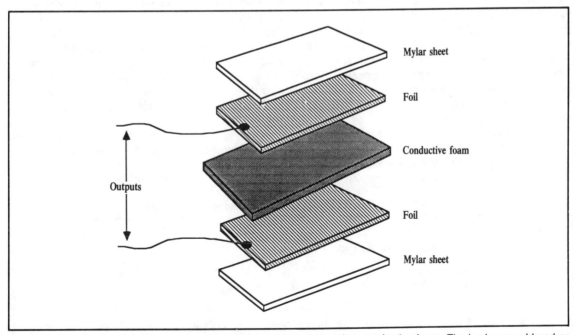

Fig. 22-3. Construction detail for a more elaborate pressure sensor using conductive foam. The leads are soldered or attached to foil (copper works best).

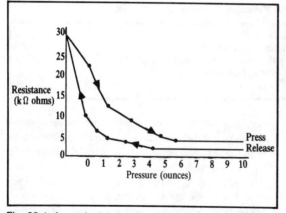

Fig. 22-4. A graph showing the response curve of the conductive foam pressure sensor. Note that the resistance varies depending on whether the foam is being pressed or released.

Strain Gauges

Obviously, the home-built pressure sensors described above leave a lot to be desired, accuracy wise. If you need greater accuracy, you should consider commercially available strain gauges. These work by registering the amount of strain—same as pressure—exerted on various points along the surface of the gauge. Strain gauges are somewhat pricey—about $10 and over in quantity of five or 10. The cost may be offset by the increased accuracy the gauges offer. You want a gauge that's as small as possible, preferably one mounted on a flexible membrane. See Appendix A, "Sources," for a list of companies offering such gauges.

Converting Pressure Data to Computer Data

The output of both the home-made conductive foam pressure transducers and strain gauges is analog—a resistance or voltage. Neither can be directly used by a computer, so the output of these devices must be converted to digital form first.

Both types of sensors are perfect for use with an analog-to-digital converter. You can use one ADC0808 chip (under $5) with up to eight sensors. You select which sensor output you want to convert to digital form. The converted output of the ADC0808 chip is an eight-bit word, which can be fed directly to a microprocessor or computer port. A circuit diagram for use with the conductive foam transducer is shown in Fig. 22-5. Notice the 10K resistor placed in series with the pressure sensor. This converts the output of the sensor from resistance to voltage. You can change the value of this resistor to alter the sensitivity of the circuit. For more information on A/D converters, see Chapter 32, "Build a Robot Interface Card."

Another method is to use the resistance values provided by the sensors (no voltage divider resistor added) in an astable multivibrator circuit. Use the common NE555 timer IC as the multivibrator. Insert the output

nesses of tubing; experiment with a few types until you find one that meets your requirements.

The output of the transducers changes abruptly when they are pressed in. The output may not return to its original resistance value (see Fig. 22-4). So in the control software, you should always reset the transducer just prior to grasping an object.

For example, the transducer may first register an output of 30K ohms (the exact value depends on the foam, the dimensions of the piece, and the distance between the wire terminals). The software reads this value and uses it as the setpoint for normal (non-grasping) level to 30K. When an object is grasped, the output drops to 5K. The difference—25K—is the amount of pressure. Keep in mind that the resistance value is relative, and you must experiment to find out how much pressure is represented by each 1K of resistance change.

The transducer may not go back to 30K when the object is released. It may spring up to 40K or go only as far as 25K. The software uses this new value as the new setpoint for the next time the gripper grasps an object.

Table 22-4. Gated CMOS Pressure Transducer Parts List.

U1	4011 CMOS Quad NAND Gate IC
R1	1 MΩ resistor
R2	470 ohm resistor
Misc.	Conductive foam pressure sensor (see text)

All resistors 5 or 10 percent tolerance, 1/4-watt; all capacitors 10 percent tolerance, rated 35 volts or higher.

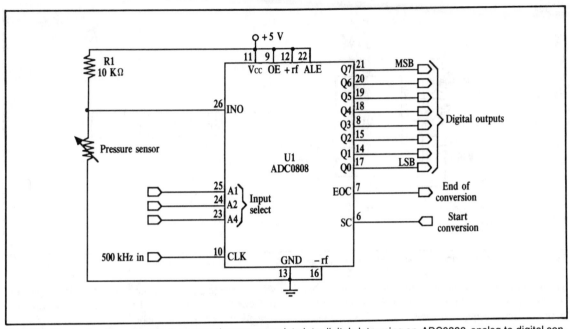

Fig. 22-5. The basic wiring diagram for converting pressure data into digital data using an ADC0808 analog-to-digital converter IC. You can connect up to eight pressure sensors to the one chip.

of the sensor in the path marked R1 in Fig. 22-6. When the pressure on the sensor is varied, its resistance varies, which in turn causes the frequency output of the 555 to increase or decrease. You can use the output in a closed-loop servo system (adapt the motor speed control servo system from Chapter 13).

Another technique is to use a gated astable multivibrator, as shown in Fig. 22-7. The amount of pressure on the sensor determines the output frequency of the multivibrator, as in the circuit above. The multivibrator is connected to one input of a NAND gate. To "pass" the output of the multivibrator through the gate, a "sampling" signal—of a fixed, timed duration—is applied to the other input of the gate. The output of the NAND gate feeds the Clock input of a counter.

At the end of the sampling period, the value of the counter represents the relative pressure on the sensor. You can use a 4040 CMOS 12-stage ripple counter as the counter circuit, or if you need a visual display, any of the BCD CMOS or TTL counters. A one-shot (monostable multivibrator) can provide a timed-duration pulse for switching on the NAND gate. The 4040 receives only those pulses from the one-shot during the sampling interval, and none of the pulses before or after. Reset the counter before taking another sample. A similar gating/sampling technique is used in Chapter 29, "Navigating Through Space." See the section on ultrasonic ranging for more information.

SOUND SENSORS

Sound makes an effective touch sensor. There are a number of inexpensive sound transducers you can use and easily attach to just about any gripper design. Three such sensors are described below.

Microphone

An electret condenser microphone is about the size of a watch battery, yet it can be used to pick up sounds from across the room. You'll use that sensitivity, but to listen for the tell-tale sounds emitted when the gripper closes around an object. The microphone concept works because most solid and near-solid objects emit the most sounds when the gripper first contacts them. The sounds are reduced as the gripper exerts more pressure. A solid gripper-object grasp should make little, if any, sound.

The makeup of the object being grasped is an important consideration. A glass salt shaker will not make a peep once the gripper has a solid grasp on it, but a piece of Styrofoam will continue to make noise no matter how much pressure is used.

Mount the microphone element on the palm of the

189

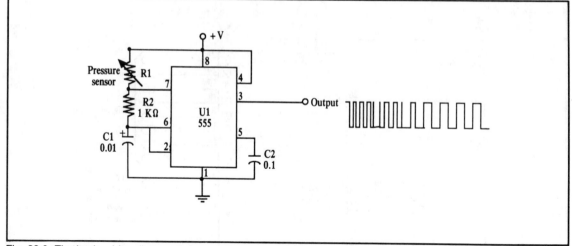

Fig. 22-6. The basic wiring diagram for converting pressure data to a variable frequency.

gripper, or directly on one of the fingers or pinchers. Place a small piece of felt directly under the element, and cement it in place using a household glue that sets up hard. Run the leads of the microphone to the sound trigger circuit, which should be placed as close to the element as possible.

A typical sound detection circuit is shown in Fig. 22-8. This circuit is adjustable so that it triggers on just about any level of sound. Experiment with the sensitivity (R1) until the circuit triggers when the proper pressure from the gripper is reached.

Note that there is limitation in this design. Sounds from the robot, such as the vibration of motors, can interfere with the sensitivity of the microphone. For best results, insulate the gripper from the rest of the robot using thick rubber pads. Insert the pads when mounting the

wrist of the gripper to the arm. You can also reduce the influence of ambient sound by using the arm fully extended away from the robot.

Ultrasonic

An ultrasonic transducer is a microphone that's sensitive only to a certain range of sounds, typically about 35 kHz to 45 kHz. These very high frequency sounds are often emitted by objects as pressure is placed on them. Use an ultrasonic transducer, like the kind found in alarm systems, in place of the electret condenser microphone. The benefit of using an ultrasonic transducer is that it is not as susceptible to ambient noise. The disadvantage is that it is not as sensitive as an electret condenser microphone. See Appendix A, "Sources," for manufacturers

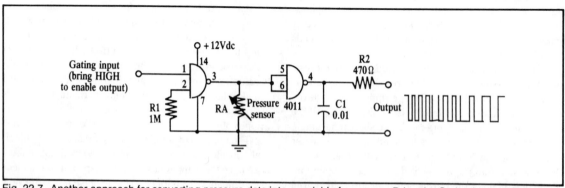

Fig. 22-7. Another approach for converting pressure data into a variable frequency. Bring the Gating Input HIGH to pass the output. The gated output can be routed to a frequency counter, which in turn displays a value that reflects the pressure on the sensor.

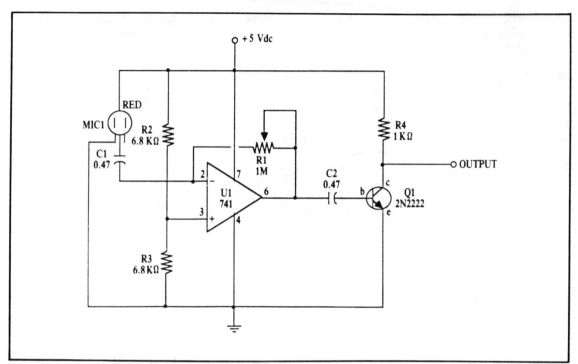

Fig. 22-8. A circuit for amplifying sound picked up by an electret condenser microphone element. The circuit is very sensitive; adjust R1 to alter the sensitivity.

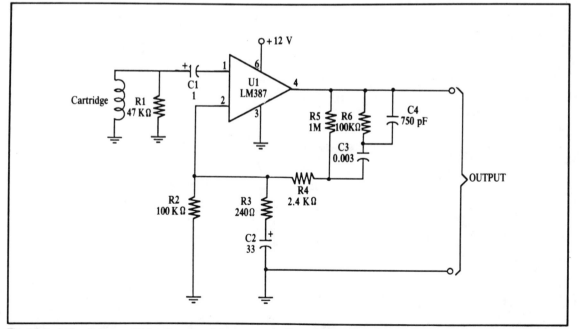

Fig. 22-9. A circuit for amplifying the sound from a magnetic cartridge. The circuit will also work with a guitar pickup, which is also suitable for sound/touch sensing.

and retailers offering ultrasonic transducers. More information for working with ultrasonics is provided in Chapters 28 and 29.

Magnetic Pickup

An inexpensive magnetic pickup cartridge from a discarded record player makes an excellent pressure and vibration sensor. You replace the needle (if the cartridge has one) with a sensing plate made of thin metal (the copper clad peeled off a blank printed circuit board is a good choice). The plate enlarges the sensing area, providing an almost fingertip-like touch pad. The output of the cartridge should be preamplified before passing the signal to the sound level detector. Use the preamplifier circuit in Fig. 22-9.

Table 22-5. Phono Cartridge Amplifier Parts List.

U1	LM387 Integrated Amplifier IC
R1	47 K resistor
R2,R6	100 K resistor
R3	240 ohm resistor
R4	2.4 K resistor
R5	1 megohm resistor
C1	1 μF electrolytic capacitor
C2	33 μF electrolytic capacitor
C3	0.003 μF ceramic capacitor
C4	750 pF ceramic capacitor

All resistors 5 or 10 percent tolerance, 1/4-watt; all capacitors 10 percent tolerance, rated 35 volts or higher.

Chapter 23

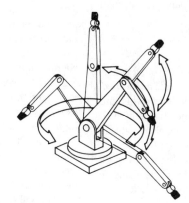

Adding a Mouth to Your Robot

Give your robot a say in things. Add a speech synthesizer and your robot can go yapping around the house like the best (or worst?) of humans! New, all-in-one voice synthesis chips make adding a voice to your robot a relatively simple and inexpensive project.

One chip, the General Instrument SPO256-AL2, available at Radio Shack and many other electronic stores, can be programmed to say any word. The SPO256 retails for under $15, and with the addition of just a few components, you can connect the synthesizer to your computer (a parallel printer port will do) and your robot can talk for hours on end.

There are several other speech synthesis chips available, such as the Votrax SC-02 and the National Semiconductor Digitalker but the SPO256 is commonly available and extremely easy to use. We'll design our robot mouth around it. Details follow:

THE FUNDAMENTALS OF SPEECH

Voice synthesis is the conversion of digital information to recognizable speech. Sounds simple, but in actual operation, the goings-on inside a voice synthesizer are considerably more complex. We humans have vocal chords, a mouth, tongue, and lips to control the various sounds

we make. A synthesizer must recreate these sounds by the use of sound and noise generators, filters, timers, and delay circuits.

In a typical speech synthesis system, you key in words on the computer keyboard or run a software program. The result is a series of speech codes. The speech synthesizer receives the codes and through the use of various period (short duration) and continuous sound generators, creates human-sounding speech. After passing through a digital-to-analog converter (or sometimes a smoothing circuit for pulse-width modulated digital output), the synthesized signal is amplified and fed to a speaker.

In a speech synthesizer system, such as the SPO256 chip, the voice you hear over the speaker is a string of individual sounds called *phonemes*. There are several dozen common phonemes in the English language and they represent the sounds created by uttering different letters and combinations of letters during regular speech. Vowels have more than one phoneme each, since they can be long or short. Combinations of letters such as "th" and "sh" have their own unique phonemes.

The ability to program a synthesizer by phoneme, instead of by pre-record voice digitization (as used in some speech synthesizers), leads to the possibility of unlimited vocabularies. English includes much of the sounds encountered in many foreign languages, so you can apply the synthesizer to speak in German, Spanish, or French as well.

Inside Phonemes

Most synthesizers employ two types of sound generators to produce speech: a *voiced sound* source and a *fricative sound* source. Voiced sounds included vowels and some consonants that have vowel sounds in them, such as "v" and "z." The letters "r" and "w" are also produced by the voiced sound generator of most synthesizers.

The fricative generator is responsible for almost all of the consonants that can be characterized by a phoneme with a "noisy" rasp to it. Fricatives include "s," "k," "f," "sh," "th," and so forth.

The output signals from the voiced and fricative generators are fed to a series of filters which are designed to mimic the human vocal tract—the mouth, tongue, and even the sinus cavities. The number of filters, and how they are used, vary from one speech chip to another. The output of the filters is usually applied to a digital-to-analog (D/A) converter for amplification. In some speech synthesizer chips, notably the SPO256, the output is a digital stream that is smoothed by a series of resistors and capacitors. The smoothed output is amplified and fed through a speaker.

Simply producing a string of vowels and consonants together without blending the transitions between the phonemes sounds awkward and stilted. *Dynamic articulation,* or the transition between sounds, must be gradual and sloped, or the resultant speech will be nothing more than a series of choppy gibberish.

Voice synthesizers employ circuitry that slope the beginning and end of each phoneme (the degree of slope depends on the individual phoneme). As an example, divide the word "computer" into its component sounds and you come up with something like "k-um-pu-t-er." Speak each of these phonetically without transitions and you have a word that's nearly impossible to understand. Smooth out the transitions and you have something like "kum-pew-ter."

While infants are learning to speak, they imitate not only the sounds of speech, but the inflection of voices as well. The rising and falling pitch of words and sentences aids in comprehension. Unfortunately, only a few voice synthesis chips and complete systems are capable of inflective speech. The SPO256 is capable of just one pitch.

BUILDING THE SPEECH SYNTHESIZER

The heart of the speech synthesizer project is the General Instruments SPO256-AL2 speech chip. This 28-pin IC requires only a few external components to operate, and is primarily intended for connection to a microprocessor. The schematic that follows uses the parallel printer port found on most personal computers. The examples given in this chapter are geared towards the IBM PC (or compatible) equipped with a parallel port board, and the Commodore 64. The IBM PC version can also be used successfully with most other computers having a standard Centronics-compatible parallel port.

Figure 23-1 shows a pin-out diagram of the SPO256 chip. Note the unusual supply pins: 1 and 7. Be sure to connect the power pins correctly and use a well-regulated 5 volt supply. Reversing the power pins, or using a higher voltage, will cause damage to the chip.

Refer to the schematic in Fig. 23-2. The SPO256 uses an odd-ball 3.12 MHz crystal as a time base. This crystal is available through Radio Shack on a special order basis, and may also be available through mail order companies specializing in crystals. Alternatively, you may have luck using a 3.2768 MHz crystal, which is much more common. Capacitors C6 and C7 are used to provide stability to the crystal. Use a 5-30 pF miniature trimmer cap for C6. Adjust the cap for proper sound output. If

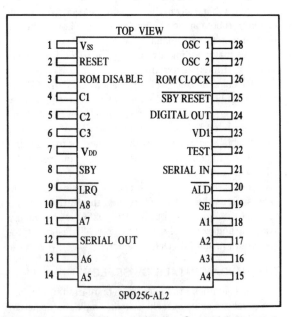

Fig. 23-1. Pinout diagram for the General Instruments SPO256-AL2 speech generator chip.

Table 23-1. Speech Processor, IBM PC and CP/M Computer Version Parts List.

U1	General Instruments SPO256-AL2 Speech Processing IC
R1	10KΩ resistor
R2	1KΩ resistor
R3	200KΩ resistor (or two 100KΩ resistors in series)
R4,R5	100KΩ resistor
R6,R7	33KΩ resistor
R8	10 K analog taper potentiometer
C1,C2	0.1 μF ceramic capacitor
C3,C4	0.022 μF ceramic capacitor
C5	1.0 μF tantalum capacitor
C6	5-30 pF trimmer capacitor
C7	10 pF ceramic or mica capacitor
D1,D2	1N914 signal diode
Q1,Q2	MPS2907 pnp transistor

All resistors 5 or 10 percent tolerance, 1/4-watt; all capacitors 10 percent tolerance, rated 35 volts or higher.

you have a frequency meter, use it to "dial in," if possible, the proper 3.12 MHz frequency with the trimmer cap.

The components connected to the SBY, $\overline{RESET}$ and $\overline{SBY RESET}$ pins provide a power-on reset and time out function. The extra circuitry is used on the IBM PC, but not the Commodore 64 version, shown in Fig. 23-3.

Resistors R6 and R7, plus capacitors C3, C4, and C5 form the output of the chip. The smoothed digital output is applied to one side of a 10K audio taper potentiometer. The wiper of the pot is connected to the input of an audio amplifier.

You can use the amplifier shown in Fig. 23-4 using the LM386 integrated amp, or the higher power ampli-

fier described in Chapter 24, "Build a Music and Sound Effects Generator." The output of the LM386 won't blast out your ears, but it should be enough for experimenting. Use a miniature 4 or 8 ohm speaker.

The speech synthesis board is best constructed using wire-wrapping techniques. All components except C5 and R8 can be placed in wire-wrap sockets. Mount the 3.12 MHz crystal as close to pins 27 and 28 as possible, and mount capacitors C6 and C7 as close to the crystal as possible. Be sure to observe proper polarity and orientation of the diodes and transistors. The polarity of C5 is not important. Tables 23-1 through 23-3 contain the parts lists for speech processors.

Table 23-2. Speech Processor, Commodore-64 Version Parts List.

U1	General Instruments SPO256-AL2 Speech Processing IC
R1	100KΩ resistor
R2,R3	33KΩ resistor
R4	10 K analog taper potentiometer
C1,C2	0.1 μF ceramic capacitor
C3,C4	0.022 μF ceramic capacitor
C5	1.0 μF tantalum capacitor
C6	5-30 pF trimmer capacitor
C7	10 pF ceramic or mica capacitor
D1	1N914 signal diode

All resistors 5 or 10 percent tolerance, 1/4-watt; all capacitors 10 percent tolerance, rated 35 volts or higher.

Table 23-3. Speech Amplifier Parts List.

U1	LM386 Audio Amplifier IC
R1	10 ohm resistor
C1	100 µF electrolytic capacitor
C2	0.05 µF ceramic capacitor
C3	10 µF tantalum or electrolytic capacitor
C4	250 µF electrolytic capacitor
SPKR1	4- or 8-ohm speaker (less than five inches diameter)

All resistors 5 or 10 percent tolerance, 1/4-watt; all capacitors 10 percent tolerance, rated 35 volts or higher.

Hooking the Board to a Computer

The SPO256 can be connected directly to the data and control lines of a standard parallel printer port or the User Port on the Commodore 64. Data lines A1 through A6 on the SPO256 connect to the D0 to D5 data output lines on the printer port (lines A6 and A7 are not used). To output a phoneme, the computer places a decimal code between 0 and 63 on the data lines of the port, then pulses the STROBE line LOW. The STROBE line triggers the ALD pin on the SPO256, signaling that there is data to be retrieved. The chip then speaks the phoneme.

After sending a phoneme, the computer quickly checks (and rechecks) the status of the BUSY line of the parallel port. The BUSY line is connected to the $\overline{LRQ}$ pin of the SPO256, which goes HIGH when the chip is speaking. When HIGH, the computer cannot send another phoneme. The line goes LOW again when the SPO256 is finished speaking. This drives the BUSY line LOW,

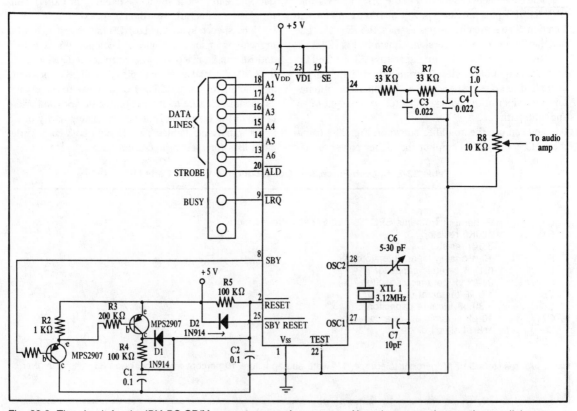

Fig. 23-2. The circuit for the IBM PC-CP/M computer speech processor. Note the connections to the parallel port.

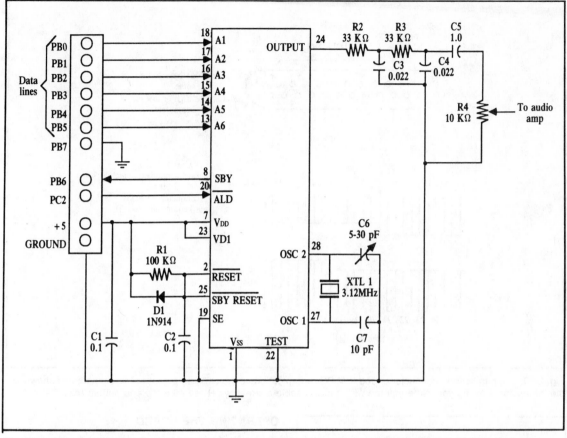

Fig. 23-3. The circuit for the Commodore 64 speech processor. Note the connections to the C-64 User Port.

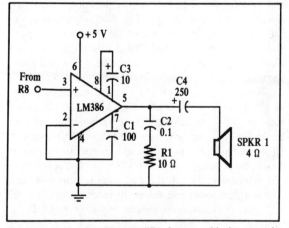

Fig. 23-4. A handy utility amplifier for use with the speech processor. Use a higher output amp (see Chapter 24) if you want more volume. The speaker can be either 4 ohm or 8 ohm.

and the computer is allowed to output the next phoneme. The process continues over again until all the phonemes are spoken.

You can connect the input and output lines of the chip to the parallel port using a 25-conductor ribbon connector terminated with a male DB-25 connector. This connector plugs directly into the parallel port on an IBM PC or compatible. Use care when wiring up the cable. Double check to make sure that the strands of the cable connect to the proper chip inputs and outputs. Refer to Fig. 23-5 for a diagram of a DB-25 connector. Note that the numbering starts from right to left (as viewed when looking at the end of the connector). When used with a ribbon connector, the pins alternate across the strands: 1, 14, 2, 15, and so forth.

Refer to Fig. 23-6 for a connection diagram when u ing a Commodore 64. Use a connector specially desig for the User Port, or cut a 22-pin connector to the pr size.

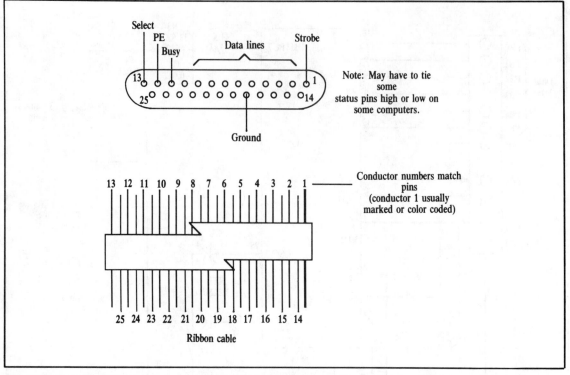

Fig. 23-5. The parallel port connector on the IBM PC (and most compatibles). The pinout of the connector is shown as the connector is facing you. Note that the wiring in the cable alternates between the top and bottom row.

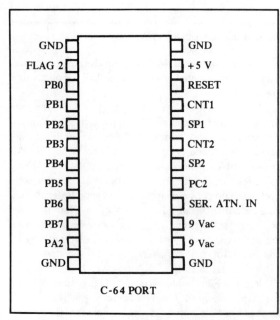

Fig. 23-6. Pinout diagram of the Commodore-64 User Port.

OPERATING THE BOARD

You can operate the board in any of several ways. Short BASIC program routines are provided in Figs. 23-7 and 23-8 for the IBM PC and Commodore 64, respectively. The program listing in Fig. 23-9 is best suited for use with computers other than IBM PC and relies fully on BASIC to output the data. The BASIC program in Fig. 23-10 is an alternative for use on the IBM PCs (and compatible).

Because of the way BASIC works, it ends each line as well as terminates execution by sending a line feed/carriage return to the printer. These codes output a sound on the SPO256 chip, which continue indefinitely until the chip receives a new phoneme code. In use, the computer program operating your robot will have the speech routine embedded within it, along with all the other routines. The program won't end unless the robot is switched off or reset, the unwanted speech probably won't be a problem.

Use the listing in Fig. 23-7 and 23-10 IF IBM PC or compatible is equipped with a parallel expansion board. If you are using the parallel port contained on the Mono-

```
5      CLS
10     Y = 0
20     READ X
30     IF X = 100 THEN GOTO 100
40     OUT 888, X
50     Y = INP (889)
60     IF Y = 143 THEN GOTO 70 ELSE 50
70     OUT 890,127
80     OUT 890,128
90     GOTO 20
100    OUT 888,0
110    OUT 890,127
120    OUT 890,128
130    REM Testing
140    DATA 13,7,55,2,13,12,44,4,4
150    REM One
160    DATA 46,15,15,11,2
170    REM Two
180    DATA 13,21,2
190    REM Three
200    DATA 29,14,19,2
210    REM Four
220    DATA 40,40,58,1,1,1,1,100
```

Fig. 23-7. Speech synthesizer talker program, for the IBM PC.

chrome Display/Parallel Printer Adapter, change the input and output values in the program as specified in Chapter 31, "Computer Control Via Parallel Port." Otherwise, use the generic BASIC version in Fig. 23-9.

```
10     REM C-64 SPEECH PROGRAM
20     POKE 56579,63
30     FOR J = 1 TO 27
40     READ A
50     POKE 56577,A
60     POKE 56577,0
70     PB = PEEK (56577)
80     F = PBAND64
90     IF F < > 64 THEN 70
100    NEXT J
110    REM I
120    DAT 24,6,0
130    REM AM
140    DATA 7,7,16,2
150    REM A TALKING
160    DATA 24,2,0,13,23,23,2,42,12,44,0
170    REM COMPUTER
180    DATA 42,15,16,9,49,22,13,51,1,4,4
190    RESTORE
170    FOR T = 1 TO 500:NEXT T: GOTO 30
1000   END
```

Fig. 23-8. Speech synthesizer talker program, for the Commodore 64.

```
10     READ X
20     IF X = 100 THEN END
30     REM READ PHONEME
40     LPRINT CHR$(X);
50     GOTO 10
60     REM TESTING
70     DATA 13,7,55,2,13,12,44,4,4
80     REM ONE
90     DATA 46,15,15,11,2
100    REM TWO
110    DATA 13,31,2
120    REM THREE
130    DATA 29,14,19,2
140    REM FOUR
150    DATA 40,40,58,4,4,4,4,4,100
```

Fig. 23-9. Speech synthesizer for CP/M computers.

IN CASE OF PROBLEMS

If your speech synthesizer does not work properly, double check your work. Be especially critical of the connections from the parallel port to the lines on the SPO256 chip. Use a logic probe to test that data as it is being received by the chip. The STROBE ($\overline{ALD}$) line should pulse on and off when the computer is outputting data. Most or all of the data lines should pulse as well.

If the chip emits a loud burst of noise, it has "crashed." Reset the chip by momentarily disconnecting the power. If the chip continues to emit nothing but noise when it should be talking, the problem probably lies in the crystal. The SPO256 seems to be extremely sensitive to fluctuations in the crystal frequency, and one of the prototype boards I constructed had to be warmed up for a few minutes so the crystal could stabilize. Adjusting trimmer cap C6 also seemed to help.

```
10     READ X
20     IF X = 100 THEN END
40     OUT 888,X
41     OUT 890, 127:OUT 890,128
42     FOR D = 1 TO 115: NEXT D
50     GOTO 10
60     REM testing
70     DATA 13,7,55,2,13,12,44,4,4
80     REM one
90     DATA 46,15,15,11,2
100    REM two
110    DATA 13,21,2
120    REM three
130    DATA 29,14,19,2
140    REM four
150    DATA 40,40,58,4,4,4,4,4,100
```

Fig. 23-10. Alternate speech program for the IBM PC.

The SPO256 will continue to make the sound of the last phoneme it has received until told otherwise. To quiet the chip, send a pause. Make it a habit to end each data line in your program with a pause to quiet the chip.

When used in a BASIC program that contains data lines, you need to add three or four pause codes to the last line. Otherwise, the chip will stop talking before all the phonemes have been sent. Note the extra pauses in the last data lines in the example program listings.

The data sheet that accompanies the SPO256 contains phoneme codes for a number of common words and phrases. Use the listings to help you construct sentences for your robot. You'll find that after using the SPO256 for a while that all 63 phonemes are committed to your own memory banks. You'll be able to write long, eloquent sentences in just a matter of minutes.

As an advanced project, you can wire up the sister chip to the SPO256, the General Instrument CTS256-AL2. This IC converts standard English text into the phoneme code used by the SPO256. The text-to-speech translator CTS256 chip can be connected to your computer via a serial or parallel port. Most outlets that offer the SPO256 also carry the companion text-to-speech translator.

Chapter 24

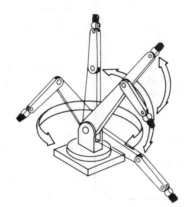

Build a Music and Sound Effects Generator

Speech synthesis is one type of sound output you can easily add to your robot creation. Others are music and sound effects. Imagine having your robot play a rock tune or two at your next party, or a few notes in the wee hours of the morning to cheerfully wake you up to the new day. Or how about an ear-piercing siren alarm at the first sign of trouble?

This chapter is devoted to making music and sound effects generators for your robot. Construction details are provided for several sound-making projects, and there's room on the sound board for a few more projects.

One advantage of music and sound effects generation is that the circuits are fairly straightforward, and the ones described here do not absolutely require a computer (although, as usual, a computer comes in handy). Sophisticated sound effects chips exist for computer control, such as the General Instrument AY3-8910A. You may wish to experiment with them for inclusion in your robot designs.

MUSIC MAKER

The UM3482A Melody Generator, as available from Radio Shack (Catalog # 276-1797) has 12 built in tunes including "Oh, My Darling Clementine," "London Bridge is Falling Down," "Happy Birthday," and "Home Sweet Home." You play them back by bringing several pins on the chip HIGH or LOW. The UM3482A is mask-ROM programmed, meaning that the music data is encoded into it at the time of manufacture, and that other versions of the chip may contain a different selection of songs. See Table 24-1 for a complete listing of songs.

Using the UM3482A is simple and straightforward. For basic play—play a song once and stop—the chip is connected as shown in Fig. 24-1. The single 2N2222 transistor is sufficient to provide a reasonable output level through a small 8 ohm speaker. You may, instead, route the sound output to the more powerful amplifiers presented later in this chapter. A pinout diagram is shown in Fig. 24-2.

When the chip is first powered up, it automatically starts at selection 1, which is "American Patrol." The song will play once and the chip will go quiet. Applying a positive-going pulse to pin 4 causes the chip to change to the next song. That's how you can change selections, like a jukebox. You can't skip back through the selections, so if you go past the selection you want, you have to change through all of them.

In a more practical circuit, you'd cut the Vcc connec-

Table 24-1. UM3482A Melody Generator Songs.

SELECTION	SONG
1	American Patrol
2	Rabbits
3	Oh, My Darling Clementine
4	Butterfly
5	London Bridge is Falling Down
6	Row, Row, Row Your Boat
7	Are You Sleeping?
8	Happy Birthday
9	Joy Symphony
10	Home, Sweet Home
11	Weigenlied
12	Melody on Purple Bamboo

tion to pin 2 and control it manually. When pin 2 is HIGH, the chip is enabled, and it will play the next selected song. When pin 2 is LOW, the chip is deselected, and though you skip through to other selections, they won't play. This is how you select a new song without playing the first few notes of all the others before it.

Refer to Table 24-2 for pin assignments and functions and Table 24-3 for a program truth table. You should have no trouble getting the chip to do all sorts of tricks for you with these two tables. Alter the connection of the other control pins to suit your requirements.

Changing Timber

You can change the timber of the sound by adjusting capacitor C3, resistor R3, or both. You may want to substitute the 150K fixed resistor, R3, for a 250K poten-

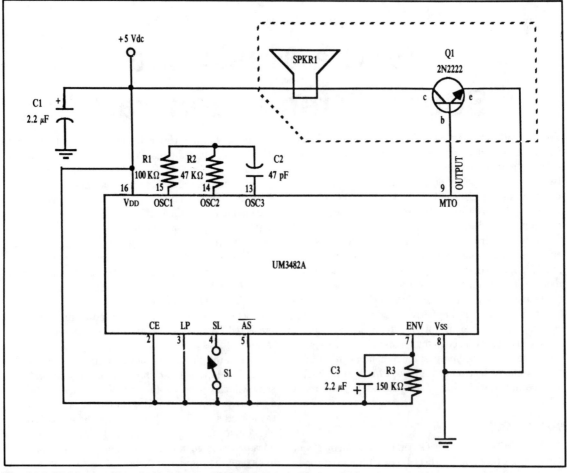

Fig. 24-1. The basic wiring diagram for the UM3482A Melody Generator IC.

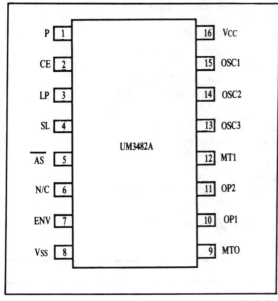

Fig. 24-2. Pinout diagram for the UM3482A Melody Generator.

tiometer. For remote control of timber, you can use the circuit shown in Fig. 24-3. This lets you select the timber via computer or automated control. Only one input of the 4066 should be on at one time. Figure 24-4 shows the effect of the sound envelope shape when altering the C3/R3 time constant.

WARBLING SIREN

If you use your robot as a security device, for detecting intruders, fire, water, or whatever, then you probably want the machine to warn you of immediate or impending danger. The warbling siren shown in Fig. 24-5 will do the trick. The circuit is constructed using two 555 timer chips (alternatively, you can combine the functions into the 556 dual timer chip). To change the pitch and speed of the siren, alter the values of R2 and R5.

For maximum effectiveness, connect the output of U2 to a high-powered amplifier. Try the 16 watt amplifier described later in this chapter, or use a more powerful integrated amplifier module. These are often available in ready-made or kit form through electronics mail order companies (check the latest editions of the electronics magazines). Some of the integrated amps put out 50 or more watts.

USING THE SPO256 VOICE CHIP FOR SOUND EFFECTS

If you built the speech synthesizer described in Chapter 23 (and you really should; it's fun!), then you probably found that it makes some pretty funny sounds when fed random data. With an IBM PC or CP/M computer, for example, you can cause all keyboard input to be sent to the parallel printer port by pressing the Control and P keys simultaneously. Start tapping keys and the chip

Table 24-2. Pin Assignments of UM3482A Melody Generator.

PIN	FUNCTION	DESCRIPTION
1	TSP	Output flap of melody autostop Leave floating for normal play
2	CE	Chip enabled if HIGH; chip disabled if LOW
3	LP	Song plays once if HIGH; all songs play if LOW
4	SL	Song trigger (positive going pulse triggers)
5	AS	Melody repeat if HIGH, melody auto-stop if LOW
6	NC	No connection
7	ENV	Envelope circuit terminal
8	GND	Ground
9	MTO	Modulated tone signal output
10	OP1	Pre-amplifier output 1
11	OP2	Pre-amplifier output 2
12	MTI	Modulated tone signal input to the preamp
13	OSC3	Oscillator tank terminal
14	OSC2	Oscillator tank terminal
15	OSC1	Oscillator tank terminal (ext. osc. signal can be applied to this pin, if desired)
16	Vcc	Positive supply power

Table 24-3. Program Truth Table for UM3482A Melody Generator.

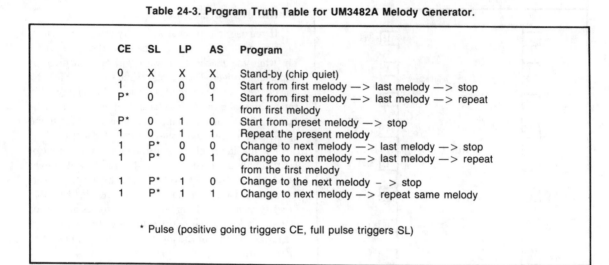

CE	SL	LP	AS	Program
0	X	X	X	Stand-by (chip quiet)
1	0	0	0	Start from first melody —> last melody —> stop
P*	0	0	1	Start from first melody —> last melody —> repeat from first melody
P*	0	1	0	Start from preset melody —> stop
1	0	1	1	Repeat the present melody
1	P*	0	0	Change to next melody —> last melody —> stop
1	P*	0	1	Change to next melody —> last melody —> repeat from the first melody
1	P*	1	0	Change to the next melody – > stop
1	P*	1	1	Change to next melody —> repeat same melody

* Pulse (positive going triggers CE, full pulse triggers SL)

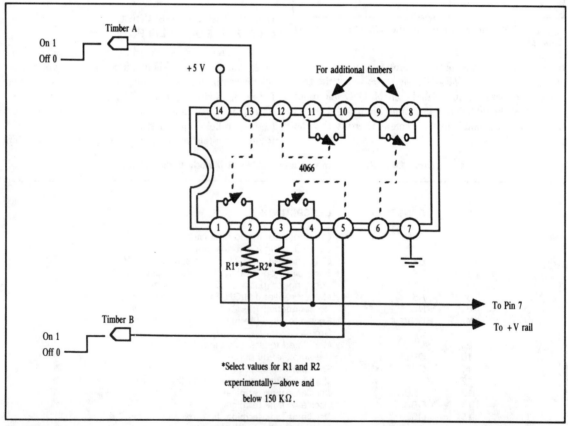

Fig. 24-3. One way to remotely vary the timber of the melody generator. Choose the values of R1 and R2 (and optionally R3 and R4 for additional timbers) experimentally. Switch the desired resistor into the circuit by applying a HIGH to its control switch input.

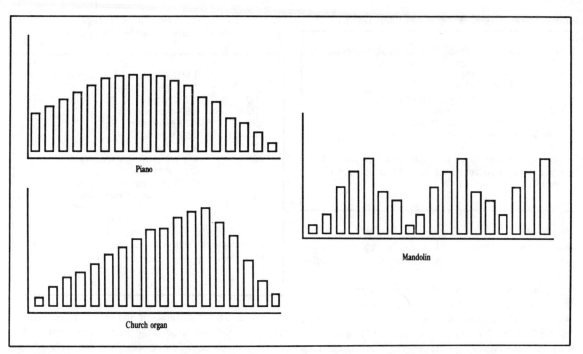

Fig. 24-4. Varying the value of R3 and C3 in the basic wiring diagram of the melody generator alters the timber, as shown graphically in these charts. You can approximate the sound from various instruments, such as a piano, church organ, or mandolin, by careful selection of these two components.

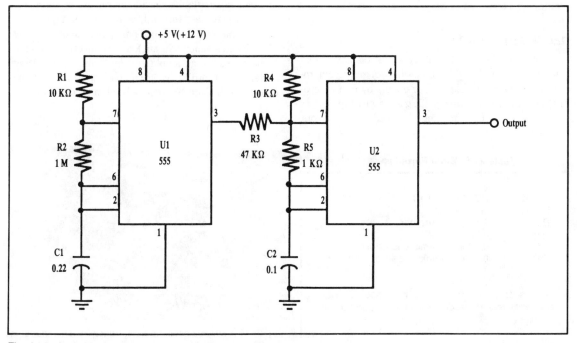

Fig. 24-5. A warbler siren made from two 555 timer ICs.

Table 24-4. Music Maker Parts List.

U1	UM3482A (Radio Shack) Music Generator IC
R1	100 K resistor
R2	47 K resistor
R3	150 K resistor
C1,C3	2.2 μF electrolytic capacitor
C2	47 pF ceramic capacitor
Q1	2N2222 npn transistor
SPKR1	4- or 8-ohm miniature speaker

All resistors 5 or 10 percent tolerance, 1/4-watt; all capacitors 10 percent tolerance, rated 35 volts or higher.

Table 24-6. Cassette Tape Head Player Amplifier Parts List.

U1	LM387 Integrated Amplifier IC
R1	300 ohm resistor
R2	12 K resistor
R3	680 ohm resistor
R4	120 K resistor
R5	360 K resistor
R6	91 K resistor
C1	0.2 μF ceramic capacitor
C2	2.2 μF tantalum capacitor
C3	3000 pF ceramic capacitor
C4	10 μF tantalum capacitor

All resistors 5 or 10 percent tolerance, 1/4-watt; all capacitors 10 percent tolerance, rated 35 volts or higher.

makes short, odd-sounding sound effects, from "oohs" and "aahs" to "booms" and "bahs." Alternate between a couple of keys and you can even get a steam train-like sound.

You can use the SPO256 for your robot by programming in sounds in a similar manner as programming speech. Lacking computer control, you can assemble a binary "random" number generator for driving the chip. Actually, the number generator may be a counter chip that counts from 1 to 64 then starts over. The "random" part comes from the seemingly random sounds emitted from the robot. In actuality, the SPO256 chip is simply reciting its phonemes in order one at a time.

RECORDED SOUND

Not all of the sounds your robot makes have to be electronically produced. You can pre-program your robot to play back music, voice, or effects by first recording the sound on cassette tape. Rig a cassette tape player in the robot and have the sound played back, on your command.

Table 24-5. Siren Parts List.

U1,U2	NE555 Timer IC
R1,R4	10 K resistor
R2,R5	1 megohm resistor
R3	47 K resistor
C1	0.22 μF ceramic capacitor
C2	0.1 μF ceramic capacitor

All resistors 5 or 10 percent tolerance, 1/4-watt; all capacitors 10 percent tolerance, rated 35 volts or higher.

You can use the transport and pre-amplifier mechanisms of an old discarded tape deck. These are plentiful on the surplus market; I've seen them for about $8, including all the circuit boards. The nicer ones, like the unit in Fig. 24-6, are solenoid controlled, which is handy for your robot designs. Instead of pressing mechanical buttons, you can actuate solenoids by remote or computer control to play, fast forward, or rewind the tape.

Most cassette decks need only power to operate the motor(s) and solenoids (if any), and connection from the playback head to an amplifier. Since you are not using the deck for recording, you don't have to worry with the erase head, biasing the record head, and all that other

Fig. 24-6. A surplus cassette deck transport. This model is entirely solenoid driven, perfect for robotics use.

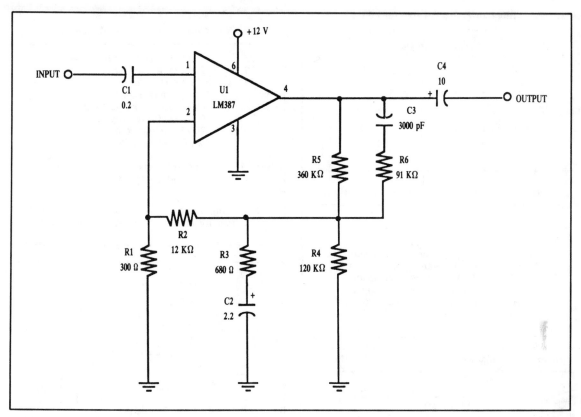

Fig. 24-7. Preamplifier circuit for use with a magnetic tape head.

stuff. If the deck already has a small pre-amplifier for the playback head, use it. It'll improve the sound quality. If not, wire up a pre-amplifier as shown in Fig. 24-7. Place the pre-amplifier board as close to the cassette deck as possible to minimize stray pickup.

SOUND CONTROL

Unless you have all of the sound making circuits in your robot hooked up to separate amplifiers and speakers (not a good idea), you'll need a way to select between the sounds. The circuit in Fig. 24-8 using a 4051 CMOS analog switch, lets you choose from among eight different analog signal sources.

Input selection is done by providing a three-bit binary word to the select lines. You can load the selection via computer, or set it manually with a switch. A binary-coded-decimal (BCD) thumbwheel switch is a good choice, or you can use a four-bank DIP switch. Table 24-7 shows the truth table for selecting any of the eight inputs.

You can also route the output of the SPO256, which

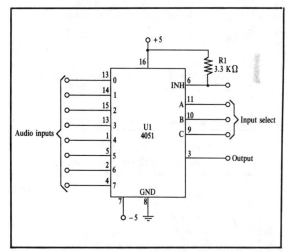

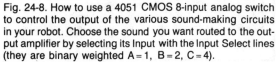

Fig. 24-8. How to use a 4051 CMOS 8-input analog switch to control the output of the various sound-making circuits in your robot. Choose the sound you want routed to the output amplifier by selecting its Input with the Input Select lines (they are binary weighted A = 1, B = 2, C = 4).

Table 24-7. Truth Table for 4051.

A	B	C	Selected Output	Pin
0	0	0	0	13
0	0	1	1	14
0	1	0	2	15
0	1	1	3	12
1	0	0	4	1
1	0	1	5	5
1	1	0	6	2
1	1	1	7	4

Table 24-8. Audio Amplifier, Gain-of-50 Parts List.

U1	LM386 Audio Amplifier IC
R1	10 ohm resistor
C1	100 μF electrolytic capacitor
C2	0.1 μF ceramic capacitor
C3	10 μF electrolytic capacitor
C4	250 μF electrolytic capacitor
SPKR1	4- or 8-ohm miniature speaker

Table 24-9. Audio Amplifier, Gain-of-200 Parts List.

U1	LM386 Audio Amplifier IC
R1	1.2 K resistor
R2	10 ohm resistor
C1	100 μF electrolytic capacitor
C2	0.1 μF ceramic capacitor
C3	10 μF electrolytic capacitor
C4	250 μF electrolytic capacitor
SPKR1	4- or 8-ohm miniature speaker

All resistors 5 or 10 percent tolerance, 1/4-watt; all capacitors 10 percent tolerance, rated 35 volts or higher.

Table 24-10. Eight-Watt Power Audio Amplifier Parts List.

U1	LM383 8-Watt Audio Amplifier IC
R1,R2	2.2 ohm resistor
R3	220 ohm resistor
C1	10 μF electrolytic capacitor
C2	470 μF electrolytic capacitor
C3	0.2 μF ceramic capacitor
C4	2000 μF electrolytic capacitor
SPKR1	4- or 8-ohm speaker (up to eight inches diameter)

Table 24-11. Sixteen-Watt Power Audio Amplifier Parts List.

U1,U2	LM383 8-Watt Audio Amplifier IC
R1,R3	220 ohm resistor
R2,R4	2.2 ohm resistor
R5	1 megohm resistor
R6	100 K audio taper potentiometer
C1,C7	10 μF electrolytic capacitor
C2,C5	470 μF electrolytic capacitor
C3,C4	0.2 μF ceramic capacitor
C6	
SPKR1	4- or 8-ohm speaker (up to eight inches diameter)

All resistors 5 or 10 percent tolerance, 1/4-watt; all capacitors 10% tolerance, rated 35 volts or higher.

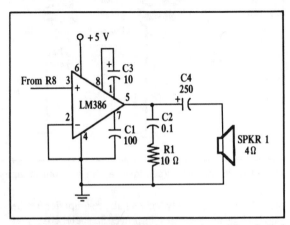

Fig. 24-9. A simple gain-of-50 integrated amplifier, based on the LM386 audio amplifier IC.

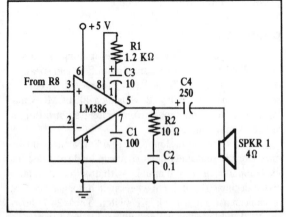

Fig. 24-10. A simple gain-of-200 integrated amplifier, based on the LM386 audio amplifier IC.

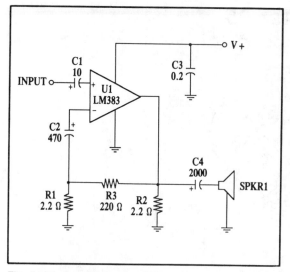

Fig. 24-11. An eight-watt amplifier designed around the LM383 power audio amplifier IC.

is probably mounted on a board of its own, through this chip. In fact, any of your sound projects can be routed through this chip, just as long as the outlet level doesn't

exceed a few milliwatts. Do not pass amplified sound through the chip. Other than likely destroying the chip, it'll cause excessive crosstalk between the channels. Also important is that each input signal should not have a voltage swing that exceeds the supply voltage to the 4051.

AMPLIFIER

Figure 24-9 shows a rather straightforward 0.5 watt sound amplifier, using the LM386 integrated amp. The sound output is minimal, but the chip is easy to get, cheap, and can be wired up quickly. It's perfect for experimenting with sound projects. The amplifier as shown has a gain of approximately 50. You can increase the gain to about 200 by making a few wiring changes, as shown in Fig. 24-10. Either amplifier will drive a small (two or three inch) four or eight ohm speaker.

If you use the siren warbler, you'll probably want a more powerful amplifier. The circuit in Fig. 24-11 should do quite nicely for most applications. The circuit is designed around an LM383 8 watt amplifier IC. The IC comes mounted in a TO-220-style transistor package, and you should use it with a suitable heat sink. Figure 24-12 shows a higher output 16 watt version using two LM383's.

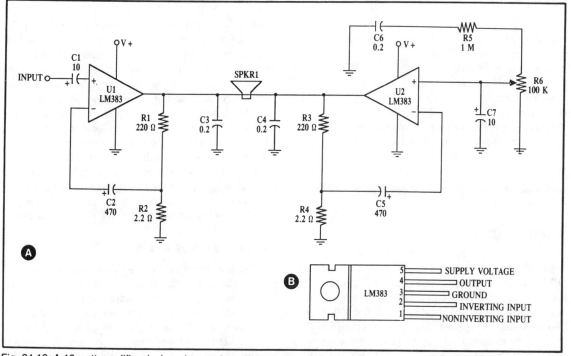

Fig. 24-12. A 16-watt amplifier designed around two LM383 power audio amplifier ICs. A. Schematic diagram; B. Pinout of the LM383.

Chapter 25

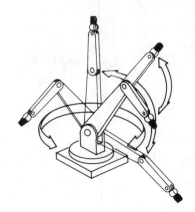

Sound Detection

Next to sight, the most important human sense is hearing. And compared to sight, sound detection is far easier to implement in robots. Simple ears you can build in less than an hour let your robot listen to the world around it.

Sound detection allows your robot creation to respond to your commands, whether they be in the form of a series of tones, an ultrasonic whistle, or a hand clap. It can also listen for the telltale sounds of intruders, or search out the sounds in the room to look for and follow its master. This chapter presents several ways to detect sound. Once detected, the sound can trigger a motor to motivate, a light to be lit, a buzzer to buzz, or a computer to compute.

MICROPHONE

Obviously, your robot needs a microphone (or mic) to pick up the sounds around it. The most sensitive type of microphone is the electret condenser, used in most higher quality hi-fi mics. The trouble with electret condenser elements, unlike crystal element mics, is that they need electricity to operate. Supplying electricity to the microphone element really isn't a problem, though, because the voltage level is low—under 4 or 5 volts.

Most all electret condenser microphone elements come with a built-in field effect transistor (FET) amplifier stage, so the sound is amplified before it is passed on to the main amplifier. Electret condenser elements are available from a number of sources, including Radio Shack, for under $3 or $4. You should buy the best one you can. A cheap microphone isn't sensitive enough.

The placement of the microphone is important. You should mount the mic element at a location on the robot where vibration from motors is minimal. Otherwise, the robot will do nothing but listen to itself. Depending on the application, such as listening for intruders, you might never be able to place the microphone far enough away from sound sources or make your robot quiet enough. You'll have to program the machine to stop, then listen.

AMPLIFIER INPUT STAGE

Use the circuit in Fig. 25-1 as an amplifier for the microphone. The circuit is designed around the common LM741 op amp, wired to operate from a single-ended power supply. Potentiometer R1 lets you adjust the gain of the op amp, and hence the sensitivity of the circuit to sound. After experimenting with the circuit, adjusting R1 for best sensitivity, you can substitute the potentiometer for a fixed value resistor. Remove R1 from the cir-

Table 25-1. Sound Detector Amplifier Parts List.

U1	LM741 Op Amp IC
R1	1 megohm potentiometer
R2,R3	6.8KΩ resistor
R4	1KΩ resistor
C1,C2	0.47 μF ceramic capacitor
MIC1	Electret condenser microphone

cuit and check its resistance with a volt-ohm meter. Use the closest standard value of resistor.

By adding the optional circuit in Fig. 25-2, you can choose up to four gain levels via computer control. The resistors are connected to the inverting input of the op amp and the inputs of a 4066 CMOS 1-of-4 analog switch. Select the resistor value by placing a HIGH bit on the switch you want to activate. The manufacturer's specification sheets for this chip recommends that only one switch be closed at a time. You can delete the 3.3K pull-down resistors if the select lines are always tied to the control computer. Otherwise, the input lines may float, and cause the 741 to behave erratically.

TIMER TRIGGER

Figure 25-3 shows a basic timer trigger that you can hook up to the output of the amplifier input stage above. The circuit is designed around a 555 single-chip IC timer, used as a monostable multivibrator. In operation, the 555 is triggered by an output signal from the 741 op amp.

Once triggered, the 555 goes into timer mode, bringing the output (pin 3) HIGH for a duration set by capacitor C1 and variable resistor R1 and fixed resistor R2. When the pot is rotated all the way to 0 ohms, the time delay is approximately 10 seconds. Rotate the pot and the time delay decreases, all the way to about 2/10ths of a second. By using a short delay period, you can intercept such sounds as clapping hands.

You can use a computer or counting circuit to count the number of claps. One clap might mean "come here," two claps might mean "go away," and so forth. At least one major motorized toy has a similar feature, and you pay a lot of money for it. You can build it into your robot for nearly nothing.

The output of the 555 can drive a relay or can connect directly to a microprocessor or computer data port.

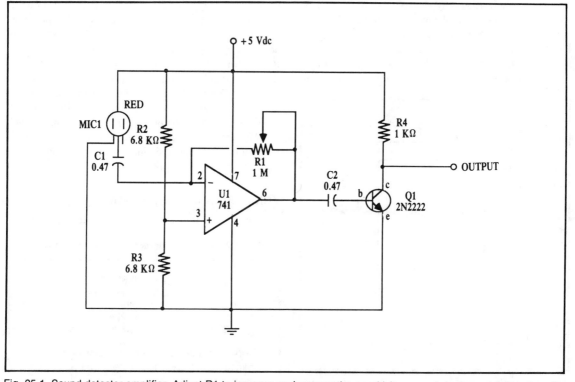

Fig. 25-1. Sound detector amplifier. Adjust R1 to increase or decrease the sensitivity, or replace the potentiometer with the circuit that appears in Fig. 25-2.

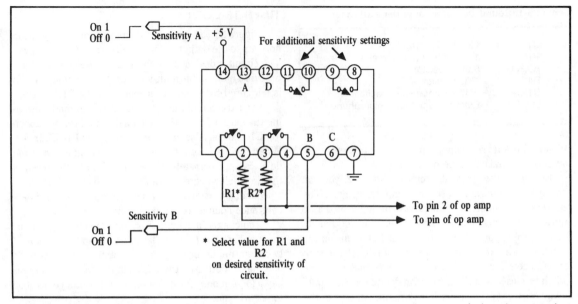

On 1 —
Off 0 — Sensitivity A +5 V

For additional sensitivity settings

A D

B C

R1* R2*

To pin 2 of op amp
To pin of op amp

Sensitivity B

On 1
Off 0

* Select value for R1 and
R2
on desired sensitivity of
circuit.

Fig. 25-2. Remote control of the sensitivity of the sound amplifier circuit. Under computer control, the robot can select the best sensitivity for a given task.

You can use just about any size of relay with the 555, since this chip can drive devices with up to 200 mA of current. But for battery conservation use a low-power relay, if possible.

Remember to add the diode across the coil terminals of the relay. This prevents the back EMF voltage generated by the relay when it turns off from flowing into the 555, and possibly destroying it. Note that the sound detection circuit is very sensitive to electrical noise generated by the robot power supply and the 555 itself. Remember to use the 0.01 μF bypass capacitors across the power supply pins of the 555 (1 and 8) as well as pin 5.

The 555 is used as an event timer, and can be omitted if you are hooking the sound detector directly to another circuit or to a computer. When connecting the 741 to CMOS or TTL circuitry, remember to add the interface components specified in Appendix C, "Interfacing Logic Families and Chips."

TONE DECODING DETECTION

The 741 op amp is sensitive to sound frequencies in

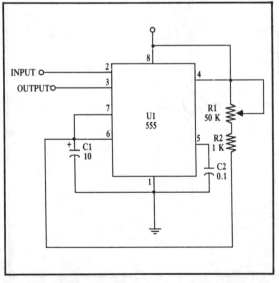

INPUT
OUTPUT

8
2
3
7
U1
555
6
+ C1
10
4
R1
50 K
5 R2
1 K
C2
0.1
1

Fig. 25-3. The latching circuit for the sound detector. Adjust R1 to vary the on-time of the 555 timer.

Table 25-2. Sound Detector Trigger Parts List.

U1	NE555 Timer IC
R1	50 K potentiometer
R2	1 K resistor
C1	10 μF tantalum capacitor
C2	0.1 μF ceramic capacitor

All resistors 5 or 10 percent tolerance, 1/4-watt; all capacitors 10 percent tolerance, rated 35 volts or higher.

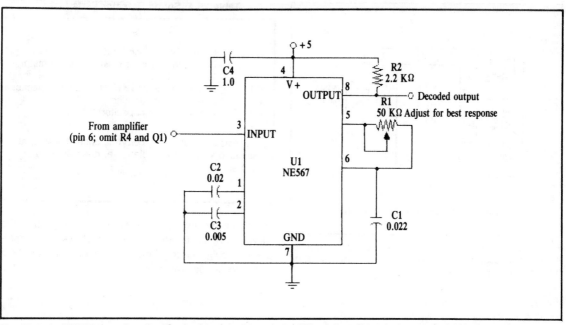

Fig. 25-4. An NE567 tone decoder IC wired to detect tones at 1 kHz. Adjust R1 to lock onto the 1kHz center frequency.

a very wide band, and can pick up everything that the microphone has to send it. You may wish to listen for sounds that occur only in a specific frequency range. An NE567 tone decoder IC can be easily added to the amplifier input stage to look for these specific sounds.

The 567 is almost like a 555 in reverse. You select a resistor and capacitor to establish an operating frequency, called the *center frequency*. Additional components are used to establish a bandwidth—the percent variance that the decoder will accept as a desired frequency (the variance can be as high as 14 percent). Figure 25-4 shows how to connect a 567 to listen to and trigger on a 1kHz tone. If you want to decode a different frequency, refer to the design equations provided in Appendix C.

Before you get too excited about the 567 tone decoder, you should know about a few minor faults. The 567 has a tendency to trigger on harmonics of the desired frequency. You can limit this effect, if it is undesirable, by adjusting the sensitivity of the input amplifier and by decreasing the bandwidth of the chip.

Another minor problem is that the 567 requires at least eight wavefronts of the desired sound frequency before it triggers on it. This reduces false alarms, but it also makes detection of very low frequency sounds impractical. Though the 567 has a lower threshold of about 1 Hertz, it is impractical for most uses at frequencies this low.

Building a Sound Source

With the 567 decoder, you'll be able to control your robot using specific tones. With a tone generator, you'll be able to make those tones so you can signal your robot via sound. Such a tone generator sound source is shown in Fig. 25-5. The values shown in the circuit generate sounds in the 12 kHz to 150 Hz range. To extend the range higher or lower, substitute a higher or lower value for C1. Design formulas and tables for the 555 are provided in Appendix C, "Interfacing Logic Families and ICs."

For frequencies between about 5 kHz and 15 kHz, use a piezoelectric element as the sound source. Use a miniature speaker for frequencies under 5 kHz and an ultrasonic transducer for frequencies over 30 kHz.

Cram all the components in a small box, stick a battery inside, and push the button to emit the tone. Be aware

Table 25-3. Tone Generator Parts List.

U1	NE555 Timer IC
R1	1 megohm potentiometer
R2	1 K resistor
C1	0.001 μF ceramic capacitor
C2	0.1 μF ceramic capacitor
SPKR1	4- or 8-ohm miniature speaker

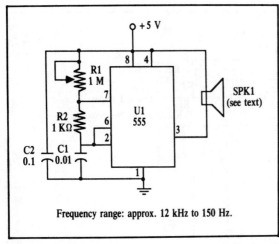

Frequency range: approx. 12 kHz to 150 Hz.

Fig. 25-5. A variable frequency tone generator, built around the common 555 timer IC. The tone output spans most of the range of human hearing.

Table 25-4. Sound Detector Parts List.

U1	NE567 Tone Decoder IC
R1	50K 3- to 15-turn precision potentiometer
R2	2.2KΩ resistor
C1	0.022 monolithic or Hi-Q ceramic capacitor
C2	0.02 ceramic capacitor
C3	0.005 ceramic capacitor

All resistors 5 or 10 percent tolerance, 1/4-watt; all capacitors 10 percent tolerance, rated 35 volts or higher.

that the sound level from the speaker and especially the piezoelectric element can be quite high. DO NOT operate the tone generator close to your ears or anyone else's ears except your robot's.

Chapter 26

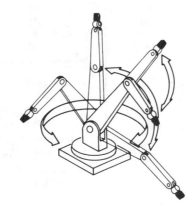

Robotic Eyes

Giving sight to your robots is perhaps the kindest thing you can do for your creations. Robotic vision systems can be simple or complex, to match your requirements and your itch to tinker. Rudimentary cyclops vision systems are used for nothing more than detecting the presence or absence of light. Despite this rather mundane task, there are plenty of useful applications for an on/off light detector. More advanced vision systems decode relative intensities of light, and can even make out patterns and crude shapes.

While the hardware for making robot eyes is rather simple, using the vision information is not. Other than the one-cell light detector, vision systems require interface to a computer to be useful. The designs presented in this chapter can be adapted to just about any computer, using a microprocessor data bus or one or more parallel printer ports.

ONE-CELL CYCLOPS

A single light sensitive photocell is all that is required to sense the presence of light. The photocell is a variable resistor that works much like a potentiometer but has no control shaft. You vary its resistance by increasing or decreasing the light. With no light present, the re-

sistance between the leads of the photocell is in the megohms. Apply light and the resistance swings sharply to just a few thousand ohms.

Connect the photocell as shown in Fig. 26-1. Note that a resistor is placed in series with the photocell and that the output tap is between the cell and resistor. This converts the output of the photocell from resistance to voltage, the latter of which is easier to use in a practical circuit.

The value of the resistor is given at 3.3K ohms but is open to experimentation. You can vary the sensitivity of the cell by substituting a higher or lower value. For experimental purposes, connect a 50K pot in place of the resistor and try using the cell at various settings of the wiper. Test the cell output by connecting a volt-ohm meter to the ground and output terminals.

So far, you have a nice light-to-voltage sensor, and when you think about it, there are numerous ways to interface this ultra-simple circuit to a robot. One way is to connect the output of the sensor to the input of a comparator (the LM339 quad comparator IC is a good choice, as shown, but you can use just about any comparator). The output of the comparator changes state when the voltage at its input goes beyond or below a certain "trip point."

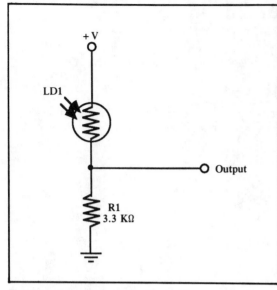

Fig. 26-1. A one-cell robotic eye.

Table 26-1. Single-Cell Robotic Eye Parts List.

U1	LM339 Quad Comparator IC
R1	3.3KΩ resistor
R2	10KΩ potentiometer
LD1	Photocell

low the set point, the output of the comparator changes state.

One practical application of this circuit is to detect light levels that are higher than the ambient light in the room. That way, your robot ignores the background light level and responds only to the higher intensity light. To begin, set the trip point pot so that the circuit just switches HIGH. Use a flashlight to focus a beam directly onto the photocell, and watch the output of the comparator change state. Another application is to use the photocell as a light detector—period. Set the pot all the way over so that the comparator changes state just after light is applied to the surface of the cell.

MULTIPLE-CELL LIGHT SENSORS

The human eye has millions of tiny light receptacles. Combined, these receptacles allow us to discern shapes, to actually see, rather than just detect light levels. A crude but surprisingly useful implementation of human sight is given in Fig. 26-3. Here, eight photocells are connected

In the circuit shown in Fig. 26-2, the comparator is hooked up so that the non-inverting input serves as a voltage reference. Adjust the potentiometer to set the *trip point*. To begin, set it mid-way, then adjust the trip point higher or lower as required. The output of the photocell circuit is connected to the inverting input of the comparator. When the voltage at this pin rises above or be-

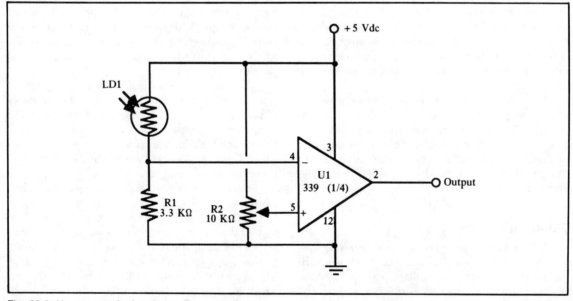

Fig. 26-2. How to couple the photocell to a comparator.

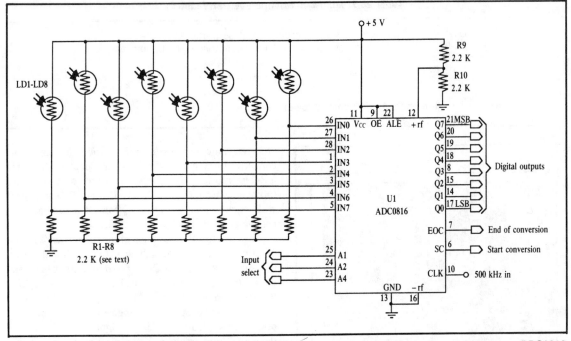

Fig. 26-3. One way to make a robotic "eye." The circuit, as shown, consists of eight photocells connected to an ADC0816 8-bit 16-input analog-to-digital converter IC. The output of each photocell is converted when selected at the Input Select lines. The ADC0816 can handle up to 16 inputs, so you can add another eight cells.

to a 16-channel multiplexed analog-to-digital (A/D) converter (the A/D converter has room for another eight cells). The A/D converter takes the analog voltages from the outputs of each photocell and one by one converts them into digital data. The eight-bit binary number presented at the output of the A/D converter represents any of 256 different light levels.

The converter is hooked up in such a way that the outputs of the photocells are converted sequentially, in a row and column pattern, following the suggested mounting scheme shown in Figs. 26-4 and 26-5. A computer hooked up to the A/D converter records the digital value of each cell, and creates an *image matrix,* which can be used to discern crude shapes, outlines, and borders.

Each photocell is connected in series with a resistor, as in the one-cell eye presented earlier. Initially, use 2.2K resistors, but feel free to substitute higher or lower values to increase or decrease sensitivity. The photocells should be identical, and for best results, should be brand-new prime components. Prior to placing the cells in the circuit, test each one with a volt-ohm meter and with a carefully controlled light source. Check the resistance of the photocell in complete darkness then again with a light shining at it a specific distance away. Reject cells that

do not fall within a 5 to 10 percent pass range.

There are many ways you can connect the converter to your computer. If the computer has a bi-directional par-

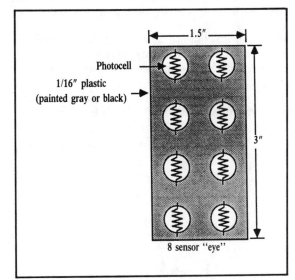

Fig. 26-4. Mounting the photocells for an eight-cell eye.

217

Table 26-2. Multi-Cell Robotic Eye Parts List.

U1	ADC0816 8-bit Analog-to-Digital Converter IC
R1-R8	2.2 K resistor (adjust value for best response of photocells)
R9,R10	2.2 K resistors
LD1-LD8	Photocell

All resistors 5 or 10 percent tolerance, 1/4-watt.

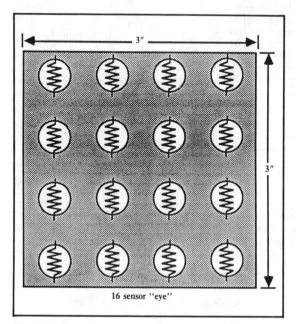

Fig. 26-5. Mounting the photocells for a 16-cell eye.

allel port (such as the Commodore 64), you can use it to retrieve the data from the converter. Another method is to convert the data to serial form, and transmit it serially through an RS-232C port. That requires a UART (Universal Asynchronous Receiver/Transmitter) as well as some programming on the computer end to assure proper transmission speed, parity, stops, bits, etc.

With a computer like an IBM PC or Apple II, or any other machine with a central bus, you can connect the converter directly to the data ports. Chapter 32, "Build a Robot Interface Card," shows how to construct a prototyping board for use in the IBM PC (or PC compatible with expansion slots). The A/D converter can easily be connected to this board.

You can test the operation of the A/D converter without a computer by manually selecting each address input and hooking up LEDs to the data output lines and ground (remember to use a 270 ohm resistor for each LED to limit current).

Note the short pulse that appears at pin 13, the End-of-Conversion Output. This pin signals that the data at the output lines are valid. You can connect this pin to the READ line in your computer (ANDing it with the specific I/O address used to access the converter, so the READ line isn't being triggered all the time!), or to one of the computer's interrupt lines (if available). Using an interrupt line lets your computer do other things while waiting for the A/D converter to signal the end of a conversion. See Chapter 32, "Build a Robot Interface Card," for more details on how to connect an A/D converter directly to the microprocessor system bus.

You can get by without using the EOC pin—it certainly is easier to implement—but you must set up a timing delay circuit or routine to do it. Simply wait long enough for the conversion to take place—a maximum of about 115 μs—then read the data. Even with a delay of 125 μs (to allow for settling, etc.), it takes no more than about 200 milliseconds to read the entire matrix of 16 cells.

Chapter 27

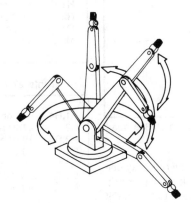

Fire Detection Systems

Everyone complains that a robot is good for nothing—except, perhaps, making a fool of itself at parties, serving drinks, or providing its master a way to tinker with gadgets in the name of science! But here's one useful and potentially life-saving application you can give your robot in short-order: fire and smoke detection. As this chapter shows, you can easily attach sensors to your robot to detect flames, heat, and smoke. Your robot becomes a kind of mobile smoke detector.

FLAME DETECTION

Flame detection requires little more than a sensor that detects infrared light and a circuit to trigger a motor, siren, computer, or other device when the sensor turns on. As it turns out, almost all phototransistors are specifically designed to be sensitive primarily to infrared or near-infrared light. You need only connect a few components to the phototransistor and you've made a complete flame detection circuit. Interestingly, the detector can see flames that we can't. Many gases, including hydrogen and propane, burn with little visible flame. The detector can spot it before you can, or before the flames light something on fire and smoke fills the room.

The simple circuit in Fig. 27-1 shows the most straightforward method of detecting flames. A phototransistor is masked from seeing anything but infrared light by the use of an opaque infrared filter. Some phototransistors have the filter built-in; with others, you'll have to add the filter yourself. Infrared filters are available at most photographic stores.

In the circuit, when infrared light hits the phototransistor, it triggers on. The brighter the infrared source, the more voltage is applied to the inverting input of the comparator. If the input voltage exceeds the reference voltage applied to the non-inverting input, the output of the comparator changes state.

Potentiometer R2 sets the sensitivity of the circuit. You'll want to turn the sensitivity down so that ambient infrared light does not trigger the comparator. You'll find that the circuit does not work when the background light has excessive infrared content. You can't, for example, use the circuit outdoors, or when the sensor is pointed directly at an incandescent light or the sun.

Another flame detector is shown in Fig. 27-2. This one works on the same principle but uses discrete components. Use any infrared-sensitive phototransistor for Q1. If the phototransistor does not have an infrared-filtered lens, add one for the light to pass through. Place

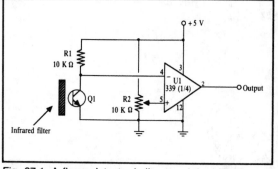

Fig. 27-1. A flame detector built around the LM399 quad comparator IC.

the transistor at the end of a small opaque tube, say one with a 1/4-inch or 1/2-inch I.D (the black tubing for drip irrigation is a good choice). Glue the filter to the end of the tube. The idea is to block all light to the transistor except that which is passed through the filter.

Connect the other components as shown. You can power the circuit with +5 volts, but you are free to use just about any voltage, as long as it's under 30 volts. Just make sure that the voltage at the output does not exceed the input voltage rating of the interface circuit you are using. If you are interfacing this circuit to a TTL AND gate, for example, don't use anything above 5 volts.

Test the circuit by connecting an LED and 270 ohm resistor from the Vout terminal to ground. Point the sensor at a wall and note the condition of the LED. Now,

wave a match in front of the phototransistor. The LED should blink on and off. You'll notice that the circuit is sensitive to all sources of infrared light, which includes the sun, strong photolamps, electric burners, even your soldering iron. If the circuit doesn't seem to be working quite right, look for hidden sources of infrared light. With the resistor values shown, the circuit is fairly sensitive;

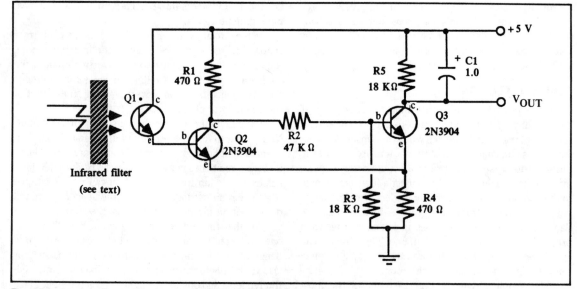

Fig. 27-2. Another flame detector circuit built around discrete transistors.

you can change it by adjusting the value of R1 and R2.

SMOKE DETECTION

"Where there's smoke, there's fire." Statistics show that more fire deaths are caused each year not by burns but by smoke inhalation. For less than $15, you can add smoke detection to your robot's long list of capabilities, and with a little bit of programming, have it wander through the house, checking each room for trouble. You'll probably want to keep it in the predominate fire-prone rooms, such as the basement, kitchen, laundry room, and robot lab.

You can build your own smoke detector using individually purchased components, but some items, such as the smoke detector cell, are hard to find. It's much easier to use a commercially available smoke detector and modify it for use with your robot. In fact, the process is so simple that you can add one to each of your robots. Tear the smoke detector apart and strip it down to the base circuit board.

To modify the detector, first put plugs in your ears, then connect a battery to the battery terminals on the board. Don't forget the plugs. The piezoelectric buzzer puts out a high-pitched, high-decibel squawk that can hurt your ears and make you dizzy.

Connect a volt-ohm meter to the terminals of the buzzer. Note the voltage level. Depress the test button and watch the meter. What does it do? Some piezoelectric elements have three terminals, and if the meter does nothing during your test, you should try another set of terminals. Unsolder the buzzer from the board (you can now take out the ear plugs) and connect the volt-ohm meter to the terminals once again. Retest the detector and note the voltages present on the meter. They should be the same as before.

You can interface the detector to your robot in a variety of ways. One simple approach is to connect the outputs of the detector to a relay (see Appendix C). Power the relay from the smoke detector battery (you can hook the detector to the robot's main power source later on if you wish). The contacts of the relay can be used by just about anything, including a motor, a siren, or a flashing light. Add a debouncing circuit (also described in Appendix C) and you can use the relay contacts with a microprocessor or computer.

Alternatively, you can use an opto-isolator. The opto-isolator bridges the gap between the detector and the robot, and serves the same general function as a relay, but does so in a smaller package and with less power consumption. The output of the opto-isolator does not need to be conditioned if you are connecting it to a microprocessor port or computer. Several opto-isolator interfacing circuits are provided in Appendix C, "Interfacing Logic Families and ICs."

You should be wary of certain limitations inherent in the application of a robot fire detector. In the early stages of a fire, smoke tends to cling to the ceilings. That's why manufacturers recommend you place smoke detectors on the ceiling, rather than on the wall. Only when the fire gets going, and smoke builds up, does it start to fill up the rest of the room.

Your robot is probably a rather short creature and it might not detect smoke that confines itself only to the ceiling. This is not to say that the smoke detector mounted on a two or three foot-high robot won't detect the smoke from a small fire, just don't count on it. Back up the robot smoke sensor with conventionally mounted smoke detection units.

HEAT SENSING

In a fire, smoke and flames are most often encountered first. Heat isn't something that's felt until the fire is going strong. But what about before the fire gets started in the first place? What about a kerosene heater that was inadvertently left on? Or an iron that's been tipped over and is melting the nylon clothes underneath?

If your robot is on wheels (or legs), and is wandering through the house, perhaps it'll be in the right place at the right time, and sense these irregular situations. A fire is brewing and before the house fills with smoke or flames, the air gets a little warm. Equipped with a heat sensor, the robot can actually seek out warmer air, and if the air temperature gets too high, it can sound an initial alarm.

Realistically, heat sensors provide the least protection in a fire. But heat sensors are easy to build, and besides, when the robot isn't sniffing out fires, it can be wandering through the house giving it an energy check, or reporting on the outside temperature, or . . . You get the idea.

Figure 27-3 shows a basic but workable circuit centered around an LM355 temperature sensor. This device is relatively easy to find and costs under $1.50. The output of the device, when wired as shown, is a linear voltage. The voltage increases 10 mV for every rise in temperature of one degree Kelvin (K).

Degrees Kelvin is the same as degrees Centigrade (C), except that the zero point is absolute zero—about −273 degrees C. One degree Centigrade equals one degree Kelvin; just the start points differ. You can use this to your advantage because it lets you easily convert

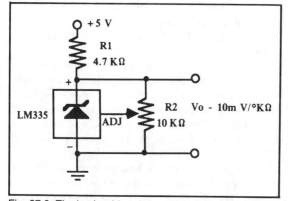

Fig. 27-3. The basic wiring diagram for the National Semiconductor LM355 temperature sensor.

Table 27-3. Basic Temperature Transducer Parts List.

R1	4.7 KΩ resistor, 1 percent tolerance
R2	10 KΩ 10-turn precision potentiometer
D1	LM335 Temperature Sensor Diode

degrees Kelvin into degrees Centigrade. Actually, since your robot will be deciding when hot is hot, and doesn't care what temperature scale is used, the conversion really isn't necessary.

You can test the circuit by connecting a volt-ohm meter to the ground and output terminals of the circuit. At room temperature, the output should be about 2.98 volts. You can figure the temperature if you get another reading by subtracting the voltage by 273 (ignore the decimal point but make sure there are two digits to the right of it, even if they are zeros). What's left is the tempera-

ture in degrees Centigrade. For example, if the reading is 3.10 volts, the temperature is 62 degrees C (310 − 273 = 62).

You can calibrate the circuit by using an accurate bulb thermometer as a reference and adjusting R2 for the proper voltage. How do you know the proper voltage? If you know the temperature, you can determine what the output voltage should be by adding the temperature (in degrees C) to 273. If the temperature is 20 degrees C, then the output voltage should be 2.93 volts. For more accuracy, float some ice in a glass of water for 30 minutes and stick the sensor in it (keep the leads dry). Wait 10 to 15 minutes for the sensor to settle and read the voltage. It should be exactly 2.73 volts.

The load presented at the outputs of the sensor circuit can throw off the reading. The schematic in Fig. 27-4 provides a buffer circuit so that the load does not interfere with the output of the 355 temperature sensor.

FIRE FIGHTING

I really can't recommend this next project to everyone, but you're free to experiment with it as you like. By attaching a small fire extinguisher to your robot, you can have the automaton put out the fires it detects.

Obviously, you'll want to make sure that the fire detection scheme that you've put into use is relatively free of false alarms, and that it doesn't overreact in non-fire situations. Having your robot rush over to one of your guests to put out a cigarette he just lit is not only bad manners, but potentially embarrassing. Besides, the Halon used in the fire extinguisher specified in the project can actually do a person harm. The special chemical fights fires by gobbling up oxygen.

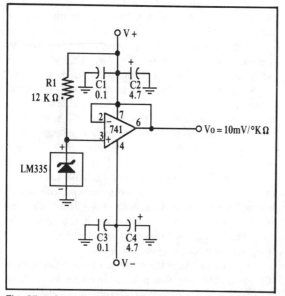

Fig. 27-4. An enhanced wiring scheme for the LM355 temperature sensor. The load of the output is buffered and does not affect the reading from the LM335.

Table 27-4. Advanced Temperature Transducer Parts List.

R1	12 KΩ resistor, 1 percent tolerance
C1,C3	0.1 µF ceramic capacitor
C2,C4	4.7 µF tantalum capacitor
D1	LM335 Temperature Sensor Diode
All capacitors 10 percent tolerance unless noted; all resistors 1/4-watt.	

Halon is a cleaner method of putting out fires than the older fashioned powder extinguishers, but it is more expensive and requires extra care. Using caution as a guide, you should probably use your firebot only when there are no people around, like a factory floor after hours or a storage warehouse.

The exact mounting and triggering scheme you use depends entirely on the design of the fire extinguisher. The bottle used in the prototype firebot is a Kidde Force-9, 2 1/2 pound Halon extinguisher. It has a diameter of about 3 1/4 inches and is refillable.

The extinguisher can be mounted in the robot using plumber's tape, that flexible strip used by plumbers to mount water and gas pipes. It has lots of holes already drilled into it for easy mounting. Use two strips to hold the bottle securely. Remember that a fully charged Halon extinguisher is heavy—in this case over 3 pounds (2 1/2 pounds for the Halon chemical and about 1/2 pound for the bottle). If you add a fire extinguisher to your robot, you must relocate other components to evenly distribute the weight.

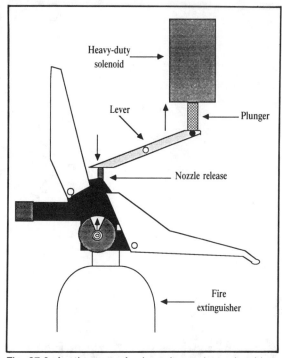

Fig. 27-6. Another way of using a heavy-duty solenoid to directly activate the release valve in a fire extinguisher.

The Halon extinguisher used in the prototype system used a standard actuating valve. To release the Halon, you squeeze two levers together. Figure 27-5 shows how to use a heavy-duty solenoid to remotely actuate the valve. You may be able to access the valve plunger itself (you may have to remove the levers to do so). Rig up a heavy-duty solenoid and lever system as shown in Fig. 27-6. A computer or control circuit activates the solenoid.

For best results, the valve should be opened and closed in quick bursts. The body of the robot should also pivot back and forth so the Halon is spread evenly over the fire. Remember that to be effective, the Halon must be sprayed at the base of the fire, not the flames. For most fires, this is not a problem because the typical robot stays close to the floor. If the fire is up high, the robot may not be able to effectively fight it.

You can test the fire extinguisher a few times before the bottle needs recharging. I was able to squeeze off four or five short blasts before the built-in pressure gauge registered that I needed a new charge. For safety's sake, experiment with an extra extinguisher. Don't use your only extinguisher for your robot experiments. Keep an extra handy in the unlikely event that you have to fight a fire yourself.

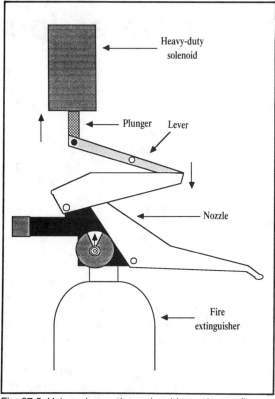

Fig. 27-5. Using a heavy-duty solenoid to activate a fire extinguisher.

Chapter 28

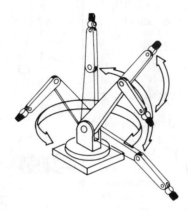

Collision Avoidance and Detection

You've spent hundreds of hours designing and building your latest robot creation. It's filled with complex little do-dads and precision instrumentation. You bring it into your living room, fire it up, and step back.

Promptly, the beautiful new robot smashes into the fireplace and scatters itself over the living room rug. You remembered things like motor speed controls, electronic eyes and ears, even a synthetic voice, but you forgot to provide your robot with the ability to look before it leaps.

Collision avoidance and detection takes many forms, and most of the systems are easy to build and use. Here, in this chapter, we present a number of passive and active detection systems for sensing when an object has been encountered, or when it is within a pre-defined danger zone. You can apply one or more of the detection schemes in a robot, using the strengths of each one to complement the others.

COLLISION DETECTION

There are three major types of collision detection systems: bumper switch, whisker, and pressure pad.

Bumper Switch

A bumper collision detector can be as simple as a spring-loaded pushbutton switch mounted on the frame of the robot, as shown in Fig. 28-1. The plunger of the switch is pushed in whenever the robot collides with an object. Obviously, the plunger must extend further than all other parts of the robot. You may need to mount the switch on a bracket to extend its reach.

The surface area of the plunger is very small. You can enlarge the contact area by attaching a plate, or bumper, to the switch plunger. A piece of rigid 1/16-inch thick plastic or aluminum is a good choice for bumper plates. Glue the plate onto the plunger.

Low-cost pushbutton switches are not known for their sensitivity. The robot may have to crash into an object with a fair amount of force before the switch makes positive contact, and for most applications, that's obviously not desirable.

Leaf switches require only a small touch before they trigger. The plunger in a leaf switch (often referred to as a Micro Switch, after the manufacturer that made them popular) is extra small, and travels only a few fractions of an inch before its contacts close. A metal-strip, or *leaf*, attached to the strip acts as a lever, further increasing sensitivity. You can mount a plastic or metal plate to the end of the leaf for an increase in surface area. If the leaf is wide enough, you can use miniature 4/40 or 3/38 hardware to mount the plate in place.

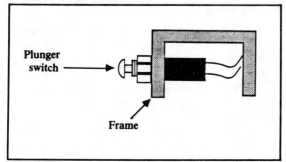

Fig. 28-1. A SPST spring loaded plunger switch mounted in the frame or body of the robot, used as a contact sensor.

Whisker

Most animal experts believe that a cat's whiskers are used to measure space. If the whiskers touch when a cat is trying to get through a hole, it knows there is not enough space for its body. We can apply a similar technique to our robot designs, whether or not kitty whiskers are actually used for this purpose.

You can use thin 20 to 25 gauge stove wire for the whiskers of the robot. Attach the wires to the end of

switches, or mount them in a receptacle so that the wire is supported by a small rubber grommet.

By bending the whiskers, you can extend their usefulness and application. The commercially-made robot in Fig. 28-2, the Movit WAO, has two whiskers that can be rotated in their switch sockets. When the whiskers are positioned so that the loop is up and down, they can detect changes in the topography, to watch for such things as the edge of a table, the corner of a rug, and so forth.

A more complex whisker setup is shown in Fig. 28-3. Two different lengths of whiskers are used for the two sides of the robots. The longer length represent a space a few inches wider than the robot. If these whiskers are actuated by rubbing against an object, but the short

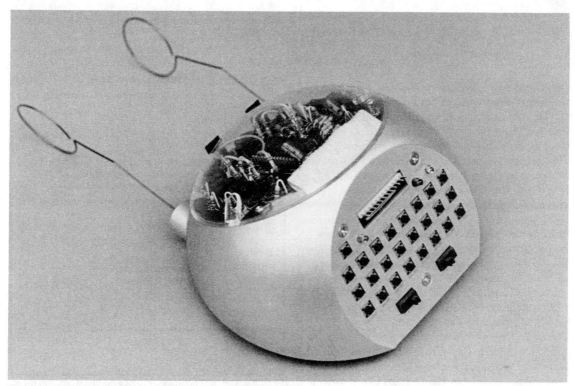

Fig. 28-2. The Movit WAO robot kit. Its two tentacles, or whiskers, allow it to navigate a space.

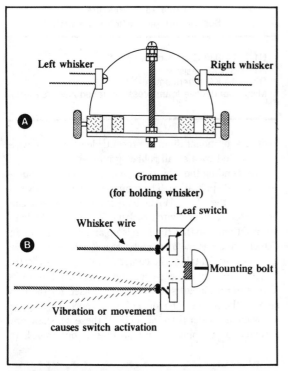

Fig. 28-3. Adding whiskers to a robot. A. Whiskers attached to the dome of the Minibot (Chapter 5); B. Construction detail of the whiskers and actuation switches.

whiskers are not, then the robot understands that the pathway is clear to travel, but that space is tight.

The short whiskers are cut to represent the width of the robot. Should the short whiskers on only one side of the robot be triggered, then the robot will turn the opposite direction to avoid the obstacle. If both sides of short whiskers are activated, then the robot knows that it cannot fit through the passageway, and it either stops or turns around.

Application of the whiskers can be fully implemented by using the optical path navigation designs described in Chapter 29, "Navigating Through Space." See that chapter if you are interested in implementing the multiple whisker design.

Before building bumper switches or whiskers into your robot, be aware that most electronic circuits will misbehave when triggered by a mechanical switch contact. The contact has a tendency to bounce as it closes and opens, so it needs to be conditioned. See the debouncing circuits in Appendix C for ways to clean up the contact closure so that switches can directly drive your robot circuits.

Pressure Pad

In Chapter 22 you learned about how to give the sense of touch to robot fingers and grippers. One of the materials used as a touch sensor was the conductive foam, packaged with most CMOS and microprocessor ICs. This foam is available in large sheets and is perfect for use on collision detection pressure pads. Radio Shack sells a nice 5-inch square pad that's ideal for the job.

Attach wires to the pad as described in Chapter 22, and glue the pad to the frame or skin of your robot. Unlike fingertip touch, where the amount of pressure is important, the salient ingredient with a collision detector is that contact has been made with something. This makes the interface electronics that much easier to build.

Figure 28-4 shows a suitable interface for use with the pad. The pad is placed in series with a 3.3K resistor between ground and the positive supply voltage to form a voltage divider. When the pad is pressed down, the voltage at the output of the sensor will vary. The output of the sensor, which is the point between the pad and resistor, is applied to the inverting pin of a 339 comparator (there are four separate comparators in the 339 package, so one chip can service four pressure pads). When the voltage from the pad exceeds the reference voltage supplied to the comparator, the comparator changes states, thus indicating a collision.

The comparator output can be used to drive a motor direction control relay, or can be tied directly to a microprocessor or computer port. Follow the interface guidelines provided in Appendix C, "Interfacing Logic Families and ICs."

COLLISION AVOIDANCE

Avoiding a collision is better than detecting it once

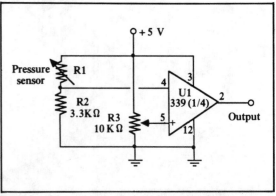

Fig. 28-4. Converting the output of a conductive foam pressure sensor to an on/off type switch output.

it has happened. Short of building some elaborate radar distance measurement system, there are two general ways of providing proximity detection for the purpose of avoiding collisions: light and sound. Let's take a look at both right now.

Infrared Light

Light may always travel in a straight line, but it bounces off nearly everything. You can use this to your advantage to build an infrared collision detection system. You can mount several infrared bumper sensors around the periphery of your robot. They can be tied together, to tell the robot that "something is out there," or they can provide specific details of the outside environment to a computer or control circuit.

The basic infrared detector is shown in Fig. 28-5. This uses the now-familiar infrared LED and infrared phototransistor. A suitable interface circuit is provided. The output of the transistor can be connected to any number of control circuits. The comparator circuit for the whisker switches will work nicely, and will provide a go/no go output to a computer. Figure 28-6 shows how the LED and phototransistor might be mounted on the top of a robot for the purpose of detecting an obstacle like a wall, chair, or person.

The set point adjustment, R2, provides a means to increase or decrease the sensitivity of the circuit. An increase in sensitivity means that the robot will be able to detect objects farther way. A decrease in sensitivity means that the robot must be fairly close to the object

Table 28-2. Infrared Proximity Switch Parts List.

R1	270 Ω resistor
R2	10 KΩ resistor
Q1	infrared sensitive phototransistor
LED1	Infrared Light Emitting Diode
Misc.	Infrared filter for phototransistor (if needed)

All resistors 5 or 10 percent tolerance, 1/4-watt; all capacitors 10 percent tolerance, rated 35 volts or higher.

before it is detected. The prototype sensor I built had an effective range of about 6 inches, even when the sensitivity of the circuit was boosted as high as it would go. If you need greater than this, use an LED-phototransistor pair with built-in lenses.

Bear in mind that all objects reflect light in different ways. You'll probably want to adjust the sensitivity so that the robot best behaves itself in a room with white walls. But that sensitivity may not be as great when the robot comes to a dark brown couch, or the coal gray suit of your boss.

The infrared phototransistor must be baffled—blocked—from both ambient room light as well as direct light from the LED. The positioning of the LED and phototransistor is very important and care must be taken to

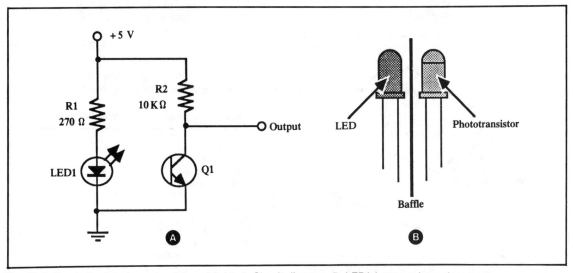

Fig. 28-5. Proximity detector using infrared light. A. Circuit diagram; B. LED/phototransistor placement.

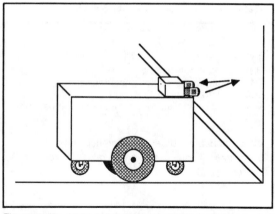

Fig. 28-6. The LED/phototransistor proximity switch mounted on a robot, detecting an obstacle.

Fig. 28-8. A commercially manufactured infrared LED/phototransistor proximity switch.

ensure that the two are properly aligned. You may wish to mount the LED-phototransistor pair in a small block of wood, as shown in Fig. 28-7. Drill holes for the LED and phototransistor.

Or, if you prefer, you can buy the detector pair already made up and installed in a similar block. The device shown in Fig. 28-8 is a TIL139, from Texas Instruments. This particular component was purchased at a surplus store for about $1.

Ultrasonic Sound

Sound can be used to detect the proximity of objects in much the same as was infrared light. Ultrasonic sound is transmitted from a transducer, reflected by a nearby object, then received by another transducer. The advan-

tage of using sound is that it is not sensitive to objects of different color and light reflective properties. Keep in mind, however, that some materials reflect sound better than others, and that some even absorb sound completely. In the long run, however, proximity detection with sound is a more fool-proof way to go.

Figures 28-9 and 28-10 provide a practical circuit you can build that provides ultrasonic proximity detection (the circuit is similar to the ones used to automatically open doors at the grocery store). The ultrasonic transducers, like the ones in Fig. 28-11, are available from a number of retail and surplus outlets, including Dick Smith Electronics. See Appendix A, "Sources," for a more complete list of electronics suppliers.

A stream of 40kHz pulses is produced by a 555 timer wired up as an astable multivibrator. The output of the 555 provides more than enough power for the transducer. The receiving transducer is positioned two or more inches away from the transmitter. For best results, place a piece

Table 28-3. Ultrasonic Transmitter Parts List.

U1	NE555 Timer IC
R1	10 KΩ resistor
R2	1 KΩ resistor
R3	1.2 KΩ resistor
R4	2.2 KΩ resistor
C1	0.1 μF ceramic capacitor
C2	0.0033 μF monolithic, mica, or ceramic capacitor
Q1	2N2222 npn transistor
TR1	Ultrasonic transducer

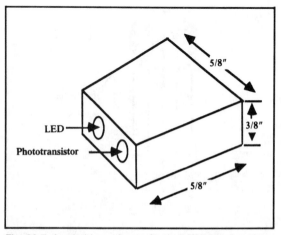

Fig. 28-7. A suitable enclosure for the LED/phototransistor pair, made from wood and cut and drilled to size.

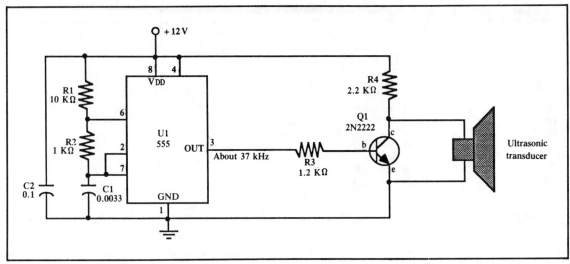

Fig. 28-9. Schematic for a 40kHz ultrasonic transmitter.

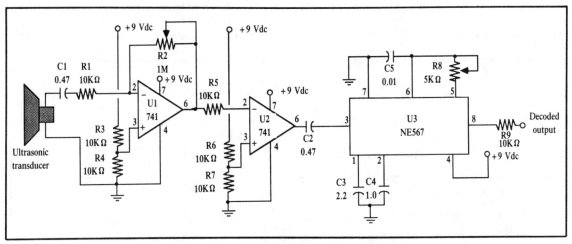

Fig. 28-10. Schematic for an ultrasonic receiver and tone decoder.

of foam between the two transducers to eliminate direct interference between the two.

The signal from the receiving transducer needs to be amplified. A 741 op amp is more than sufficient for the job. As it is, the amplified output can be directly connected to a comparator, as with the infrared light detector described above. But the advantage of using ultrasonics is that you can be particular about the frequency of sound. Specifically, you want to limit the sensitivity of the circuit to 40 kHz, the same as the output of the transmitter. A 567 tone decoder IC is connected to the output of the 741 amp.

The output of the tone decoder alternates between

Fig. 28-11. Sample ultrasonic transducers.

229

Table 28-4. Ultrasonic Receiver Parts List.

U1	LM741 Op Amp IC
U2	NE567 Tone Decoder IC
R1,R3,R4	10 KΩ resistor
R5,R6,R7,R9	
R2	1 megohm potentiometer
R8	5 K 10- or 15-turn precision potentiometer
C1,C2	0.47 μF ceramic capacitor
C3	2.2 μF tantalum capacitor
C4	1.0 μF tantalum capacitor
C5	0.01 μF Monolithic or Hi-Q ceramic capacitor
TR1	Ultrasonic transducer

All resistors 5 or 10 percent tolerance, 1/4-watt; all capacitors 10 percent tolerance, rated 35 volts or higher.

LOW and HIGH as the sensor is moved closer to or further away from an object. Connect the output to a gate or other circuit. Remember that the logic levels are at 0 and about 9 volts. If you use the circuit with a computer or microprocessor, you must follow the interfacing guidelines provided in Appendix C.

Once you get the circuit up and working, adjust potentiometer R2, on the op amp, to vary the sensitivity of the

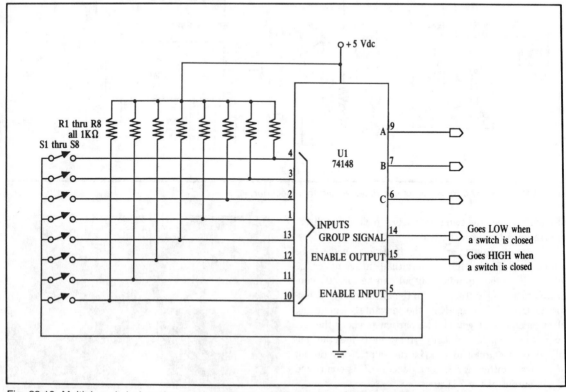

Fig. 28-12. Multiple switch detection using the 74148 priority encoder IC. Basic wiring diagram.

circuit. You will find that, depending on the quality of the transducers you use, that the range of this sensor is quite large. When the gain of the op amp is turned all the way up, the range may be as much as eight to 10 feet (the op amp may ring, or oscillate, at very high gain levels, so use your logic probe to choose a sensitivity just below the ringing threshold).

By the way, this basic setup is used in the ultrasonic ranging system described in detail in Chapter 29, "Navigating Through Space."

Multiple Switch Contacts

What happens when you have many switches or proximity devices scattered around the periphery of your robot? You could connect the output of each switch to the computer, but that's a waste of interface ports. A better way to do it is to use a priority encoder or multiplexer. Both schemes allow you to connect several switches to a common control circuit. The robot's microprocessor or computer queries the control circuit, instead of the individual switches or proximity devices.

The circuit in Fig. 28-12 uses a 74148 priority encoder

IC. Switches are shown at the inputs of the chip; you can use the output of a comparator or tone decoder as well. When a switch is closed, its binary equivalent (see the truth table) appears at the A-B-C output pins. For example, if switch #7 is closed, then binary number at the output pins of the chip is 100. If switch #4 is closed, then the binary number at the output pins of the chip is 001. With a priority encoder, only the highest value switch is indicated at the output. In other words, if switch 4 and 7 are both closed, the output will only reflect the closure of pin 4.

Another method is shown in Fig. 28-13. Here, a 74150 multiplexer IC is used as a switch selector. To read if a switch is closed or not, the computer or microprocessor applies a binary weighted number to the input select pins. The state of the desired input is shown in inverted form at the Output, pin 10. For example, if a binary 011 (decimal 3) is applied to the input lines, and the #3 switch is closed, then the output is HIGH. The advantage of the 74150 is that the state of any switch can be read at any time, even if a number of switches are closed. The disadvantage is that the switches must be addressed.

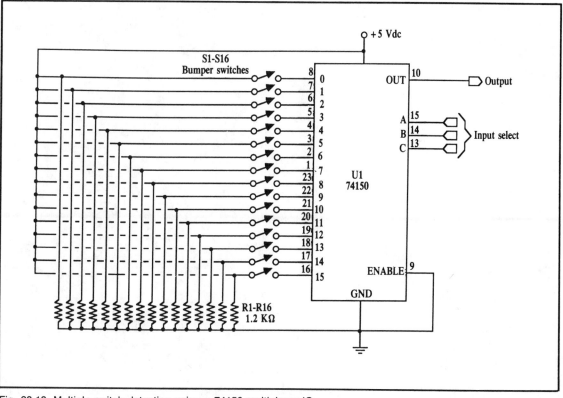

Fig. 28-13. Multiple switch detection using a 74150 multiplexer IC.

**Table 28-5. Priority
Bumper Switch Detector Parts List.**

U1	74148 8-input Priority Encoder IC
R1-R8	1 KΩ resistor
S1-S8	Bumper switches, SPST, momentary contact, normally open

**Table 28-6. Multiplexer
Bumper Switch Detector Parts List.**

U1	74150 1-of-16 Multiplexer IC
R1-R16	1.2 KΩ resistor
S1-S16	Bumper switches, SPST, momentary contact, normally open

All resistors 5 or 10 percent tolerance, 1/4-watt.

Chapter 29

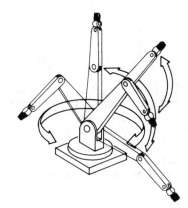

Navigating Through Space

The projects in this chapter deal with navigating your robot through space—not the outer-space, but the space between two chairs in your living room, or the space between your bedroom and the hall bathroom, or the space outside your yard by the pool. Robots suddenly become useful once they can master their surroundings, and being able to wend their way through their surrounds is the first step towards that mastery.

The techniques used to provide the navigation are varied: path-track systems, infrared beacons, ultrasonic rangers, even controllers for a pre-programmed player-bot, that make your invention trace the same spot in the room over and over again.

LINE TRACING

Robots that work in factories or deliver mail in a large building use tracks to keep them on course. The track can be a groove in the floor, a strip of reflective tape stretched out on the concrete, or a wire buried in the carpet. The reflective tape method is preferred because the track can easily be changed without ripping up the floor.

You can incorporate a tape track navigation system in your robot. The line tracing feature can be the robot's only means of semi-intelligent action, or it can be just a part of a more sophisticated machine. You could, for example, use the tape to help guide a robot back to its battery charger nest.

In a line tracing robot, a piece of white or reflective tape is placed on the floor. For best results, the floor should be hard, like wood, concrete or linoleum, not carpeted. One or more optical sensors are placed on the robot. These sensors incorporate an infrared LED and an infrared phototransistor. The transistor turns on and it sees the light from the LED reflected off the tape. Obviously, the darker the floor the better, because the tape shows up against the background.

In a working robot, mount the LED and phototransistors in a suitable enclosure, as described more fully in Chapter 28, "Collision Detection and Avoidance." Or, use a commercially available LED-phototransistor pair (again, see Chapter 28). Mount the detectors on the bottom of the robot, as shown in Fig. 29-1. Two detectors are shown, placed a little further apart than the width of the tape. I used 1/4-inch art tape in the prototype and placed the sensors 1/2-inch apart from one another.

Figure 29-2 shows the basic sensor circuit, and how the LED and phototransistor are wired. Feel free to experiment with the value of R2; it determines the sensi-

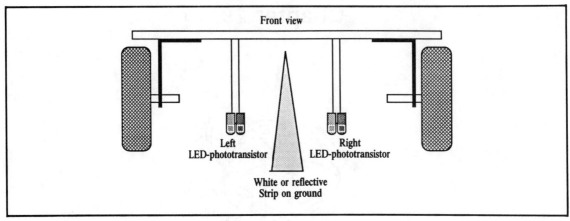

Fig. 29-1. Placement of the left and right phototransistor/LED pair for the line tracing robot.

tivity of the transistor. Figure 29-3 shows the sensor and comparator circuit that forms the basis for the line tracing system. Refer to this figure often because this circuit is used in many other applications.

Here is how the circuit works: When no light is reflected off the tape, the transistor is off, so the voltage at the inverting input of the comparator (here, 1/4 of an LM339 IC) is lower than the threshold, or set point, voltage, as adjusted by R3.

Conversely, when light is reflected off the tape, the transistor switches on, and if the set-point pot has been adjusted properly, the voltage at the inverting input of the comparator will exceed the threshold, or reference voltage (see Fig. 29-4 for a graphic illustration of this). The comparator will then trip over and trigger the motor relay. The DPDT relay (one for each motor) is con-

nected so that when actuated, the motor turns in reverse.

You can use the schematics in Fig. 29-5 and Fig. 29-6 to build a complete line tracing system. You can build the circuit using just three IC packages: an LM339 quad comparator, a 7486 quad Exclusive OR gate, and a 7403 open collector quad NAND gate. Before using the robot, block the phototransistors so that they don't receive any light. Rotate the shaft of the set point pots until the relays kick in, then back off again. You may have to experiment with the settings of the set point pots as you try out the system.

The line tracer robot is designed so that when both phototransistors see the background, the wheels roll straight ahead. If the robot strays off to the left, the right sensor sees the tape, so the right motor reverses to place the robot back on course.

Depending on the motors you use, and the switching speed of the relays, you may find your robot waddling its way down the track, over-correcting for its errors every time. You can help minimize this by using faster acting relays. Another approach is to vary the gap between

Table 29-1. Infrared Tape Guideway Circuit Parts List.

U1	LM339 Quad Comparator IC
U2	7486 Quad Exclusive OR Gate IC
U3	7403 Quad NAND Gate IC
R1,R4	270 Ω resistor
R2,R5	10 KΩ resistor
R3,R6	10 KΩ potentiometer
Q1,Q2	Infrared sensitive phototransistors
LED1,2	Infrared Light Emitting Diode
Misc.	Infrared filter for phototransistor (if needed)

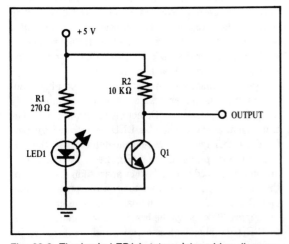

Fig. 29-2. The basic LED/phototransistor wiring diagram.

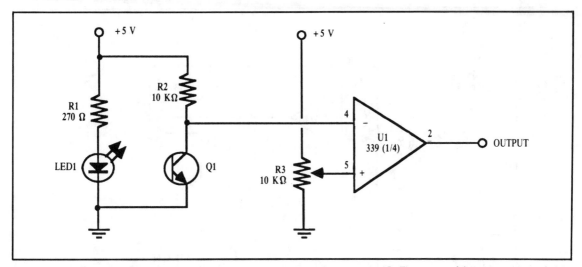

Fig. 29-3. Connecting the LED and phototransistor to an LM339 quad comparator IC. The output of the comparator switches between HIGH or LOW depending on the amount of light falling on the phototransistor.

the two sensors. By making it wide, the robot won't be turning back and forth as much to correct for small errors. I have also found that this so-called *overshoot* effect can be minimized by careful adjustment of the set point pots.

The circuit in Fig. 29-5 can easily be used to detect a black line on a large piece of white paper. Switch the connections to the comparators and the system will work in reverse. You can also affect the same change by reversing the connections on the relays.

Also try reversing the connections to just one of the comparators and placing the sensors so that one works normally when it sees the line, and the other works nor-

mally when it doesn't see the line. When the set point potentiometers have been set up correctly, the robot will turn in a circle when it loses the line but straighten out when the line is picked up again (the logic of this approach is reversed from the one before: straddling the line is considered the error condition and missing the line is considered the normal condition!). This setup has the effect of picking up a lost line after a few spins around the floor. With the previous method, the robot just goes on its way after it loses the guide path.

The line sensing robot is one approach to optical navigation. Another approach is shown in Fig. 29-7. Here, two white strips are placed any distance apart. You could,

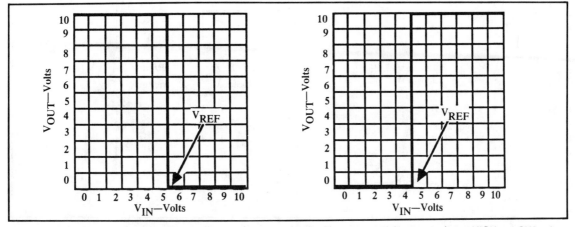

Fig. 29-4. A graphic representation of the output of the comparator. The output changes state (goes HIGH or LOW) when the input exceeds the reference voltage (also called the setpoint or threshold).

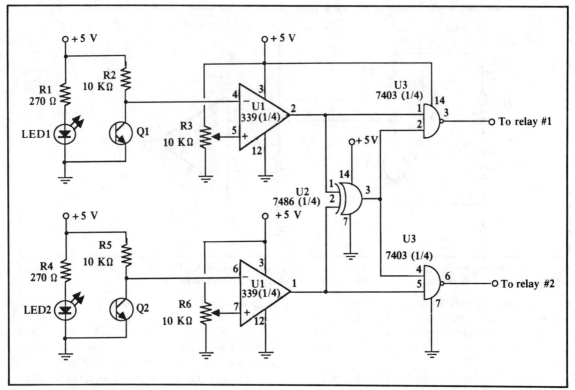

Fig. 29-5. Wiring diagram for the line tracing robot. The outputs of the 7403 are routed to the relays shown in Fig. 29-6.

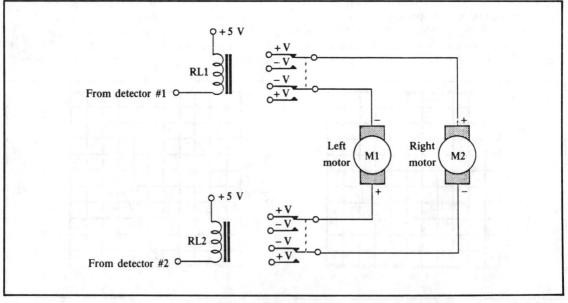

Fig. 29-6. Motor direction and control relays for the line tracing robot. You can substitute the relays for purely electronic control; refer to Chapter 13.

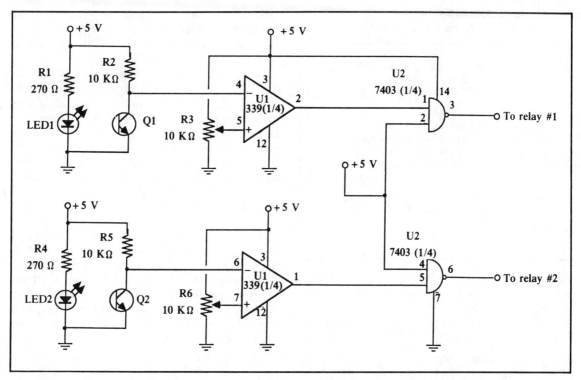

Fig. 29-7. Wiring diagram for the guidefence robot. The outputs of the 7403 are routed to the relays shown in Fig. 29-6.

Table 29-2. Relay Control Parts List.

RL1,RL2	DPDT fast-acting relay, contacts rated 2 amps or more

Table 29-3. Infrared Tape Guidefence Circuit Parts List.

U1	LM339 Quad Comparator IC
U2	7403 Quad NAND Gate IC
R1,R4	270 Ω resistor
R2,R5	10 KΩ resistor
R3,R6	10 KΩ potentiometer
Q1,Q2	Infrared sensitive phototransistors
LED1,2	Infrared Light Emitting Diode
Misc.	Infrared filter for phototransistor (if needed)

All resistors 5 or 10 percent tolerance, 1/4-watt; all capacitors 10 percent tolerance, rated 35 volts or higher.

for example, construct a six or seven inch wide roadway for your robot. When one of the lines is struck, the robot counters by moving in the opposite direction. You can also use this system to keep your robot from entering a protected area.

You'll hardly ever see a railroad track that has a turn

Table 29-4. Audio Induction Receiver Parts List.

U1	LM3900 Quad Comparator IC
R1,R3	1 KΩ resistor
R2	100 KΩ resistor
R4	50 KΩ potentiometer
R5	3.3 KΩ resistor
R6	68 KΩ resistor
R7	33 KΩ resistor
R8	2.2 KΩ resistor
C1,C2,C3	1.0 μF electrolytic capacitor
L1	Coil (see text)

All resistors 5 or 10 percent tolerance, 1/4-watt; all capacitors 10 percent tolerance, rated 35 volts or higher.

**Table 29-5. Ultrasonic
Ranger Timing Circuit Parts List.**

U1	4011 CMOS Quad NAND Gate IC
U2	4040 CMOS Counter
U3	4017 CMOS Counter
R1	220 KΩ resistor
R2	5.1 megohm resistor
R3	3.6 KΩ resistor
C1	5-20 pF trimmer capacitor
C2	22 pF ceramic capacitor
XTL1	5.0688 MHz crystal

All resistors 5 or 10 percent tolerance, 1/4-watt;
all capacitors 10 percent tolerance, rated 35 volts or
higher.

tighter than about eight degrees. There is good reason for this. If the turn is made tighter, the train cars can't stay on the track, and the whole thing derails. There is a similar limitation in line tracing robots (this includes the other methods given below). The lines cannot be tighter than about 10 to 15 degrees, depending on the turning radius of the robot, or the thing can't act fast enough when it crosses over the line. The robot will skip the line and go off course.

The actual turn radius will depend entirely on the robot. If you need your robot to turn very tight, small corners, build it small, like the commercially-available robot kit shown in Fig. 29-8, the Movit Line-Tracer. This robot responds to a thick black line drawn on a piece of white paper. It can turn corners as tight as about 18 inches, but not always without losing the line. Corners greater than 24 inches in diameter are handled with ease. A close-up of its light sensing electronics—a single infrared LED and two phototransistors, is shown in Fig. 29-9.

If your robot has a brain, whether it be a computer or central microprocessor, you can use it instead of the direct connection to the relays for motor control. The output of the comparators, when used with a +5 volt supply, is compatible with computer and microprocessor circuitry, as long as you follow the interface guidelines provided in Appendix C. The two sensors require only two bits of an eight bit port.

Fig. 29-8. The Movit Linetracer robot kit. It uses a similar technique as described in the text for tracing a black line drawn on a white piece of paper.

Fig. 29-9. The underside of the Movit Linetracer, showing the sensing electronics. One LED is used, flanked by two phototransistors.

WIRE TRACING

You may not be able to stick a piece of white tape on a shag carpet, but you can still devise a guide path navigation system for your robot. A wire tracing robot tracks a cable buried underneath the carpet. Visual contact isn't necessary; the robot picks up signals from the wire through rf energy.

The motor control circuit for use with the optical line

tracing robot can be used for a wire tracing robot as well. Apply the signal from the output of the receiver/amplifier to the inverting input of the comparator. Refer to the schematic diagram shown in Fig. 29-10.

You can interface the comparators directly to a microprocessor or to a computer, as long as you follow the design suggestions given in Appendix C, "Interfacing Logic Families and ICs." One possible way to connect the sensors to your personal computer is to use the two status bits of a parallel printer port. The eight data output ports can then drive the relays. Refer to Chapter 31, "Computer Control Via Parallel Port," for more information.

You can select only a certain range of frequencies (rejecting other sources of rf energy like the 60 Hz hum found in electrical wiring) by applying the output of the amplifier to a 567 tone decoder IC. Design specifications for the 567 chip can be found in Chapter 28, "Collision Detection and Avoidance," as well as Appendix C.

The design of the wire tracing coils and the pickup electronics is rather loose, and is open to a lot of experimentation. Initially, try wrapping 100 to 500 turns of 26 AWG magnet (solenoid) wire around a 1- or 2-inch coil form (you can also use a plastic spool or bobbin). Make two such coils. Position the coils on the underside of the robot as shown in Fig. 29-11. Spacing is important here as it was with the white tape above. You may want to mount the coils on sliding tracks so that you can experiment with the spacing.

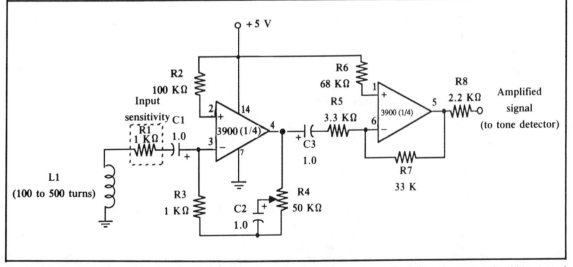

Fig. 29-10. The pickup amplification circuit for the wire tracing robot. For maximum flexibility, the output should be routed to an NE567 tone decoder IC, tuned to a center frequency of 1 to 10 kHz. You can vary the value of R1 to change the input sensitivity.

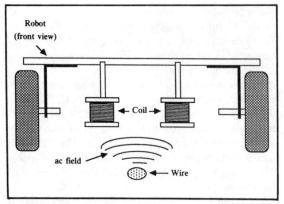

Fig. 29-11. Placement of the left and right wire coils for the wire tracing robot.

The wire you use can be insulated or noninsulated, but you'll have the best results if the insulation is very thin. Coated 26 AWG magnet wire is a good choice. Connect one end of another long piece of wire to the transmitter and place the wire along the path you want to trace.

Loop the free end of the wire back around to the transmitter and hook it up to ground.

The wire tracing system works when the coils mounted in your robot sense the magnetic field induced in the wire by the transmitter. This field is created by passing a 1 kHz to 10 kHz ac signal through the wire. You can use just about any transmission source that provides a reasonable amount of power, but a good makeshift system is to use the amplifier in your home stereo.

The voice and music output from the speaker terminals is well within the desired range, and you can easily control the volume (and hence the magnetic field induced through the wire) by turning the volume control up or down. Connect a long length (20 or 30 feet is a good start) between the speaker terminals of your hi-fi. Position the loop on the floor over the path you want the robot to trace.

ULTRASONIC DISTANCE MEASUREMENT

Police radar systems work by sending out a high frequency radio beam which is reflected off nearby objects, such as your car as you are speeding down the road. The

Table 29-6. Ultrasonic Ranger Trigger Circuits Parts List.

Microprocessor Trigger with 4528:

U1	4528 CMOS Monostable Multivibrator IC
R1	220 KΩ resistor, 1 percent tolerance
C1	100 pF Hi-Q ceramic or mica capacitor

Self-Running/Timed Duration

U1	NE555 Timer IC
U2	4528 CMOS Monostable Multivibrator IC
R1	5 KΩ resistor
R2	74 KΩ resistor
R3	22 KΩ resistor
R4	220 KΩ resistor
C1	0.33 μF ceramic capacitor
C2	0.1 μF ceramic capacitor
C3	100 pF Hi-Q ceramic or mica capacitor

Self-Running/Non-Timed Duration

U1	NE555 Timer IC
U2	4011 CMOS Quad NAND Gate IC (from Timing Section)
R1	1 MΩ resistor
R2	1 KΩ resistor
C1	1 μF ceramic capacitor
C2	0.1 μF ceramic capacitor

All resistors 5 or 10 percent tolerance, 1/4-watt unless noted; all capacitors 10 percent tolerance.

difference between the time when the transmit pulse is sent and when the echo is received denotes distance. Speed is calculated using the Doppler effect: the time between the sending pulse and echo increases or decreases proportionally depending on how fast you are going.

Radar systems are complex, expensive, and require certification by the Federal Communications Commission. All three make them unsuitable for homebrew robotics. High frequency sound can be used instead to measure distance, and with the right circuitry, can even provide a rough indication of speed.

Ultrasonic ranging is, by now, an old science. Polaroid has been using it for years as an automatic focusing aid on their instant cameras. Other camera manufacturers use a similar technique (some use infrared ranging, which won't be covered here). The doppler effect caused when something moves toward or away from the ultrasonic unit is used in home burglar alarm systems.

Certainly, the popularity of ultrasonics does not detract from its usefulness in robot design. The system given here can be used with or without a computer, but you'll enjoy the greatest flexibility if the project is connected to some form of brain. The transducers used in this project are available from a variety of sources, including Dick Smith Electronics. You can also salvage the transducers from a discarded home alarm.

To measure distance, a short burst of ultrasonic sound— more specifically 40 kHz sound—is sent out through a transducer (a specially built ultrasonic speaker). The sound bounces off an object and the echo is received by another transducer (this one a specially built ultrasonic microphone). A circuit then computes the time it took between the transmit pulse and the echo and comes up with distance. Yes, it is that simple, but there is a little more to a working ultrasonic ranger than this.

Facts and Figures

First some statistics. At sea level, sound travels at a speed of about 1,085 feet per second or 13,030 inches per second (this time varies depending on atmospheric conditions). The time it takes for the echo to be received is in microseconds if the object is within a few inches or even a few feet of the robot. The short duration is really no problem, however, for fast-acting CMOS and TTL ICs. The overall time between transmit pulse and echo is divided by two, to compensate for the round-trip travel time between the robot and the object.

Refer to the block diagram in Fig. 29-12. The ultrasonic ranger is composed of many small blocks (each block is in itself a simple circuit with few individual components). First, a 5.0688 MHz crystal generates a precise timing signal that, after being divided down into more reasonable frequencies, is used by the transmitting transducer as an ultrasonic sound source, and by the timing counter to determine distance. If your robot has a computer, you can use its crystal as the time base, but you may have to refigure the division to come up with the proper frequencies.

The 5.0688 MHz reference is divided to 39.6 kHz by a 4040 ripple counter IC. This IC can divide up to 4,096, but the divide-by-128 output is perfect for this application. The 39.6 kHz output is just 400 Hertz from the ideal ultrasonic frequency of 40 kHz.

The output of the 4040 is sent two ways. Let's look at the transmitter leg first. The 39.6 kHz signal is applied to one half of a NAND gate, used as a pulse gate. The other half of the NAND gate is connected to a pulse generator trigger. This generator sends out a timed pulse that allows the NAND gate to pass the 39.6 kHz tone for a specific amount of time (roughly 400 μs). The output of the NAND gate is not powerful enough to drive the transducer directly, so the signal is first boosted by a driver transistor. The result is a 400 μs burst of ultrasonic sound from the transducer.

When the generator sent out its pulse, it also triggered a counter to reset and start counting. This counter is clocked at a precise speed by a 4017 IC hooked up to divide the 39.6 kHz ultrasonic tone by six—or to 6.6 kHz. This frequency is roughly half that of the speed of sound—in inches—at sea level (13,030 times per second). Since the frequency is halved before it gets to the counter, you don't have to divide the resulting time difference later on to account for the round trip of the ultrasonic burst. Each tick of the counter represents one inch, which means that this system is fundamentally limited to an accuracy of no more than one inch. This is far more than enough for most applications.

The counter would go on forever if not stopped, and it is the entire idea of this circuit to stop the counter before things go too far. The echo pulse, if there is one, is captured by the receiver transducer. The signal is very weak and must be amplified. A 567 tone decoder IC is wired up to listen only for sounds that occur at about 40 kHz. So, when the echo pulse is received, the tone decoder latches onto it, and changes its output. At this time, the counter is stopped, and the distance, in inches is displayed.

You don't have to use the counter if you have a computer controlling your robot. You can use the computer as a timer, if you can access the timer interrupt lines on the system bus. Or you can use just the counting chips

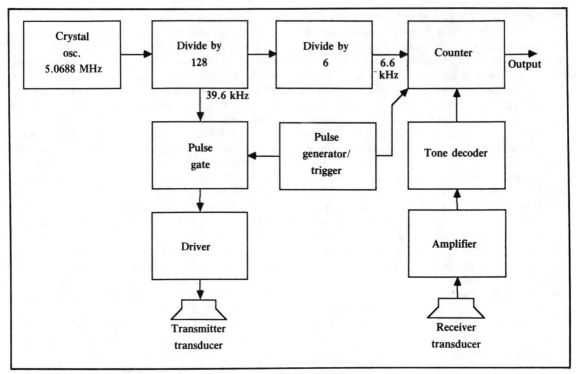

Fig. 29-12. Block diagram of the ultrasonic ranging system.

and present the result, which is in binary form, to a microprocessor port. You can count in BCD or binary, to suit your tastes and requirements.

Building the Circuit

Figures 29-13 through 29-17 show the circuits for the various blocks described above. The circuits are shown independently to make it easier for you to understand their function, and to allow you to more easily modify the ranger to suit your needs. You can use just about any counter circuit, as long as you can control the reset, start count, and stop count functions. You may need to add inverters to change the logic levels so you can make the

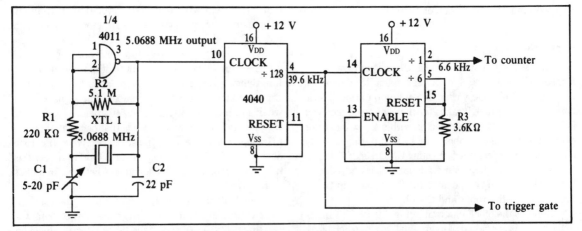

Fig. 29-13. The timing section of the ultrasonic ranging system. Outputs are 39.6kHz and 6.6kHz.

242

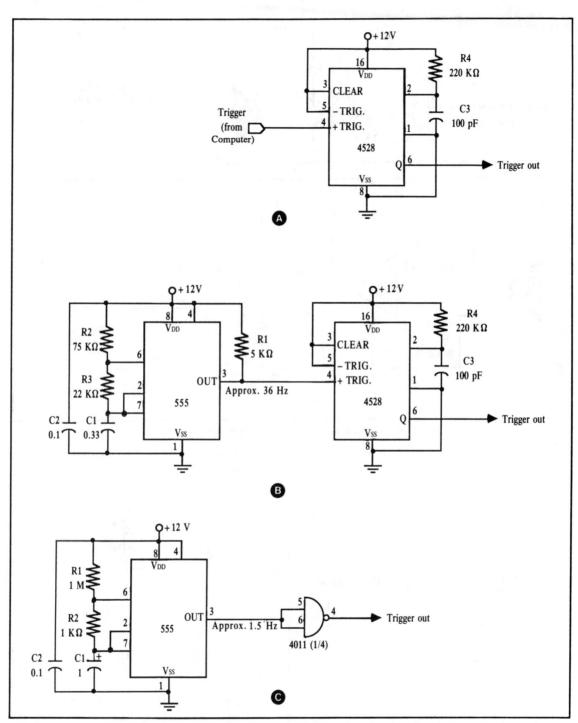

Fig. 29-14. The gate trigger section of the ultrasonic ranging system. A. One-shot timer for microprocessor/computer actuation; B. Free-running actuation, with timed gating interval (preferred); C. Free-running actuation, with non-timed gating interval.

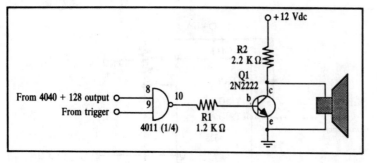

Fig. 29-15. The gate and output section of the ultrasonic ranging system.

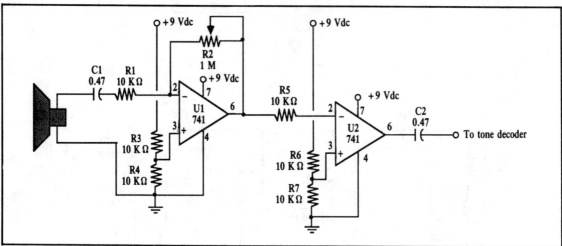

Fig. 29-16. The receiver and amplifier section of the ultrasonic ranging system.

Fig. 29-17. The 40kHz tone decoding section of the ultrasonic ranging system.

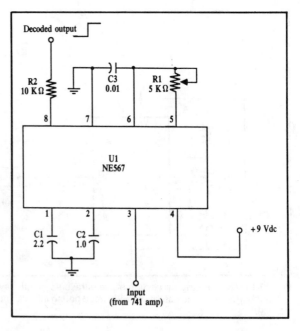

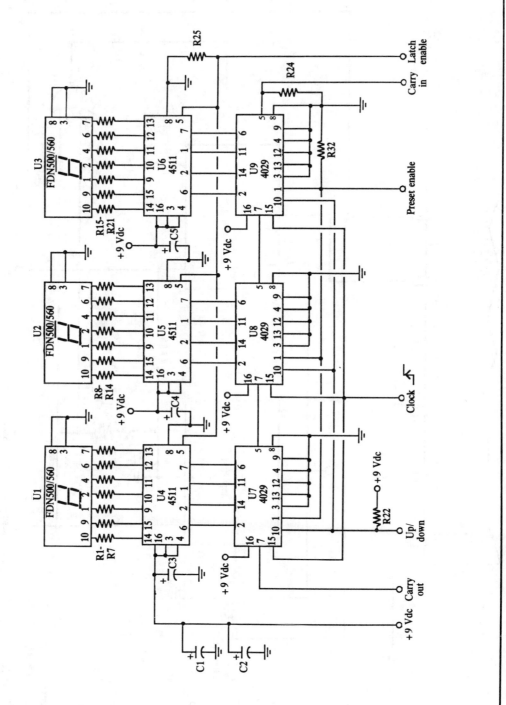

Fig. 29-18. A versatile 3-digit counter, suitable for use with the ultrasonic ranging system.

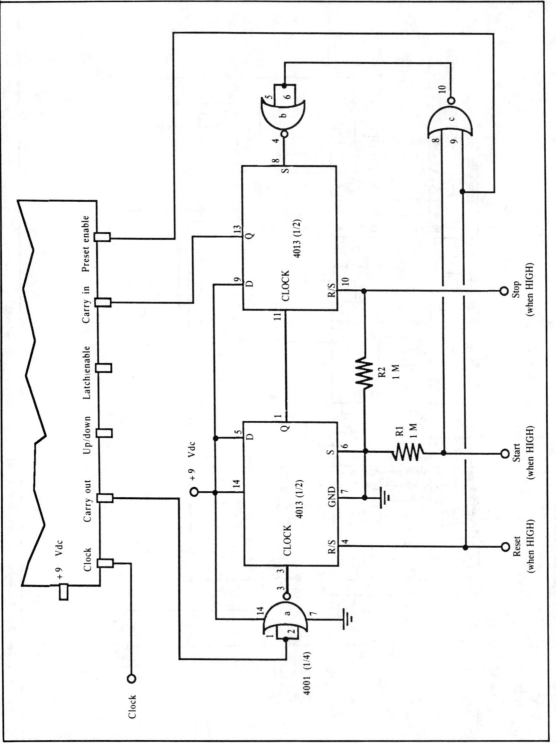

Fig. 29-19. Push-button actuation of the 3-digit counter.

**Table 29-7. Ultrasonic
Ranger Gate/Driver Circuit Parts List.**

U1	4011 CMOS Quad NAND Gate IC (from Timing section)
R1	1.2 K resistor
R2	2.2 K resistor
Q1	2N2222 npn transistor
TR1	Ultrasonic transducer

All resistors 5 or 10 percent tolerance, 1/4-watt.

**Table 29-8. Ultrasonic
Ranger Receiver/Amplifier Parts List.**

U1,U2	LM741 Op Amp IC
R1,R3,R4 R5,R6,R7	10 KΩ resistor
R2	1 MΩ potentiometer
C1,C2	0.47 μF ceramic capacitor
TR1	Ultrasonic transducer

All resistors 5 or 10 percent tolerance, 1/4-watt.

**Table 29-9. Ultrasonic
Ranger Echo Decoding Parts List.**

U1	NE567 Tone Decoder IC
R1	5K 10- or 15-turn precision potentiometer
R2	10 KΩ resistor
C1	2.2 μF tantalum capacitor
C2	1.0 μF tantalum capacitor
C3	0.01 μF Hi-Q ceramic or monolithic capacitor

All resistors 5 or 10 percent tolerance, 1/4-watt.

Table 29-10. Three-Digit Counter Parts List.

U1-U3	FND500/560 7-segment LED display (or equiv).
U4-U6	4511 CMOS LED Driver/Latch IC
U7-U9	4029 CMOS Counter IC
R1-R21	470 Ω resistor
R22,R23 R24,R25	1 MΩ resistor
C1,C2	100 μF electrolytic capacitors
C3-C6	0.47 μF ceramic capacitors

Table 29-11. Counter Gating Circuit Parts List.

U1	4013 CMOS "D" Flip-Flop IC
U2	4001 CMOS Quad NOR Gate IC
R1,R2	1 MΩ resistor

All resistors 5 or 10 percent tolerance, 1/4-watt;
all capacitors 10 percent, rated 35 volt or higher.

**Table 29-12. Beacon
Transmitter/ with Keypad Parts List.**

U1	Plessey SL409A Remote Control Transmitter
R1	100 Ω resistor
R2	50 KΩ potentiometer
R3	2.2KΩ resistor
R4	15 KΩ resistor
C1	0.68 μF ceramic capacitor
C2	150 μF electrolytic capacitor
C3	0.22 μF ceramic capacitor
C4	4.7 μF electrolytic capacitor
Q1	2N3906 pnp transistor
Q2	2N3904 npn transistor
LED1,2	TIL906 high-output Light Emitting Diode
Misc.	Keypad, infrared filter for diodes

All resistors 5 or 10 percent tolerance, 1/4-watt;
all capacitors 10 percent tolerance, rated 35 volts
or higher.

**Table 29-13. Beacon
Transmitter/ with DIP Switch Parts List.**

U1	Plessey SL409A Remote Control Transmitter
R1	100 Ω resistor
R2	50 KΩ potentiometer
R3	2.2 KΩ resistor
R4	15 KΩ resistor
C1	0.68 μF ceramic capacitor
C2	150 μF electrolytic capacitor
C3	0.22 μF ceramic capacitor
C4	4.7 μF electrolytic capacitor
Q1	2N3906 pnp transistor
Q2	2N3904 npn transistor
LED1,2	TIL906 high-output Light Emitting Diode
SW1	Four-position DIP switch
Misc.	Infrared filter for diodes

All resistors 5 or 10 percent tolerance, 1/4-watt;
all capacitors 10 percent tolerance, rated 35 volts or
higher.

counter work for you. A suitable counter circuit that provides a three digit visible display is shown in Figs. 29-18 and 29-19.

Note that the duration of the gated 39.6 kHz pulse has been set to accommodate the 567 tone decoder IC. At a frequency of 40 kHz (give or take a 1 kHz or so), there are 40,000 wavefronts each second, or one every 25 μs. The 567 needs at least eight wavefronts of its target signal to latch on; 15 or 16 to make it reliable. Multiply 25 μs times 16 and you get 400 μs. The component values for the one-shot triggers in Fig. 29-14 were chosen to provide a timed duration as close to 400 μs as possible. You can trigger the pulse generator/trigger manually, through a low-frequency astable multivibrator (as shown), or by microprocessor command.

Adjusting and Troubleshooting the Circuit

Of all the circuits for this book, the ultrasonic distance measurement system gave me the most trouble. Not because it was complex, but because it's hard to see exactly what's happening. Possible trouble points include the crystal oscillator, driver, amplifier, tone decoder, and counter. If the circuit isn't working right, make sure that the crystal oscillator is indeed working (check with a logic probe that accepts pulsing data), and that it is putting out the proper frequency. A frequency meter or oscilloscope is required to check this.

The driver may not be putting out enough power for the transducers you are using. Increase the gain by using a bigger transistor or connecting two transistors together Darlington style (of course, you can always use an npn Darlington transistor, like a TIP120).

The amplifier will probably give you the most trouble. By necessity, the op amp used in the amplifier is operated at very high gain. At high gain levels, the op amp may ring, thus masking any signal that it might otherwise amplify. Set the gain potentiometer, R2, below the point where the amp oscillates.

Wiring the 567 tone decoder is straightforward and there is little that can go wrong, but you'll want to be sure to use high-quality components to minimize the effects of frequency drift. The decoder responds to a wide range of frequencies several thousand Hertz higher and lower than 40 kHz, providing some margin for error.

The greatest difficulty with the tone decoder comes when using the circuit where the object being measured moves quickly. If you move your hand swiftly towards the receiver transducer during a measurement interval, the circuit may not appear to work. What's happening is that the frequency of the 39.6 kHz output pulse is compressed (thanks to the Doppler shift effect) as you move your hand, so the frequency reaching the receiving transducer is increased. Testing with a scope or frequency meter may reveal a change of 10 kHz to 20 kHz if your hand is moving fast enough. This is beyond the bandwidth of the tone decoder, so it will likely ignore the echo.

The effects of Doppler shift are often minimized by a design flaw of the 567 IC. The chip may also trigger on harmonics of the 39.6 kHz fundamental frequency. The compressed or expanded frequencies you get when the object is in motion during a ranging period may well be a harmonic of the fundamental tone, so you may get a reading after all.

By adding a phase-locked loop (PLL) or similar circuit, you can use the Doppler shift effect to measure speed. Be sure the circuit you use can cope with the wide range of frequencies. PLL's can only track within a defined band of frequencies. Another approach is to convert the frequency to voltage, using a frequency-to-voltage converter IC, or a circuit of your own design. As the frequency changes, so does the voltage. Sample the voltage and convert it to digital form, if necessary, to provide you with a speed measurement.

INFRARED BEACON

Unless you confine your robot to playing just within the laboratory, you'll probably want to provide it with a means to distinguish one room in your house from the next. This is particularly important if you've designed the robot with even a rudimentary form of object and area mapping. This mapping can be stored in the robot's memory, and used to steer around objects and avoid walls.

For less than a week's worth of groceries, you can construct an infrared beacon system that your robot can use to identify when it has passed from one room to the next. The robot is equipped with a receiver; in each room is a transmitter. The transmitters send out a unique code, which the robot interprets as a specific room. Once it has identified the room, it can retrieve the mapping information previously stored for it, and use it to navigate through its surroundings.

The beacon system that follows is designed around a set of television/VCR remote control chips sold by Plessey. The chips are reasonably inexpensive, but must be obtained through Plessey or one of their distributors. It is not carried by most electronics specialty stores. The unique design aspect of the Plessey chips is that they require very little additional components to operate. This is unlike some other remote control chips, such as the Motorola MC14457 and 14458, described in lucid detail

Table 29-14. Beacon Timer Parts List.	

U1	NE555 Timer IC
U2	4066 CMOS Analog Switch IC
R1	1 MΩ potentiometer
R2	47 KΩ resistor
R3	3.3 KΩ resistor
C1	10 μF tantalum capacitor
C2	0.1 μF ceramic capacitor

All resistors 5 or 10 percent tolerance, 1/4-watt; all capacitors 10 percent tolerance, rated 35 volts or higher.

Table 29-15. Beacon Receiver.	

U1	Plessey SL486 Infrared/Ultrasonic Preamplifier
U2	Plessey ML926 Remote Control Receiver
R1	50 Ω resistor
R2	200 Ω resistor
R3	56 KΩ resistor
R4	100 KΩ potentiometer
R5-R8	330 Ω resistor
C1,C2	6.8 μF electrolytic or tantalum capacitor
C3	0.047 μF ceramic capacitor
C4	0.22 μF ceramic capacitor
C5	0.15 μF ceramic capacitor
C6	33 μF electrolytic or tantalum capacitor
C7	22 μF electrolytic or tantalum capacitor
Cn	0.022 to 0.22 μF Hi-Q ceramic capacitor
D1	Infrared phototransistor or diode
LED1-4	Light Emitting Diode

All resistors 5 or 10 percent tolerance, 1/4-watt; all capacitors 10 percent tolerance, rated 35 volts or higher.

in Chapter 30, ''Remote Control Systems.''

You can, of course, use just about any wireless remote control system you desire. The only requirements is that you must be able to set up different codes for each transmitting station, and that the system works with infrared light. You'll experience too much interference if you use radio control or ultrasonics.

Figure 29-20 shows the pinouts of the three Plessey chips used in the project. The basic transmitter, equipped with a matrix keypad, is shown in Fig. 29-21. For a beacon system, you'd replace the keypad with a four-position DIP switch bank, as shown in Fig. 29-22. The switches set the room code. Closing a switch is like pushing a button on the remote control keypad. When a particular button is pressed, its unique code (in this case, a 5-digit serial binary word) is transmitted through the LEDs. With a four-position switch, you can have up to four rooms. If you have more rooms you want to code, use DIP switches

with more positions (they come six, eight, and 10 position packages). Alternatively, you can hard wire the transmitters, but this doesn't allow you to easily experiment or change room codes.

The transmitters are pulsed by a 555 timer so that they emit a beacon signal once every three to five seconds. The beacon timer is shown in Fig. 29-23. You can power each transmitter from batteries, but you'll find that the batteries will go dead after a day or two. A better way is to invest in surplus ac adapters. A typical 12-volt 300 mA adapter (which is enough for this circuit) costs less than $4 on the surplus market.

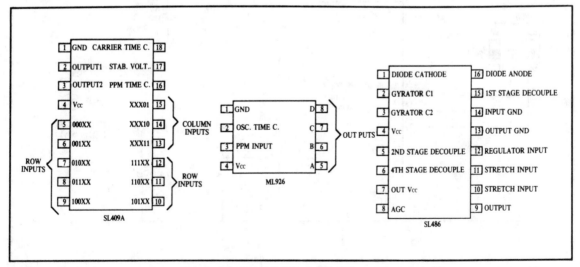

Fig. 29-20. Pinout diagrams for the Plessey ML409A, ML926 and SL486 remote control chips.

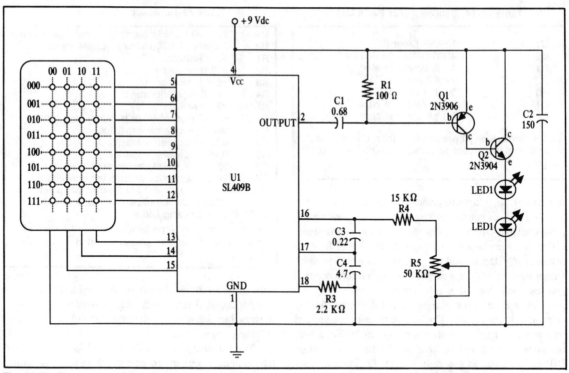

Fig. 29-21. The basic wiring diagram for the ML409A remote control transmitter. Shown with matrix keypad.

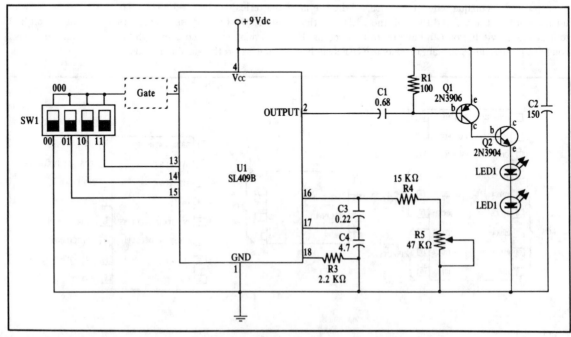

Fig. 29-22. Wiring diagram for the ML409A remote control transmitter used as a room beacon.

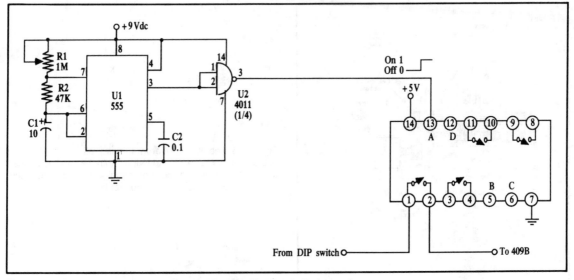

Fig. 29-23. Gate timer circuit to pulse the ML409A IC at periodic intervals.

The schematic diagram in Fig. 29-24 shows the receiver for use in the robot. An additional IC, for amplifying the incoming infrared code, is used to increase the distance. In my experiments, I found that with the amplifier I could reliably use the transmitters up to 20 feet from the receiver. The decoded output of the receiver is connected to the ports on the robot's computer. Capacitor Cn is chosen to set the desired frequency of the

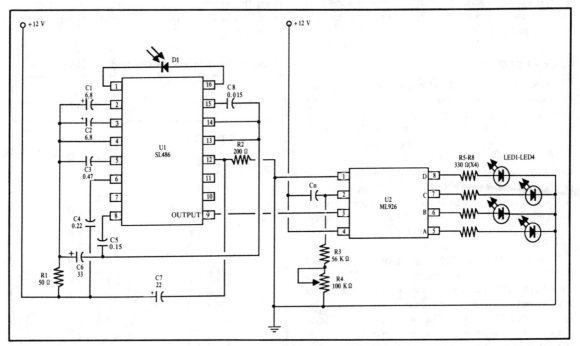

Fig. 29-24. The basic wiring diagram for the SL486 preamplifier IC and ML926 remote control receiver. Adjust R4 for best response.

251

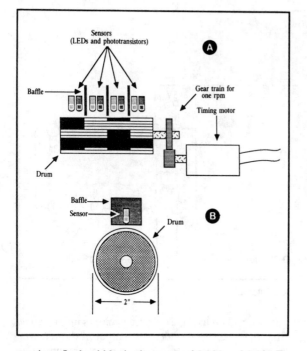

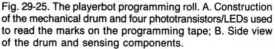

Fig. 29-25. The playerbot programming roll. A. Construction of the mechanical drum and four phototransistors/LEDs used to read the marks on the programming tape; B. Side view of the drum and sensing components.

receiver. It should be in the range of 0.022 and 0.22 μF.

To use the transmitter/receiver, adjust the transmitter frequency pot, R5 (see Fig. 29-22) while sending light pulses to the receiver. Keep adjusting the pot until the receiver kicks in.

PLAYER ROBOT

Here's a bonus project that you may want to try some rainy day. It uses the techniques covered earlier in this chapter on line tracing to preprogram the movement of a robot. I call it the Playerbot, because it's fashioned somewhat after the old-time player pianos.

In a player piano, a roll of paper is punched with holes; each hole represents a note (additional holes control the pedals). Vacuum is used to trigger a set of valves. If there is a hole at a particular location across the length of the paper, the vacuum is released, so the valve for that hole opens. The valve, in turn, triggers the corresponding key on the piano.

The Playerbot uses light instead of a vacuum, and there are only four "valves" to worry about: The valves are infrared (IR) LED-phototransistor sensors connected to comparators. Sensors control the power and direction of the right and left motors. The roll of paper in the Player is a cylinder that's driven by a timing motor. The basic Playerbot setup is shown in Fig. 29-25.

To create a program, you mark on a strip of paper

with black ink, as shown in Fig. 29-26. You control the functions of the motors by leaving the squares blank or filling them in with ink. Figure 29-27 shows a block diagram of how the comparators activate the various control relays.

Table 29-16. Transmitter Codes and Output Signals.

Transmitter Code	ML926
E D C B A	D C B A
0 0 0 0 0	0 0 0 0
0 0 0 0 1	0 0 0 1
0 0 0 1 0	0 0 1 0
0 0 0 1 1	0 0 1 1
0 0 1 0 0	0 1 0 0
0 0 1 0 1	0 1 0 1
0 0 1 1 0	0 1 1 0
0 0 1 1 1	0 1 1 1
0 1 0 0 0	1 0 0 0
0 1 0 0 1	1 0 0 1
0 1 0 1 0	1 0 1 0
0 1 0 1 1	1 0 1 1
0 1 1 0 0	1 1 0 0
0 1 1 0 1	1 1 0 1
0 1 1 1 0	1 1 1 0
0 1 1 1 1	1 1 1 1

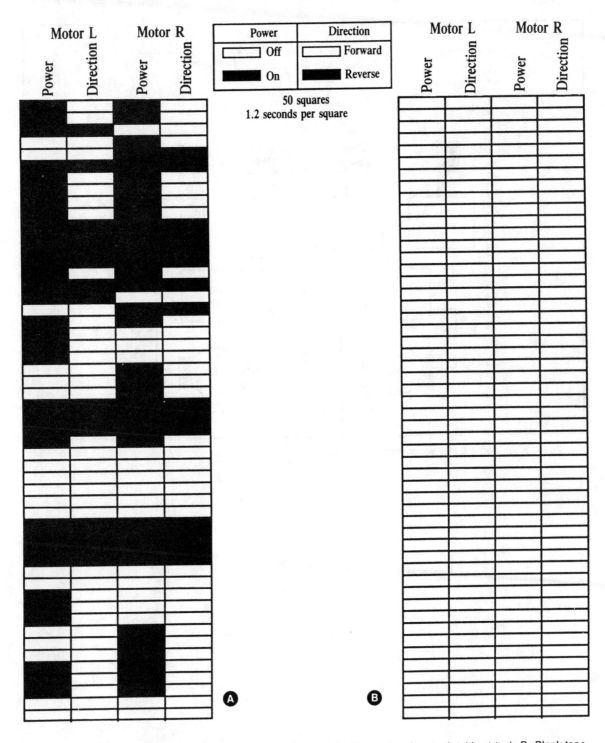

Fig. 29-26. Player roll tape. A. Sample program for both left and right drive motors (see truth table at top); B. Blank tape for your own use. Make copies of the blank and fill it in to program the playerbot.

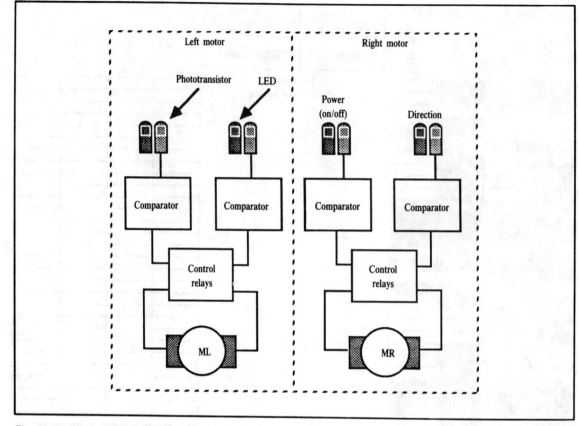

Fig. 29-27. Block diagram of playerbot control circuit.

Chapter 30

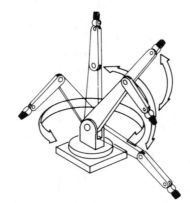

Remote Control Systems

The most basic robot designs, which are really just a step above motorized toys, use a wired control box where you flip switches to move the robot around the room, or activate the motors in a robotic arm and hand. The wire link can be a nuisance and acts as a tether preventing your robot from freely navigating through the room. You can cut the physical umbilical cord and replace it with a fully electronic one with a remote control receiver and transmitter.

This chapter details several popular ways of achieving wireless links between you and your robot. You can use the remote controller to activate all of the robots functions, or with a suitable on-board computer working as an electronic recorder, you can use the controller as a teaching pendant. You manually program the robot through a series of steps and routines, then play it back under the direction of the computer. Some remote control systems even let you connect your personal computer to your robot. You type on the keyboard, or use a joystick for control, and the invisible link does the rest.

INFRARED PUSHBUTTON REMOTE CONTROL

The Motorola MC14457 and MC14458 chips form the heart of a useful remote control receiver-transmitter pair.

The chips are available through Motorola distributors as well as several mail order outlets and retail stores, such as Dick Smith Electronics. Price is under $15 for the pair. Alternatively, you may want to use the receiver/transmitter pair from Plessey, described in Chapter 29, "Navigating Through Space." The 14457 is the transmitter, and can be used with up to 32 pushbutton switches (we'll be using 20). Pushing a switch commands the chip to send a binary serial code through a set of high-output infrared LEDs (ultrasonic transducers will also work). Decoded output pins on the 14458 receiver chip can be connected directly to a controlled device, such as a relay or LED, a counter, or a microprocessor port.

Figure 30-1 shows the pin-out diagrams for the two chips. The 14457 comes in a small 16-pin DIP package, and with all the other components added in, takes up a space of less than two inches square. The chip uses CMOS technology to conserve battery power, and when no key is pressed, the entire thing shuts down. Battery power is used only when a key is depressed.

Transmitter

The basic hookup diagram for the 14457 transmitter circuit is shown in Fig. 30-2. Note the oscillator and

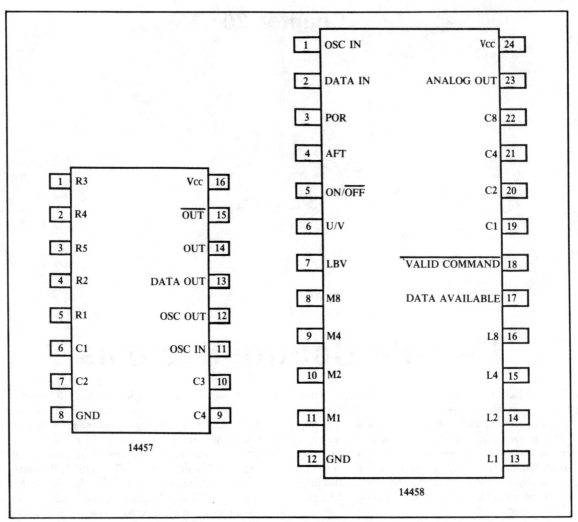

Fig. 30-1. Pinouts for the Motorola 14457 remote control transmitter IC and the 14458 remote control receiver.

Table 30-1. Remote Control Transmitter Output Parts List.

U1	Motorola MC14457 Remote Control Transmitter IC
R1	330 Ω resistor
R2	15 Ω resistor
R3	10 mΩ resistor
R4	680 Ω resistor
C1	50 μF electrolytic capacitor
C2	100 μF electrolytic capacitor
C3	1000 μF electrolytic capacitor
Q1	2N2222 npn transistor
D1	1N4001 diode
D2,D3	1N914 signal diode
LED1-3	TIL906-1 high output Light Emitting Diode
XTAL1	300 to 650 kHz ceramic resonator or crystal

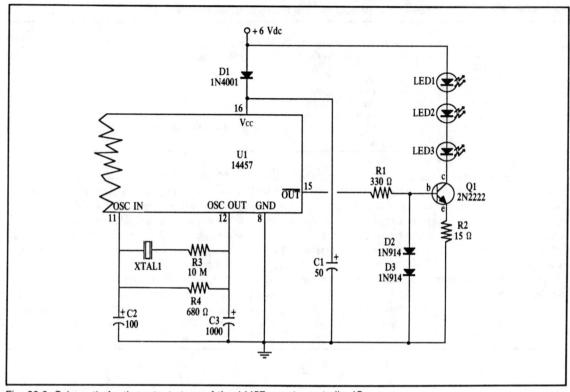

Fig. 30-2. Schematic for the output stage of the 14457 remote controller IC.

**Table 30-2. Remote Control
Transmitter Keypad Parts List.**

Q1-Q4	2N2222 npn transistor
S1-S20	SPST momentary switches, normally open

All resistors 5 or 10 percent tolerance, 1/4-watt; all capacitors 10 percent tolerance, rated 35 volts or higher.

tank circuit connected to pins 11 and 12. The oscillator required by the 14457 (and a matching one for the 14458 receiver) is a hard to find ceramic resonator, not a standard crystal. The ceramic resonator works just like a crystal, but comes in frequencies under 1 MHz. The exact value of the resonator isn't critical, I found, as long as it is within a range of about 300 to 650 kHz, and that the resonators for both chips are identical. I successfully used a 525 kHz resonator in the prototype circuit.

See Appendix A, "Sources," for a listing of mail order firms that deal with ceramic resonators. Note that one prime source, Halted Specialties (Sunnyvale, CA) has

a $10 mail order minimum, but the last time I checked, the 525 kHz resonators were only 49 cents each! You'll have to either buy a lot of resonators or fill up the order with other goodies.

**Table 30-3. Remote Control
Receiver Amplifier Parts List.**

U1	4069 CMOS Hex Inverter IC
R1	250 KΩ potentiometer (adjust for best response)
R2,R5	1 megohm resistor
R3	100 KΩ resistor
R4,R6	680 KΩ resistor
R7	68 KΩ resistor
C1,C2, C3,C6	0.001 µF ceramic capacitor
C4,C5	0.0047 µF ceramic capacitor
D1,D2	1N914 signal diode
LED1	Infrared sensitive phototransistor or diode

All resistors 5 or 10 percent tolerance, 1/4-watt; all capacitors 10 percent tolerance, rated 35 volts or higher.

Table 30-4. Remote Control Receiver Module Parts List.

U1	Motorola MC14458 Remote Control Receiver IC
U2	4049 CMOS Hex Inverter IC (from Amplifier section)
R1	10 MΩ resistor
R2	680 ohm resistor
C1	10 pF ceramic capacitor
C2	100 pF ceramic capacitor
C3	0.47 μF ceramic capacitor
XTAL1	300 to 640 kHz ceramic resonator or crystal (must match transmitter)

The inverted output of the 14457 (OUT) is boosted by a 2N2222 npn transistor. Three high power IR LED's, TIL906-1 or equivalent, put out a great deal of light, making the effective range of the system at least 10 feet. You can substitute all the components connected to the output pin with an ultrasonic transducer (connect the leads of the transducer between pins 14 and 15 of the chip). You'd use another ultrasonic transducer on the receiving end, of course.

So much for the output stage of the transmitter; how about the input stage? Figure 30-3 shows how to connect a series of 20 pushbutton switches to the column and row inputs of the 14457. You can use separate switches, but a cheaper way is with a surplus keypad. Membrane keypads that are already engineered in row and column format as shown are available for under $2 or $3. Note the DOWN and UP buttons. These two buttons have interesting possibilities, as you'll soon see.

You can use just about any wiring technique to construct the 14457 transmitter, but because you'll probably want to make the unit handheld, stay away from wire-wrapping. The stems of wire-wrapping posts and

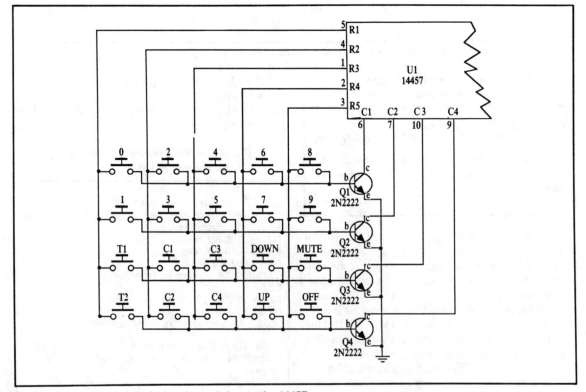

Fig. 30-3. How to add up to 20 function switches to the 14457.

**Table 30-5. Remote Control
Receiver Output Parts List.**

U2	4028 CMOS 1-of-10 Demultiplexer IC

Table 30-6. Up/Down Counter Output Parts List.

U1	4029 CMOS Up/Down Counter IC
U2	4001 CMOS Quad NOR Gate IC

All resistors 5 or 10 percent tolerance, 1/4-watt; all capacitors 10 percent tolerance, rated 35 volts or higher.

sockets are too long and will fatten the controller considerably. Use a set of four "AA" batteries to power the transmitter, a 9-volt transistor battery doesn't provide enough for the high-output LEDs.

Receiver

You must amplify the received signal before it can be applied to the 14458 receiver chip. The amplifier shown in Fig. 30-4 is a good choice, because it uses up some of the remaining gates in an IC that is required by the receiver chip.

This amplifier has extremely high gain and it is prone to oscillation if you construct it haphazardly or build it on an experimenter's breadboard. If you continue to have trouble with oscillation after the circuit is built (you can test it with a logic probe), reduce the value of the 1MΩ resistors until the ringing stops. Both the phototransistor, as shown, and the ultrasonic transducer can use this circuit. When working with the transducer, connect its leads to the input capacitor and ground.

You can vary the value of R1 to set the overall sensitivity of the amplifier. Initially, install a 250K potentiom-

eter as a variable resistor and test the circuit at various settings. The prototype amplifier worked well with a value of about 70K.

The basic wiring diagram for the 14458 receiver chip is shown in Fig. 30-5. Once again, the ceramic resonator is used as a timing reference for the IC. One inverter from a 4069 is used to provide an active element in the oscillator circuit.

All that's required now is to connect the devices to be controlled to the output lines, shown in Fig. 30-6. Note the various sets of outputs and the VC function pin. The $\overline{VC}$ line goes HIGH when all but the number keys are pressed. The pin is used in some advanced decoding schemes as a function bit. The chart in Table 30-7 shows what happens when the 20 keys are pressed (the chip can accommodate another 12 pushbuttons; see the manufacturers data sheet for more information).

For most routine application, you need only to connect the controlled device to pins 19 through 22 (labeled C1, C2, C4, and C8). These are binary weighted and by

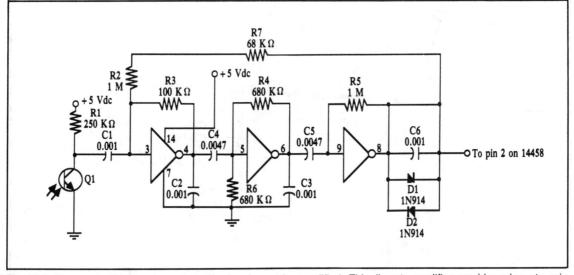

Fig. 30-4. The received infrared or ultrasonic signal must be amplified. This discrete amplifier provides adequate gain for a range of over 10 feet.

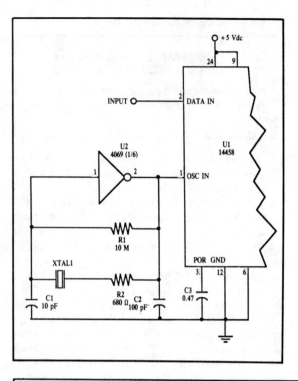
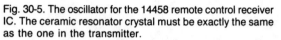

Fig. 30-5. The oscillator for the 14458 remote control receiver IC. The ceramic resonator crystal must be exactly the same as the one in the transmitter.

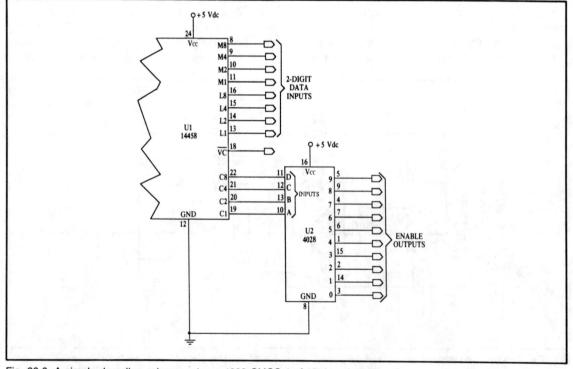

Fig. 30-6. A simple decoding scheme using a 4028 CMOS 1-of-10 decoder chip.

Table 30-7. Motorola Receiver Output Codes (First 20).

Key	Row	Colum	FB	C8	C4	C2	C1	$\overline{VC}$ Pulse
0	1	1	0	0	0	0	0	
1	1	2	0	0	0	0	1	
2	2	1	0	0	0	1	0	
3	2	2	0	0	0	1	1	
4	3	1	0	0	1	0	0	
5	3	2	0	0	1	0	1	
6	4	1	0	0	1	1	0	
7	4	2	0	0	1	1	1	
8	5	1	0	1	0	0	0	
9	5	2	0	1	0	0	1	
Toggle1	1	3	1	0	0	0	0	X
Toggle2	1	4	1	0	0	0	1	X
CONT1	2	3	1	0	0	1	0	X
CONT2	2	4	1	0	0	1	1	X
CONT3	3	3	1	0	1	0	0	X
CONT4	3	4	1	0	1	0	1	X
DOWN	4	3	1	0	1	1	0	X
UP	4	4	1	0	1	1	1	X
MUTE	5	3	1	1	0	0	0	X
OFF	5	4	1	1	0	0	1	X

connecting a 4028 one-of-ten decoder IC to the receiver, you can individually control up to 10 control devices or functions.

The receiver is wired to accept a single keypress on the transmitter as a complete command. The receiver can also be made to wait until *two* keys are pressed (this is used because in television and VCR applications, you are able to dial in multi-digit channels). The two-digit data outputs are used when the chip is in two-digit mode. To change from one- to two-digit mode, disconnect the power leads to pins 9 and 6.

Figure 30-7 shows how you can use the UP and DOWN functions with a counter to provide variable step control. Note the DOWN ENABLE and UP ENABLE lines. When depressing the UP or DOWN buttons on the transmitter, the UP and DOWN codes are sent continually, rather than just one per each push, as with the other buttons.

The enable outputs from the 4028 can only handle 10 functions, so the UP and DOWN functions are integrated with the number functions. That is, when DOWN button is pressed, the output at the 4028 chip is the same as if you depressed the number 6 button (binary code 0110). However, the $\overline{VC}$ pin is toggled HIGH, which can be used in further decoding. Similarly, when the UP button is pressed, the output of the 4028 chip is the same as if you depressed the number 8 button (binary code 1000).

To enable you to count the number of UP and DOWN pulses, you connect the UP ENABLE input of the circuit shown in the figure to the number 7 output of the 4028 and the DOWN ENABLE input to the number 6 output. The $\overline{VC}$ pin acts to gate the circuit so that the counter doesn't count when numbers 7 and 6 are pressed.

WIRELESS JOYSTICK

This easy project does not require you to assemble a remote control circuit out of ICs and components, and is perfect for use when you want to control five or less functions. The wireless joystick shown in Fig. 30-8 is a commercially available product that is sold through many department, discount, and computer stores. Cost is around $25. The joystick is made to work with all computers that accept Atari-type (switch contact) joysticks. It is not for use with the Apple II or IBM PC, which use potentiometer-type joysticks.

The joystick comes in two parts: a transmitter and receiver. The transmitter looks just like a standard joystick except that it has a whip antenna on it. It is powered by a 9-volt battery. The receiver connects to the joystick port, and is powered by four "AA" cells. During my tests, I found that the effective range of the units is about 20 feet. Not great but useful for most purposes.

Figure 30-9 shows the functions of the pins on the joystick receiver. The pins serve the same purpose as on a wired joystick. To interface the receiver to your robot,

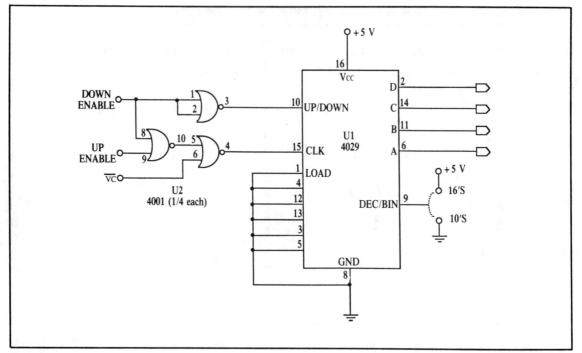

Fig. 30-7. A suitable up/down counter for use with the Up and Down keys on the transmitter. The 4029 IC can count in decimal or binary.

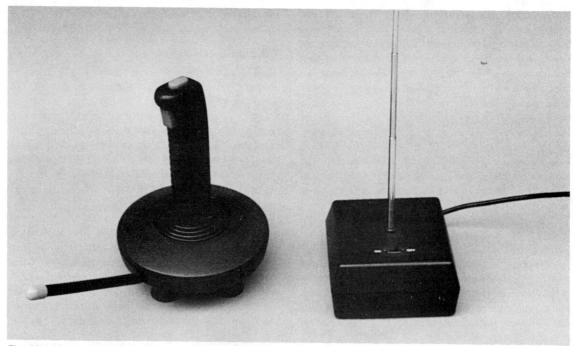

Fig. 30-8. A commercially available wireless joystick.

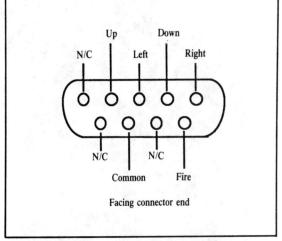

Fig. 30-9. Pinout diagram of an Atari-type joystick connector. The connector is shown facing you.

wire the pins as shown in Fig. 30-10. You can interface to TTL or CMOS gates, but for any application where you want to drive heavy loads, use relays or opto-isolators. An opto-isolator setup is shown in the figure.

If you can't find a suitable ready-made opto-isolator, construct one yourself using an infrared LED and pho-

totransistor. Butt the LED/phototransistor pair end to end, as shown in Fig. 30-11. Use heat-shrink tubing or a length of dowel drilled out to accept the components.

RADIO CONTROL SYSTEMS

Walkie-talkies are primarily intended for use in voice communications, but you can also use them to transmit codes, even digital data through the airwaves. In addition to walkie-talkies, you can use radio control (RC) receiver-transmitters designed for model airplanes and miniature racing cars. These are perfectly suited for controlling robots, especially over long distances (up to a mile or so with the good units).

You may, if you wish, build your own radio controlled receiver and transmitter using off-the-shelf parts, but it's generally more trouble than its worth. Rf circuitry is far more involved than digital and even analog circuitry, and you to need to find (or build) special coils, tuning slugs, and other inductive components. You also must keep the power output under 100 mW, stick a little "Experimental" sticker on the thing, and comply with a variety of other FCC regulations.

In one of the systems you'll build, you have to tear into a toy walkie-talkie and make a slight modification to it. According to the FCC rules, making changes to a

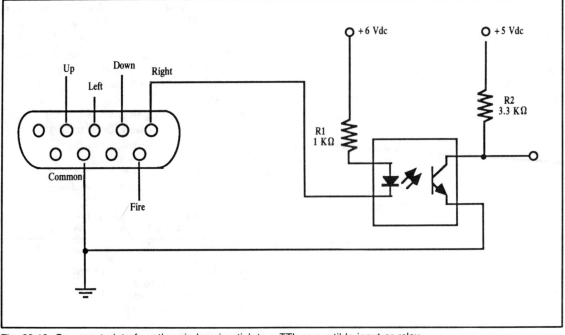

Fig. 30-10. One way to interface the wireless joystick to a TTL compatible input or relay.

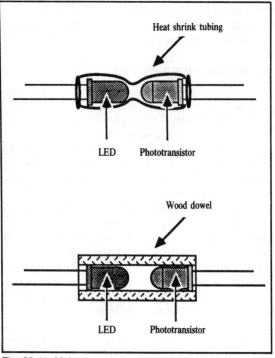

Fig. 30-11. Make your own opto-isolator: from one infrared LED and one infrared phototransistor.

commercial product, whether or not it is already certified by the Commission, is also a no-no. But since the walkie-talkie used in the project has a power output of 100 mW or under, the contraption can be classified as an experimental device.

Walkie-Talkie Code Switch Conversion

A number of toy walkie-talkies are on the market that cost $10 a piece or less. A reasonably good transmitter-receiver pair shouldn't set you back more than $25; even less if you shop around. The bulk of the very inexpensive toy transceivers, like the model in Fig. 30-12, have a "code" button that, when pressed, sends a tone to the receiving station. You're supposed to tap on this button to send Morse Code.

As it happens, the tone generator used in these walkie-talkies can be used as an audio generator. A decoding circuit attached to the receiving end picks up the beep, distinquishing it from the other noise and signals on the line. The system works fairly well, considering its simplicity.

You can use the code switch as is, and press it to activate the decoder in the receiver. You'll probably want

Fig. 30-12. Toy walkie-talkies are ideally suited for use as wireless remote control systems.

to adapt the switch so it can be operated by a computer or other circuit. Modify the transmitter by clipping the leads going to the code switch and route them to the interface electronics.

Tape the transmit button down so that the walkie-talkie is continually transmitting. In the prototype, I took the switch apart and removed the spring, but I had a hard time getting the switch back together again. Use the tape method; it's safer.

On the receive walkie-talkie, remove the back cover

Table 30-8. Walkie-Talkie Transmitter Tone Generator Parts List.

U1	4011 CMOS Quad NAND Gate IC
R1	100 K resistor
R2	470 Ω resistor
R3	1 megohm resistor

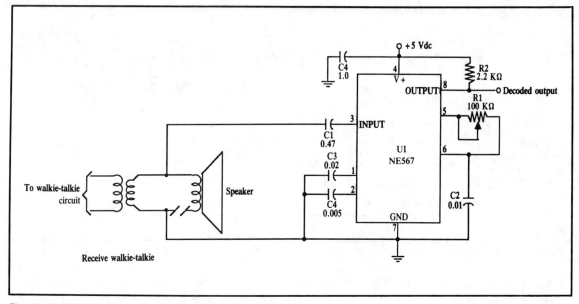

Fig. 30-13. Walkie-talkie tone decoder circuit. Adjust R1 for best reponse.

and locate the speaker wires. Next, cut the leads to the speaker (unless you want to use it as a monitor source), and route them to the circuit shown in Fig. 30-13.

At the heart of this circuit is an NE567 tone decoder IC. The values shown are for decoding a 1 kHz tone. Note that the code tone oscillator used in some walkie talkies may have a higher or lower frequency, so if the circuit doesn't seem to work at first, adjust R1 to change the center frequency of the 567. If the tone decoder doesn't seem to lock onto the incoming tone, you may need to add the signal conditioning circuit, shown in Fig. 30-14, immediately before the input of the 567.

Depress or trigger the code switch on the transmitter walkie-talkie and the output of the 567 should repeatedly go from LOW to HIGH. You can connect the output to an interface circuit to control a motor, a relay, or other device. See Appendix C for information on interfacing the tone decoder to other circuitry.

If the walkie-talkie you have lacks a tone switch, you can easily add a tone generator circuit. Follow the schematic in Fig. 30-15. The output of the tone generator is 1kHz.

Adding More Tones

The Morse Code switch or tone generator provides just one tone (or channel), so it only can control one device with this system. There are a number of ways you can add more channels to the basic walkie-talkie.

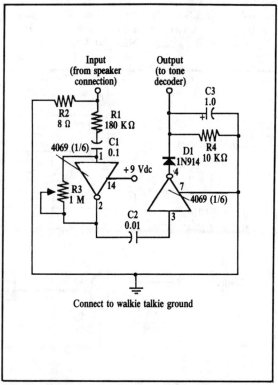

Fig. 30-14. A signal conditioner circuit for use between the walkie-talkie output stage and the input of the 567 decoder chip.

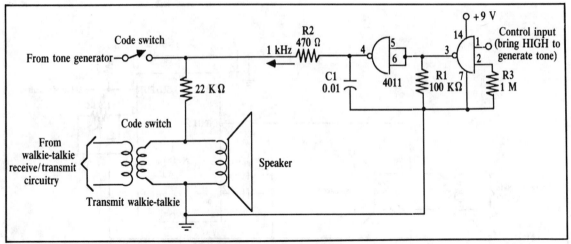

Fig. 30-15. A 1kHz tone generator circuit. Bring the control input HIGH to send a tone.

One of the easiest is to rig up multiple tone generators to the audio transformer on the receiver, as shown in Fig. 30-16. Each generator uses two gates of a 4011 CMOS NAND gate package. Calculating the frequency is easy: multiply the resistance of R1 (in KΩ) by the capacitance of C1 (in μF). With 100K for R1 and 0.01 for C1, the output is 1,000 Hz. With 150 K for R1, the fre-

quency is increased to 1,500 Hz. The walkie-talkie broadcasts the tones the same as it does voices. You pass the tone you want by bringing its gate input HIGH.

On the receiver end, you must add a separate 567 for each tone you want to decode. Vary the component values on each decoder according to the manufacturer's specifications (or check Appendix C) to receive only the desired

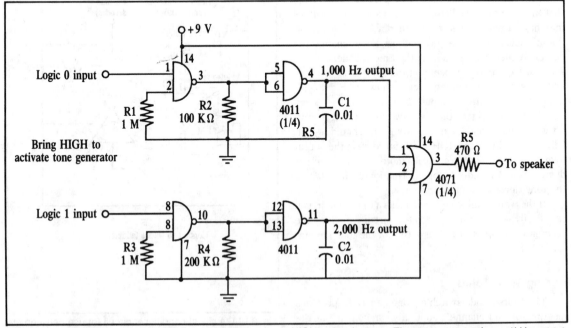

Fig. 30-16. A two-tone generator. Bring one or the other input HIGH to send a tone. The top generator has a 1kHz output; the bottom generator has a 2kHz output.

266

**Table 30-9. Walkie-Talkie
Receiver Tone Decoder Parts List.**

U1 NE567 Tone Decoder IC
R1 100 K resistor (adjust to decode 1 kHz input signal)
R2 2.2 K resistor
C1 0.47 µF ceramic capacitor
C2 0.01 µF ceramic capacitor
C3 0.02 µF ceramic capacitor
C4 0.005 µF ceramic capacitor
C4 1.0 µF tantalum capacitor

All resistors 5 or 10 percent tolerance, 1/4-watt; all capacitors 10 percent tolerance, rated 35 volts or higher.

frequency. A typical two-channel system is shown in Fig. 30-17.

If you need more than a couple of channels, you may want to consider going to a specialty IC, like a telephone tone dialing generator and decoder. These chips, available through sources like Radio Shack, can decode 12 channels and do not require you to carefully select components.

Two way communications is possible, but you need to either control the receive/transmit button of the walkie-talkies or use two sets of transceivers. Two of the walkie-talkies must operate at one frequency, and the other set must operate on another.

Adapting Radio Control Modules

Radio control modules designed for use with model

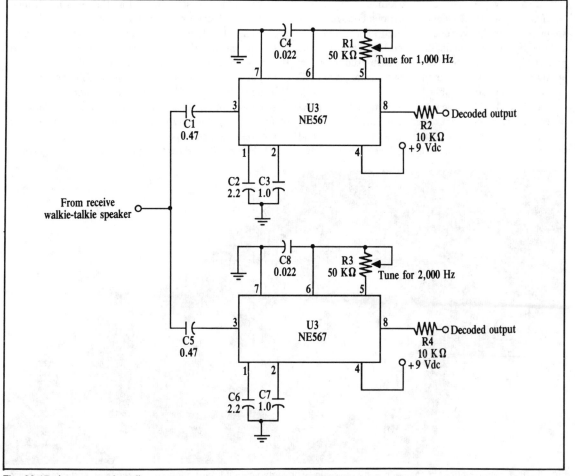

Fig. 30-17. A two-tone decoding circuit, using two independent NE567 ICs. Adjust R1 to receive the 1kHz tone; adjust R3 to receive the 2kHz tone.

U1	4069 CMOS Hex Inverter IC
R1	100 K resistor
R2	8 Ω resistor
R3	1 megohm potentiometer (adjust for best response)
R4	10 K resistor
C1	0.1 μF ceramic capacitor
C2	0.01 μF ceramic capacitor
C3	1.0 μF tantalum capacitor
D1	1N914 signal diode

Table 30-10. Walkie-Talkie Tone Shaper Parts List.

All resistors 5 or 10 percent tolerance, 1/4-watt; all capacitors 10 percent tolerance, rated 35 volts or higher.

airplanes, helicopters, cars, boats, and other toys are also perfectly suited for controlling a robot. You can purchase an RC transmitter and receiver set without the model, or you can buy the whole thing together.

The RC kits are designed for serious hobbyists and are priced accordingly. For example, $150 buys a moderate quality two-function digital proportional transmitter, receiver, and two servos. The servos are used to power or steer the model. The range of an average priced RC kit is about one-quarter mile.

Pre-assembled kits are geared towards kids and playful adults who are more interested in the thrill of throttling a miniature Porsche 911 down the street than the art and science of radio controlled modeling. The cost for the toy

is between $20 to $75, depending mostly on the model, not the RC electronics. Most of the better toys are two-function digital proportional, and have built into them a drive motor and steering motor. Range is much more modest—a block or two at the most.

The advantage of buying an RC kit is that it is easier to experiment with it. The manufacturer provides detailed instructions on how to use it in your model, and you can just as easily use this information to apply the system to a robot. With an RC toy, you get no manufacturers instructions other than how to make the toy go down the street. Using it in your robot designs requires some tinkering, testing, and experimenting on your part.

Figure 30-18 shows a typical RC model race car that

Fig. 30-18. A radio control toy car, with body removed. The single circuit board contains the receiver and motor control circuitry. Note the four power transistors located at the bottom left corner of the board.

has the body removed. Clearly visible is the receiver and control circuit. This particular model has two-channel proportional speed control and steering. Proportional control lets you choose in-between speed and steering settings, rather than full-ahead, full-left, and full-right.

In the model car, the drive motor is controlled by four power transistors, obviously connected in the familiar H arrangement as discussed in Chapter 13, "Robot Locomotion with DC Motors." The transistors look as if they can handle up to one amp, which is fine for a small turtle robot. By examining the circuit board traces on the other side, it's possible to backtrack and find the input to the transistors. Once located, new and more powerful transistors could be added, in a similar arrangement, to provide power to even larger motors.

The servo mechanism for the steering is a little more complex, because it has a feedback system. It can't be easily traded in for a larger motor. The servo could, however, be used as-is on a small Minibot-like robot with no modification.

DATA TRANSMISSION

The walkie-talkie conversions described above can be used for manual remote control as well as digital data transmission. Instead of sending a control signal as a discrete tone, you send a series of tones, timed at just the right intervals to create a serial data train. You can use wireless data transmission to connect two computers together, but the ideal use is linking your desktop computer to your robot.

The circuit in Fig. 30-19 shows how to connect the

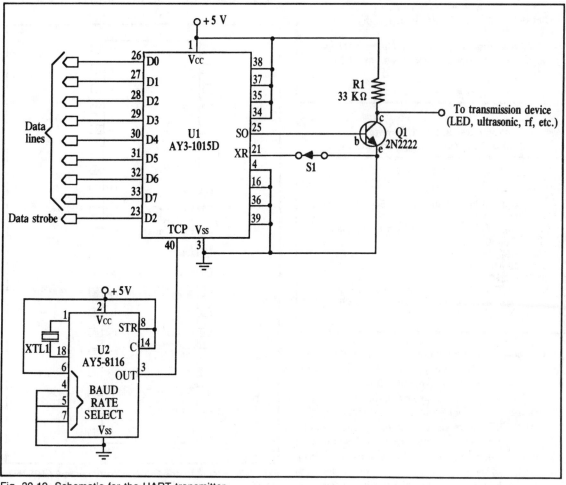

Fig. 30-19. Schematic for the UART transmitter.

269

Table 30-11. Tone Generator Parts List.

U1	4011 CMOS Quad NAND Gate IC
U2	4071 CMOS Quad OR Gate IC
R1,R3	1 megohm resistor
R2	100 K resistor
R4	200 K resistor
R5	470 ohm resistor
C1,C2	0.01 μF ceramic capacitor

Table 30-12. Dual Tone-Decoder Parts List.

U1,U2	NE567 Tone Decoder IC
R1,R3	50 K 15-turn precision potentiometers
R2,R4	10 K resistor
C1,C5	0.47 μF ceramic capacitor
C2,C6	2.2 μF tantalum capacitor
C3,C7	1.0 μF tantalum capacitor
C4,C8	0.022 μF ceramic capacitor

All resistors 5 or 10 percent tolerance, 1/4-watt; all capacitors 10 percent tolerance, rated 35 volts or higher.

Table 30-13. UART Remote Control Data Transmitter Parts List.

U1	General Instrument AY3-1015D (or equiv) UART IC
U2	General Instrument AY5-8116 (or equiv) Baud Rate IC
XTL1	3.59 MHz crystal
R1	33 K resistor
Q1	2N2222 npn transistor
S1	SPST momentary switch, normally closed

parallel port of your computer to a UART (Universal Asynchronous Receive/Transmit) chip. This chip converts parallel to serial and serial to parallel. It's much more involved than a shift register, which simply converts parallel data to pure serial form, or vice versa.

The UART allows you to send data to devices like printers, plotters, and modems, and yet be assured that all the information you are sending is getting there intact. Built into the chip are provisions for sending and receiving at the same time, for adding parity bits and stop

bits to the serial data train, and more. A pin-out diagram for the IC is shown in Fig. 30-20. Note that a number of other UARTs will work as well, and that these chips may even have the same pin-outs. The functions of the pins are listed in Table 30-14.

For all their sophistication, however, UART chips are surprisingly inexpensive—under $5 or $6. They require accurate timing, however, which requires the addition of a crystal and a baud rate generator (the generator can be replaced by other circuits, but in the long run, the generator is a better choice). With all the components added, a UART system costs about $15. You place one on your computer (or other controlling device) and one on the robot. We'll show the receiver in a bit.

The Transmitter

In the schematic, 8-bit parallel data from your com-

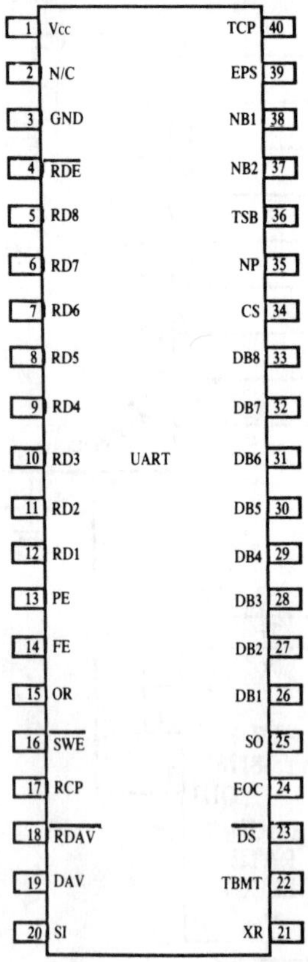

Fig. 30-20. Pinout for the General Instruments AY3-1015D UART (many other UART devices share the same pinout).

270

Table 30-14. UART Pin Functions.

Pin No.	Name	Function
1	Vcc	Power (+ Vdc).
2	N/C	No Connect.
3	GND	Ground.
4	RDE	Receive Data Enable. A LOW places the received data onto the output lines.
5-12	RD8-RD1	Received Data Bits. Data output lines. The least significant bit (LSB) appears on RD1. Tri-state outputs: TTL output when RDE is LOW, high-Z when RDE is HIGH.
13	PE	Parity Error. Goes HIGH if received data does not agree with selected parity. Tri-state.
14	FE	Framing Error. Goes HIGH if the received data has no valid stop bit. Stri-state.
15	OR	Over Run. Goes HIGH if previously received character is not read (DAV line not reset) before the present character is transferred to the receiver holding register. Tri-state.
16	SWE	Status Word Enable. Forcing LOW places the status word bits (PE, FE, OR, DAV, TBMT) onto the output lines. Tri-state.
17	RCP	Receiver Clock. Clock frequency 16X desired baud rate.
18	RDAV	Forcing LOW resets the DAV line.
19	DAV	Data Available. Goes HIGH when entire character has been received and is ready to be read.
20	SI	Serial Input. Incoming serial data stram. A mark (HIGH) to space (LOW) transition is required for initiation of data reception.
21	XR	External Reset. Forcing HIGH resets chip except for control bits.
22	TBMTR	Transmitter Buffer Empty. Goes HIGH when data bits holding register may be loaded with a new character.
23	DS	Data Strobe. Forcing LOW enters data bits on to be transmitted.
24	EOC	End of Character. Goes HIGH each time a full character is transmitted. Remains HIGH until start of next transmission.
25	SO	Serial Output. Serial output line. Stays HIGH (mark) when no data is being sent.
26-33	DB1-DB8	Data Bit inputs.
34	CS	Control Strobe. Forcing HIGH enters control bits (EPS, NB1, NB2, TSB, NP) into the control bits holding register). Can be strobed HIGH or tied HIGH.
35	NP	No Parity. Forcing HIGH eliminated parity bit from transmitted and received character (no PE indication). The stop bit(s) will immediately follow the last data bit. If not used, must be tied LOW.
36	TBS	Number of Stop Bits. Selects number of stop bits sent with each character, appended after parity bit. Forcing LOW inserts one stop bit; forcing HIGH inserts two stop bits.
37-38	NB2, NB1	Number of Bits/Character. Setting pins HIGH or LOW internally decodes number of data bits per character:

NB2	NB1	Bits per Character
0	0	5
0	1	6
1	0	7
1	1	8

Pin No.	Name	Function
39	EPS	Odd/Even Parity Select. HIGH or LOW determines type of parity, either odd or even, sent with each character. Forcing LOW inserts odd parity; forcing HIGH selects even parity.
40	TCP	Transmitter Clock. Clock frequency 16X desired baud rate.

Table 30-15. UART Tone Switch Parts List.

U1	4066 CMOS 4-Way Analog Switch

puter (like the parallel printer port) is routed to the data lines on the UART. When the computer is ready to send the byte, it pulses the STROBE line high (the line may be called DATA READY or something similar). The UART converts the data to serial format and sends it through the Serial Output (SO) pin. The speed of the data leaving the output is determined by the baud rate generator. The COM8116 dual baud rate generator sets the speed of the transmission and reception, and it is hooked up here to be rather slow—about 300 baud. This means that the UART sends serial data at the rate of roughly 300 bits per second (equivalent to 30 bytes per second).

You can connect any number of transmission devices to the output of the UART. One approach is to use a series of high output LEDs, as shown in Fig. 30-21. The receiver is equipped with an IR photodetector. If you could see infrared light, you'd see the LED flash on and off very rapidly as the data were sent. The on and off periods are equal to 0's and 1's, or spaces and marks as they are called in serial communications.

You could also use the one- or two-tone walkie-talkie described earlier. You may have to slow the UARTs down to accommodate the NE567 tone decoder. When used to decode low frequency tones, the 567 takes longer to react. Figure 30-22 shows the basic walkie-talkie transmitter setup. The 4066 CMOS analog switch is used in place of the code switch on the walkie-talkie.

A better way to transmit the data when using a walkie talkie or other rf source is by using FSK, which stands for frequency shift keying. Instead of sending tone and

Table 30-16. UART LED Output Parts List.

R1	330 Ω resistor
R2	15 Ω resistor
C1	50 μF electrolytic capacitor
D1,D2	1N914 signal diode
LED1-3	TIL906-1 high-output Light Emitting Diode

All resistors 5 or 10 percent tolerance, 1/4-watt; all capacitors 10 percent tolerance, rated 35 volts or higher.

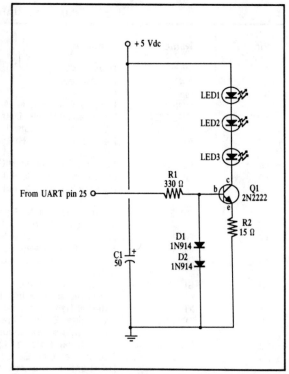

Fig. 30-21. An LED output stage for the UART transmitter.

no tone, you send two tones; one for digital 0's and one for digital 1's. The circuits in Figs. 30-16 and 30-17, show how to rig the transmit and receiving stations in FSK fashion.

The Receiver

The receiving UART is connected almost in reverse to the transmitting UART (see Fig. 30-23). The transmitter uses a baud rate generator that is operating at the same frequency as the receiver. The serial transmitted data is received and amplified if necessary.

Figure 30-24 shows a pair of amplifiers to use with an infrared phototransistor. Alternatively, you could use a receiving walkie talkie and 567 tone decoder, as described in the sections above, to receive data over the airwaves.

The amplified output is applied to the serial data pin on the UART. When an entire word is received, the UART places it on the parallel data output pins, then pulses the DATA AVAILABLE pin. In this circuit, a short time delay is used to automatically reset the UART so it processes the next word. The reconverted parallel data can now be used by your robot.

272

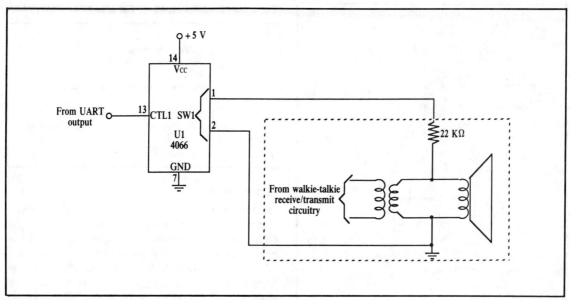

Fig. 30-22. A walkie-talkie tone switch for use with the UART transmitter.

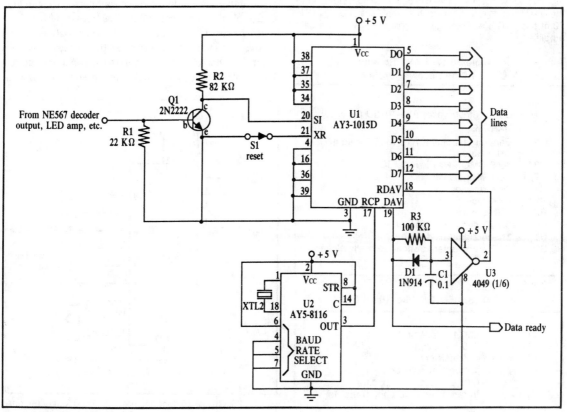

Fig. 30-23. Schematic for the UART receiver.

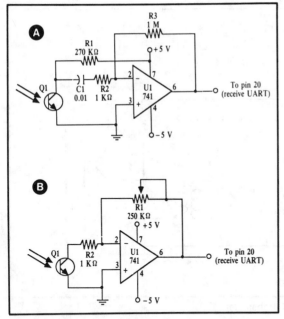

Fig. 30-24. Two LED amplifiers to boost the light signal from the UART transmitter. Connect the amplifiers directly to Pin 20 of the UART. A. Fixed gain amplifier; B. Variable gain amplifier.

Table 30-17. UART Remote Control Data Receiver Parts List.

U1	General Instrument AY3-1015D (or equiv) UART IC
U2	General Instrument AY5-8116 (or equiv) Baud Rate IC
U3	4069 CMOS Hex Inverter IC
XTL1	3.59 MHz crystal
R1	22 K resistor
R2	82 K resistor
R3	100 K resistor
C1	0.1 μF ceramic capacitor
D1	1N914 signal diode
S1	SPST momentary switch, normally closed

Table 30-18. LED Amplifier Parts List.

U1	LM741 Op Amp IC
R1	270 K resistor
R2	1 K resistor
R3	1 megohm resistor
C1	0.01 μF ceramic capacitor
Q1	Infrared sensitive phototransistor

Table 30-19. LED Amplifier, Variable Gain, Parts List.

U1	LM741 Op Amp IC
R1	250 K potentiometer
R2	1KΩ
Q1	Infrared sensitive phototransistor

All resistors 5 or 10 percent tolerance, 1/4-watt; all capacitors 10 percent tolerance, rated 35 volts or higher.

Applications

The most basic application of the UART receiver-transmitter is as a wireless remote control. With no encoding of the output pins on the receiving UART, you can directly control up to eight separate functions in your robot. For most robot designs, this is more than enough. Encode some or all of the lines, as depicted in the basic schematic in Fig. 30-25, and you can control many more (though not always at the same time).

Another application is to use your personal computer as the robot's control computer. This frees you from committing a separate computer just for the robot. To be effective, however, you should make the link two-way: the computer should be able to talk to the robot and the robot should be able to talk to the computer. You can, with

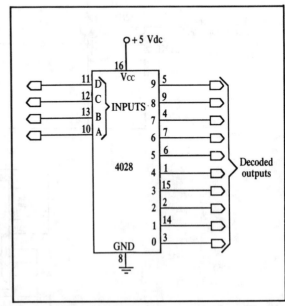

Fig. 30-25. One-of-10 decoded outputs, using four of the parallel output lines from the UART receiver.

some modification, convert the basic UART circuits above so that they receive and transmit, at the same time if necessary. The COM8116 baud rate generators let you select different speeds for each way, which can be a help. The transmission system for such a scheme must be one that operates at two different frequencies.

Chapter 31

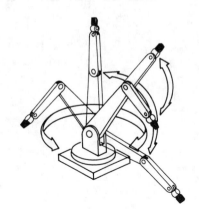

Computer Control Via Printer Port

In the "Wizard of Oz," the Scarecrow laments "If I only had a brain." He imagines the wondrous things he could do, and how important he'd be had he more than straw filling his noggin! In a way, your robot is just like the Scarecrow. Without a computer to control it, your robot can only be so smart. Hard-wiring functions into the robot is a suitable alternative to computer control, and you should always look to the simpler approaches than immediately connecting all of the parts of your robot to a super Cray-2 computer. Yet there are plenty of applications that cry out for computer control; some tasks like image and voice recognition require a computer. You can always build a computer from scratch for use in your robot, using various IC building blocks to make a reasonably smart brain for your automaton. There are numerous advantages to constructing your own computer for your robot, but a far easier and more immediately rewarding way is to use the personal computer that's sitting on your desk at home or in the office.

With little circuitry, you can use your personal computer to control your robot. If the computer is small enough, like the the old Timex/Sinclair 1000 or Commodore 64, you may even be able to install it temporarily on your robot.

Most personal computers come with, or have as an option, a parallel printer port. This interface port is intended primarily for connecting the computer to printers, plotters, and some other computer peripherals. With a few ICs and some rudimentary programming, it can also be used to directly control your robots. If your computer has several parallel ports, you can use them together to make a very sophisticated control system.

Despite its many advantages, using the computer's parallel port has some disadvantages, too. You are limited to controlling only a handful of functions on your robot, although with some creative design work, you can effectively increase that number without giving up too much control. Also, most parallel ports are designed primarily to get data out of the computer, not into it (one notable exception is the Commodore 64). Although the average parallel port does have input lines, their number is small.

If your robot must speak to the computer heavily, you should probably bypass the parallel port and go right for the microprocessor system bus. Some computers, like the IBM PC, Apple II, and TRS-80 Model 100 (the lapheld portable) provide easy access to the bus.

This chapter deals primarily with using the parallel port on an IBM PC or compatible. Why the IBM PC? It's a common computer, often found in businesses and homes, and provides a great deal of flexibility. And, as you'll read in Chapter 33, "Adding an On-Board Computer," ready-made IBM PC compatible *motherboards* are available that you can use as easily as slipping the board into your robot.

If you don't have an IBM PC or compatible, you're still in luck, because the information presented here also applies to other computers. If you have a different computer, read this chapter, then refer to the technical reference manual for your machine to learn how to make the designs work for you.

THE FUNDAMENTAL APPROACH

In the IBM PC, expansion input/output (I/O) boards, like a parallel printer or RS-232 serial board, are plugged into the main motherboard. Slots in the motherboard accept these auxiliary cards and you are relatively free to put any type of card in any slot (see Chapter 32, "Build a Robot Interface Card," for more details on this subject).

The PC addresses, or accesses, its various I/O ports by using a unique address code, as shown in Table 31-1. Each device or board in the computer has an address unique to itself, just as you have a home address that no one else in the world shares with you.

The parallel port that comes on the monochrome display adapter has a starting address of 956. This address is in decimal, or base-10 numbering form; some programming languages require that the address be given in hexadecimal, or base-16, form. In hex, the starting address is 3BCH (the address is really 3BC; the H means that the number is in hex). The parallel port contained on an I/O expansion board, like a multifuction card, has a decimal address of 888 (or 378H hex) or 632 (278H). Usually, you specify the address of the port when you install the board.

Parallel ports in the IBM PC are given the *logical* names LPT1:, LPT2:, and LPT3:. Every time the system is powered up or reset, the ROM BIOS (Basic Input/Output System) chip on the computer motherboard automatically looks for parallel ports at these I/O addresses, 3BCH, 378H, and 278H, in that order (it skips 3BCH if you don't have a monochrome card/printer port installed). The logical names are assigned to these ports as they are found.

Table 31-2 is a chart that shows the port addresses

Table 31-1. PC Input/Output Map.

I/O Address Map											
HEX RANGE	9	8	7	6	5	4	3	2	1	0	Device
00-0F	0	0	0	0	0	Z	A3	A2	A1	A0	DMA CHIP 8237-2
20-21	0	0	0	0	1	Z	Z	Z	Z	A0	INTERRUPT 8259A
40-43	0	0	0	1	0	Z	Z	Z	A1	A0	TIMER 8253-5
60-63	0	0	0	1	1	Z	Z	Z	A1	A0	PPI 8255A-5
80-63	0	0	1	0	0	Z	Z	Z	A1	A0	DMA PAGE REGS
AX	0	0	1	0	1						NMA MASK REG
CX	0	0	1	1	0						RESERVED
EX	0	0	1	1	1						RESERVED
3F8-3FF	1	1	1	1	1	1	1	A2	A1	A0	TP RS-232C CD
3F0-3F7	1	1	1	1	1	1	0	A2	A1	A0	5 1/4-INCH DRV ADAPTER
2F8-2FF	1	0	1	1	1	1	1	A2	A1	A0	RESERVED
378-37F	1	1	0	1	1	1	1	Z	A1	A0	PARALLEL PRINTER PORT
3D0-3DF	1	1	1	1	0	1	A3	A2	A1	A0	COLOR/GRAPHICS CARD
278-27F	1	0	0	1	1	1	1	Z	A1	A0	RESERVED
200-20F	1	0	0	0	0	0	A3	A2	A1	A0	GAME I/O ADAPTER
3B0-3BF	1	1	1	0	1	1	A3	A2	A1	A0	MONO CARD/PAR. ADAPTER

Notes: Z = Don't care; not necessary for decoding

Table 31-2. Parallel Port Addresses.

Adaptor	Data Output	Status	Control
Mono Card (PC,XT,AT)	3BCH, 956D	3BDH, 957D	3BEH, 958D
PC/XT Printer Adapt.	378H, 888D	379H, 889D	37AH, 890D
PC. jr. Printer Adapt.			
AT S/P card (as LPT1:)			
AT S/P card (as LPT2:)	278H, 632D	279H, 633D	27AH, 634D

"H" Suffix = Hex
"D" Suffix = Decimal

for all the possible parallel ports in the IBM PC and AT. The logical port names are often used by applications software instead of the actual addresses. In robotics, there is more flexibility using the addresses, so we'll stick with them.

The parallel port on the IBM PC is a 25-pin connector, often referred to as a DB-25 connector. Cables and mating connectors are in abundant supply, making it easy for you to wire up your own peripherals. You can buy connectors that crimp onto 25-conductor ribbon cable, or connectors that are designed for direct soldering.

Figure 31-1 shows the pin-out designations for the connector (shown with the end of the connector facing you). Note that only a little more than half of the pins are in use. The others are not connected inside the computer, or are grounded to the chassis. Table 31-3 shows the meaning of the pins.

Take a closer look at the chart in Table 31-2, above. Notice that not one address is given, but three. The so-called starting address is used for *data output*. The data output is comprised of eight binary weighted bits, something on the order of 01101000, as shown in Fig. 31-2. There are 256 possible combinations of the eight bits. In a printer application, this means that the computer can

send out the specific code for up to 256 different characters. The data output pins are numbered 2 through 9. The bit positions and their weights are shown in Table 31-4.

The other two addresses are for *status* and *control*, some of which are shown in Fig. 31-3. The function of the status and control bits is shown in Table 31-5.

The most important control pin is pin number 1. This is the STROBE line, used to tell the peripheral (printer, robot) that the parallel data on lines 2 through 9 is ready to be read. The STROBE line is used because all the data may not arrive at their outputs at the same time. It is also used to signal a change in state. The output lines are latched, meaning that whatever data you place on them stays there until you change it or turn off the computer. During printing, the STROBE line toggles HIGH to LOW and then HIGH again. You don't have to use the STROBE line when commanding your robot, but it's a good idea if you do.

Table 31-3. Parallel Port Pinout Functions.

Pin #	Function (Printer Application)
1	STROBE
2	DATA BIT 0
3	DATA BIT 1
4	DATA BIT 2
5	DATA BIT 3
6	DATA BIT 4
7	DATA BIT 5
8	DATA BIT 6
9	DATA BIT 7
10	ACKNOWLEDGE
11	BUSY
12	PE (OUT OF PAPER)
13	PRINTER ON LINE
14	AUTO LINEFEED AFTER CARRIAGE RETURN
15	PRINTER ERROR
16	INITIALIZE PRINTER
17	SELECT/DESELECT PRINTER
18-25	UNUSED OR GROUNDED

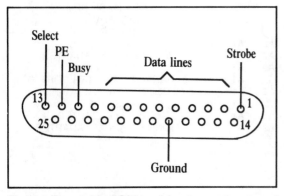

Fig. 31-1. Pinout of the DB25 parallel port connector, as used on the IBM PC and most compatibles.

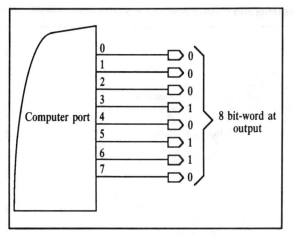

Fig. 31-2. The parallel port outputs eight bits at a time.

Other control lines:

- Auto form-feed (not always implemented).
- Select/deselect printer.
- Initialize printer (not always implemented).
- Printer interrupt (not always implemented).

The status lines are the only ones that feed back into the computer. There are five status lines, but not all parallel ports support every one. They are:

- Printer error (not always implemented).
- Printer not selected.
- Paper error (not always implemented).
- Acknowledge.
- Busy.

The acknowledge and busy lines are really the same as far as printers are concerned, but you can use the two separately in your own programs (one helpful tidbit: when the BUSY line is LOW, the ACK line is HIGH). Later

Bit Position		Weight
D7	=	128
D6	=	64
D5	=	32
D4	=	16
D3	=	8
D2	=	4
D1	=	2
D0	=	1

**Table 31-4.
Bit Position Weights.**

in this chapter we will look at creative ways to use the control and status lines for purposes other than their intended use.

EXPERIMENTER'S INTERFACE

The experimenter's interface, shown in Fig. 31-4, can be built in a matter of minutes and requires very few components. The interface uses a solderless experimenters breadboard so you can create circuits right on the interface. The input and output buffering is provided by the 74367 hex buffer driver (buffering is a good idea to protect the parallel port; it can be damaged through careless use). Three such chips are used to provide 18 buffered lines, which is more than enough.

Build the interface in a suitable enclosure that is large enough to hold the breadboard and the wire-wrapping socket. Make a cable using a male DB25 connector and a four or five foot length of 25-connector ribbon cable. Solder the data output, status, and control line conductors to the proper pins of the 74367 ICs, as shown in the schematic. Route the outputs to the bottom of the wire-wrap socket. A finished interface should look something like the one in Fig. 31-5.

Using the interface requires you to provide a +5Vdc source. Use a length of 22 AWG solid conductor wire to connect the signals at the wire-wrap socket to whatever points on the breadboard you desire.

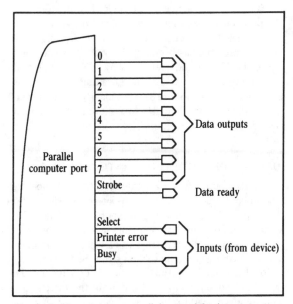

Fig. 31-3. The minimum parallel port: eight data outputs, a STROBE (Data Ready) line, and inputs from the printer or plotter, including Select, Printer Error, and Busy.

279

Table 31-5. Parallel Port Status and Control Bits.

Control Bits	
Bit	**Function**
0	LOW = normal; HIGH = output of byte of data
1	LOW = normal; HIGH = auto linefeed after carriage return
2	LOW = initialize printer; HIGH = normal
3	LOW = deselect printer; HIGH = select printer
4	LOW = printer interrupt disabled; HIGH = enabled
5-7	Unused
Status Bits	
Bit	**Function**
0-2	Unused
3	LOW = printer error; HIGH = no error
4	LOW = printer not on-line; HIGH = printer on-line
5	LOW = printer has paper; HIGH = out of paper
6	LOW = printer acknowledges data sent; HIGH = normal
7	LOW = printer busy; HIGH = printer ready

The first order of business is to connect the interface as shown in Fig. 31-6. Connect the cable to the parallel port of your computer (some of the LEDs will light) and start BASIC (GW-BASIC or similar for clones). Type the program shown in Fig. 31-7. The program is written assuming that the LPT1: port in your computer is on an I/O board. If it is on a monochrome display adapter, see the table above to convert the addresses.

The LED connected to the STROBE line (pin 1 of the port connector), should flash on and off very rapidly a few times. If it does not, recheck your wiring and make sure the program has been typed correctly. Some other LEDs should flash and a few will stay on. The LEDs that are on represent a logic 1 state; those that are off represent a logic 0 state. Figure 31-8 is a blank dotted-line version of the interface. Feel free to use it to sketch out your own designs.

USING THE PORT TO OPERATE A ROBOT

You can use the experimenter's interface to operate your robot or you can build another interface, using 74367 ICs especially for connecting to your robot. The 74367 cannot sink or source very much current, so you can't operate a motor directly from it. However, it can drive a low-power relay or a power transistor. See Chapter 13, "Robot Locomotion with DC Motors," for some popular ways to bridge the low-level output of the interface to control a real-world robot.

The simplest way to operate your robot via computer is to connect each of the data output lines to a suitable transistor or relay. You can control up to eight motors or other devices in this fashion. Let's say that you only have three motors connected to the interface, and that you are using lines 0, 1, and 2 (pins 2, 3, and 4, respectively). To turn on motor #1, you must activate the bit for line 0; that is, make it HIGH. To do this, output a *bit pattern* number to the port using the BASIC OUT command. You'll see how to do this in a bit.

The OUT command is a lot like the BASIC POKE command, but it is used to send data to an I/O port, not to memory. The OUT command consists of just two variables: port address and value. The two are separated by a comma. To send data to the data output, type:

OUT 888, x

In place of x you put the decimal value of the binary bit pattern. Table 31-7 shows the first ten binary numbers, and the bit pattern that makes them up.

Normally, you initialize the port at the beginning of the program by outputting decimal 0, or all logic 0's. The line of code for this is:

OUT 888, 0

You'll note that in every other decimal number, bit 0 (on the right hand side) changes from 0 to 1. To activate just the #1 motor, choose a decimal number where only the first bit changes. There is only one number that meets the criteria: it is decimal 1, or 00000001. So type:

OUT 888,1

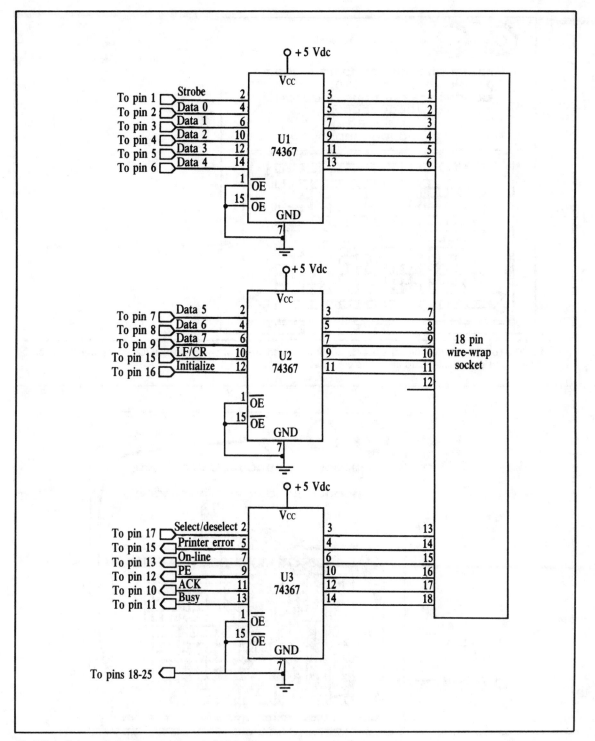

Fig. 31-4. Schematic for the experimenter's interface.

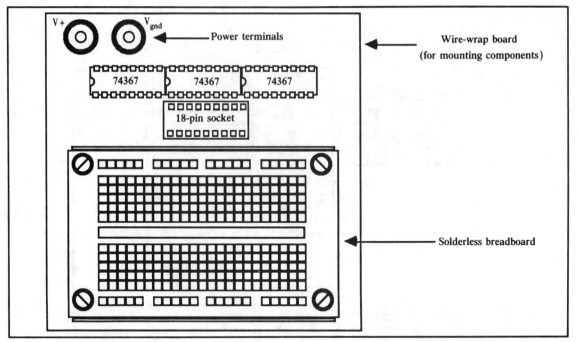

Fig. 31-5. The completed experimenter's board. Mount the breadboard, wire-wrap socket, ICs, and power terminals on a perf board, and secure the board into a project case.

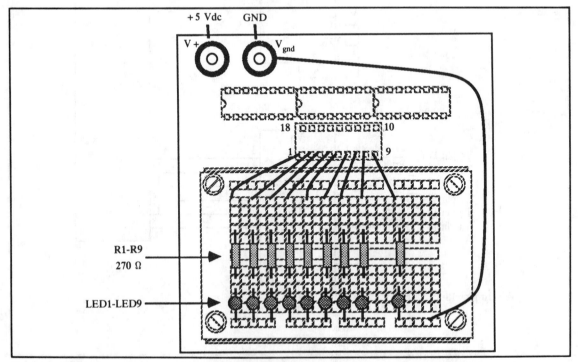

Fig. 31-6. Component arrangement for testing the experimenter's interface.

```
10      READ X
20      IF X = 100 THEN END
40      OUT 888,X
41      OUT 890,127:OUT 890,128
42      FOR D = 1 TO 100: NEXT D
50      GOTO 10
60      DATA 1,2,3,4,5,6,
70      DATA 12,13,14,15,16,17
80      DATA 100,101,102,103,104
90      DATA 200,210,221,232,244,255
```

Fig. 31-7. Sample program for testing the experimenter's interface. To make the test go slower, choose a number over 100 in line 42.

The motor turns on. To turn it off, send a decimal 0 to the port. Now, how might you turn on both motor #1 and #2? Look for the binary bit pattern where the first and second bits are 1 (it's decimal 3), and output it to the port.

Table 31-6. Experimenter's Board Parts List.

U1-U3   74367 TTL Hex Inverter/Buffer IC
Misc   18-pin Wire-Wrap socket, solderless
        experimenter board, binding posts
        (for power connection), enclosure

Controlling More Than Eight Devices

One parallel port has only eight data output lines, but you can control more than eight robotic devices in either of three ways (a fourth way is to use a second parallel port, but that won't be covered here).

The most straightforward method is to make use of some or all of the control lines. You send bits to the control lines in exactly the same way as you send bits to the data output lines, except that you use a different address. For LPT1: (expansion board), the decimal address is 890. Only the first five bits of the address are used in the port, which means the decimal numbers you use will be between 0 and 31. You should consider bit 0 reserved, be-

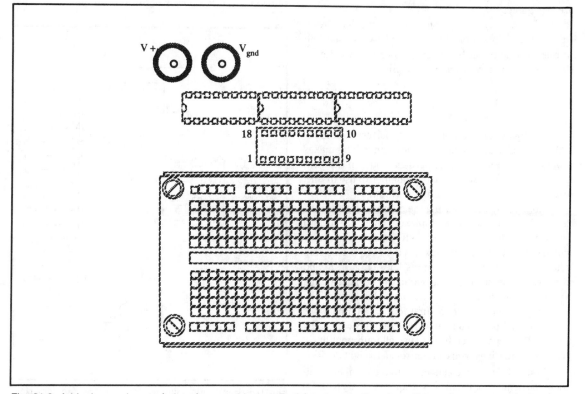

Fig. 31-8. A blank experimenter's interface card layout. Feel free to copy it and use it to make your own designs.

Table 31-7. Binary and Decimal Equivalents.

Decimal	Binary	BCD
0	0000	0000 0000
1	0001	0000 0001
2	0010	0000 0010
3	0011	0000 0011
4	0100	0000 0100
5	0101	0000 0101
6	0110	0000 0110
7	0111	0000 0111
8	1000	0000 1000
9	1001	0000 1001
10	1010	0001 0000
11	1011	0001 0001
12	1100	0001 0010
13	1101	0001 0011
14	1110	0001 0100
15	1111	0001 0101

cause this bit controls the STROBE line. You'll probably want to save its use for special applications, such as voice synthesis, as discussed in Chapter 23. Bits 1 through 4 are fair game.

Let's say you are using bit 2 of the control address. In a printer application, bit 2 is used to initialize the printer. You turn that bit on (and no others) by entering the following program line into BASIC:

OUT 890, 4

Note that you can output a binary pattern to address 890 and it will not affect the data output lines. This makes programming a little easier, because you don't always have to think about what decimal number to use to turn drive motor #1 off and arm motor #3 on.

Another way of increasing the number of controlled devices is to use a data demultiplexer. There are several types in both the TTL and CMOS IC families. A popular data demultiplexer (or "demux") is the 74154. This chip takes four binary weighted inputs lines (1, 2, 4, 8) and provides 16 outputs. Only one output can be on at a time. See the schematic in Fig. 31-9 on how to hook it up. The IC is shown connected to the first four data output lines of the parallel port. You can actually connect it to any four, and you don't even have to use all four input lines. With just three lines, you can control up to eight devices.

To select the device connected to the number 3 output of the demux, for example, you apply a binary 3 (0011) to its input lines. Write the line as follows:

OUT 888, 3

A limitation of the demux is that you can't control more than one device connected to it at the same time. You can't, for example, attach both drive motors to the demux outputs and have them on at the same time. There will be many times, when your robot will only be doing one thing (such as triggering an ultrasonic ranger), and in these cases, the demux is perfect.

External Addressing

As mentioned earlier, inside the computer, on the microprocessor bus, are all sorts of data and control lines. There is also a set of special purpose lines, the address lines, that are used to pass data to specific devices and expansion boards. The data output lines of the parallel port are addressed by sending out the address 888.

The address for the parallel port triggers just the parallel port, but with some ingenuity (and no extra components) you can wire up a sub-address scheme so that the one parallel port can fully control a very large number of devices. This is the third and most sophisticated way of sapping all the power out of the parallel port.

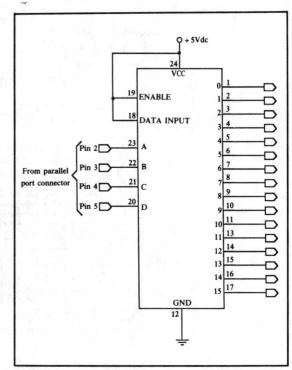

Fig. 31-9. Basic wiring diagram for the 74154 demultiplexer chip.

The 74367 hex buffer IC, used to link the port to the outside world, can be disabled. In the experimenter's interface, the ENABLE lines of the chip, pins 1 and 15, are held low, so data is passed from the input to the output. When the ENABLE pins are brought HIGH, the outputs are driven to a high-impedance state, and no longer pass digital data. In this way, the 74367 acts as a kind of valve. The two ENABLE lines control different input/output pairs, as shown in Fig. 31-10. The high impedance disabled state is engineered so that many 74367 chips can be paralleled on the same data lines, without loading the rest of the circuit.

You can use the ENABLE pins of the 74367 and a few of the unused control lines in the parallel port to make yourself an electronic data selector switch. In operation, you output a binary word onto the data output lines. You then send the word to the desired device by addressing it with the control lines.

Here is an example: Let's say that you have connected three sub-address ports to the parallel printer port, as shown in Fig. 31-11. Control lines 1, 2, and 3 are connected to the ENABLE pins of the 74367. The inputs of the three 74367s are connected together. The outputs of each feeds to the specific device.

To turn on bits 0 and 1 on device #2, enter the following lines into BASIC and run the program:

```
10    OUT 888, 3
20    OUT 890, 2
30 .  END
```

Program line 10 outputs a decimal 3 to the data output lines. That places the binary bit pattern 00000011 on the parallel port data output lines. Program line 20 enables device #2, because it turns on the second 74367.

Inputting Data

The five status lines (may be less with some parallel ports) can be used to send data back into the computer. To read data from the port, you use the BASIC command statement INP (for input). The input command is used as follows:

$$Y = INP(x)$$

In place of x you put the decimal address of the port you want to read. In the case of the LPT1: expansion board parallel port, that address is 889. The Y is a variable and is used to store the value for future use in the program. Most often, you want to PRINT the value of Y. When printed, the decimal equivalent of the binary bit pattern is shown on the screen.

If you built the speech synthesizer from Chapter 23, you used Y to test for a certain condition. In the case of the speech synthesizer, that condition was the status of pin 11, the BUSY line.

Normally, the decimal equivalent of the binary pattern of the status lines, when connected to the speech synthesizer, is 143 (when a printer is connected the bit pattern

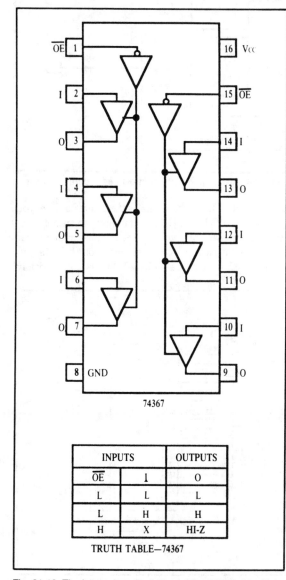

74367

INPUTS		OUTPUTS
$\overline{OE}$	I	O
L	L	L
L	H	H
H	X	HI-Z

TRUTH TABLE—74367

Fig. 31-10. The internal configuration of the 74367 chip. Note the two independent enable lines, on pins 1 and 15.

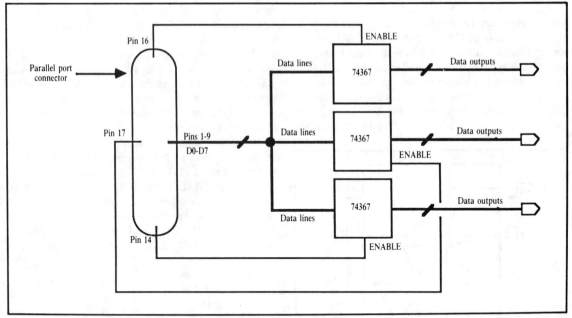

Fig. 31-11. Block diagram for a selectable parallel port, using three 74367 ICs to independently control three separate devices.

is usually 223). Decimal 143 means that the BUSY line is LOW, and that the speech chip is ready to take data. At that point, and that point only, the program sends another byte of speech data. Even if you didn't build the speech synthesizer, you should study the simple program provided for it. It is a good example of how to use the status lines to report a condition external to the computer.

There are a maximum of five status lines, which means that you can't use the parallel port for applications such as analog-to-digital (A/D) conversion. An A/D converter translates analog information into eight-bit (or more) bytes. You can use the A/D converter with the port, of course, but you'll get only the first 32 values.

You can, however, use the status bits for the robot's various sensors, like whiskers, line tracing detectors, heat and flame detectors, and so forth. The simple on/off nature of these sensors make them ideal for the parallel port. Normally, you can have up to five sensors attached to the computer, but by using the ENABLE pins of the buffers, as you've seen before, it is possible to select the input from a wide number of sensors. For example, using just four control lines with a 74150 data selector, means that up to 16 sensors can be routed to the parallel port. See Fig. 31-12 for a simplified schematic.

OTHER SOFTWARE APPROACHES

There are plenty of other ways to get data to the

printer port. If LPT1: is set up as the printer device, you can use the LPRINT CHR$(x) statement to output data. As usual, in place of x you put the decimal number that equals the bit pattern you want placed on the data output lines. Note that the STROBE line is automatically

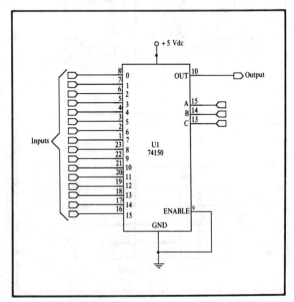

Fig. 31-12. Basic wiring diagram of the 74150 multiplexer chip.

activated for each byte and that you don't have direct control of the status and control lines of the port.

Also, the LPRINT command ends each line with a carriage return and line feed, which may be interpreted by your robot as something like "kill master" or "wreck living room." You can prevent carriage returns and line feeds from occurring by placing a semicolon after each LPRINT statement line. When the program ends, however, BASIC sends out a carriage return and line feed.

Another approach is to use the OPEN command, as in:

OPEN "LPT1" as #1

You can then use the PRINT# command statement. Again, a carriage return/line feed pair is sent when using PRINT#, but you can suppress it by ending each PRINT# statement line with a semicolon.

Chapter 32

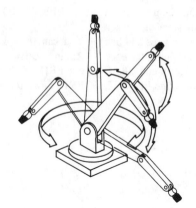

Build a Robot Interface Card

The previous chapter detailed how to use the parallel printer port of a computer—specifically the IBM PC or compatible—to control a robot. While a parallel port is ideally suited for many robotic applications, and is easy to use, it is not perfect for them all. For greater flexibility and function, you must tap into the microprocessor system bus of your personal computer. Once connected directly to the microprocessor, there is practically no limit to what your robot can be made to do.

This chapter is devoted to the theory, design, and construction of a prototyping card designed to fit into the IBM PC. The card will also work in IBM PC compatible computers that use the standard PC bus structure. A number of other computers allow direct connection to the system bus, including the Apple *II*e and the Radio Shack Model 100 laptop portable. If you own one of these other computers, you can apply the techniques covered in this chapter to build an interface card designed for your machine.

The IBM PC was chosen over the less expensive computers like the Commodore 64 and Apple *II*e partly because of its immense popularity and partly because you can easily buy all the component parts to make your own PC clone. The make-it-yourself approach is ideally suited for robotics, since you can purchase a ready-made com-

puter motherboard and begin to use it almost immediately in your robot. Chapter 33, "Adding an On-Board Computer," goes into more detail on using an IBM PC clone motherboard in your robot.

A word of caution before going on. In previous chapters, you have built mechanisms and circuits that, if something didn't work or you loused up, it was of no dire consequence—that is, other than upsetting your day and possibly making you start all over again. Until now, you haven't run the risk of damaging something valuable to you. Although there is really little that can go wrong when building and using the robot interface card, you should be aware that there is a possibility that your computer could be damaged if you hook up something wrong or try to drive a two horsepower motor from it.

If you feel the least bit unsure of your electronic expertise, pass up this project for a little while until you gain more experience. Double—no!—triple-check your work for errors, and never hook up anything to the interface card before you've sat down and analyzed what effect it will have on your robot, the interface card itself, and most importantly, the computer. Be sure that you build the circuit as shown, and don't try any shortcuts unless you know what you are doing. The interface card provides protective buffering, to limit the chances of damage to the computer by something connected through the card. Enough said. Let's get on with it.

THE IBM SYSTEM BUS

The IBM PC (or compatible) motherboard contains five or more 62 pin expansion slots. Each of the pins is connected to a vital signal line generated by the various electronics on the motherboard. These signals include the data output/input lines, address lines for accessing memory and ports, control lines to manipulate circuits and devices, and interrupt lines (the latter are used to immediately grab the attention of the Intel 8088 microprocessor in the computer). Figure 32-1 shows one slot and the func-

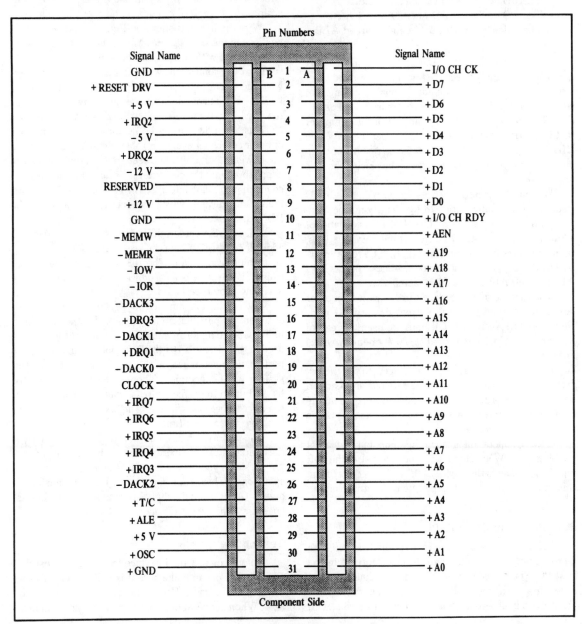

Signal Name	Pin Numbers	Signal Name
GND	B 1 A	− I/O CH CK
+ RESET DRV	2	+ D7
+5 V	3	+ D6
+ IRQ2	4	+ D5
− 5 V	5	+ D4
+ DRQ2	6	+ D3
− 12 V	7	+ D2
RESERVED	8	+ D1
+ 12 V	9	+ D0
GND	10	+ I/O CH RDY
− MEMW	11	+ AEN
− MEMR	12	+ A19
− IOW	13	+ A18
− IOR	14	+ A17
− DACK3	15	+ A16
+ DRQ3	16	+ A15
− DACK1	17	+ A14
+ DRQ1	18	+ A13
− DACK0	19	+ A12
CLOCK	20	+ A11
+ IRQ7	21	+ A10
+ IRQ6	22	+ A9
+ IRQ5	23	+ A8
+ IRQ4	24	+ A7
+ IRQ3	25	+ A6
− DACK2	26	+ A5
+ T/C	27	+ A4
+ ALE	28	+ A3
+5 V	29	+ A2
+ OSC	30	+ A1
+ GND	31	+ A0

Component Side

Fig. 32-1. Pinout of the internal expansion slot inside the IBM PC.

tions of each pin. Note that the slot is oriented so that the front of the motherboard is nearest you.

Of all the pins on the slot, we are interested in just a handful of them. Specifically, these are:

- The data output lines, labeled D0 through D7.
- The first ten address lines, labeled A0 through A9.
- The I/O device WRITE line, marked IOW.
- The I/O device READ line, marked IOR.
- The Address Latch Enable line, marked ALE.

The function of the data output lines is, as you may guess, to pass data between the microprocessor and other circuits. The data lines are bi-directional, meaning that signals can flow into the microprocessor as well as out of it. The microprocessor tells the rest of the computer that it is writing (sending) data by placing the IOW line HIGH. Likewise, the processor tells the rest of the computer that it is reading (receiving) data by placing the IOR line HIGH.

The microprocessor chooses which memory or I/O location to send information to by identifying the location with a specific address. The IBM PC is capable of addressing up to 1,048,576 different memory locations, which requires 20 address lines (20^2). The robot interface card uses just the first ten address lines; address lines A10 through A19 are for high memory.

Note that the robot interface card, as with most other devices other than memory, is treated as an I/O port. That's why the IOW and IOR lines are used, instead of the MEMR (memory read) and MEMW (memory write) lines.

The ALE line is used as final trigger, to tell the addressed device that the bits on the data bus are ready to be used. If the ALE line is never pulsed, the data on the data bus is never used by the device. It's like mail that's never retrieved from the mailbox.

A READ operation is much the same, but instead of strobing the IOW line, the microprocessor pulses the IOR line. This prevents the device at the addressed port from thinking its supposed to accept data rather than send it.

BUILDING THE INTERFACE

Building the interface card requires that you purchase a prototyping wire-wrap board designed for use in the IBM PC. You should be warned right now that the board is not cheap. The lowest I've seen a full-length (13-inch) PC prototyping board is about $30. The better ones, with gold contacts, cost $50 to $75.

Use wire-wrapping techniques instead of point-to-point soldering. You'll probably want to make changes as you go; wire-wrapping joints can be readily altered if need be. Bear in mind that wire-wrapping sockets and terminals extend a full half-inch beyond the back surface of the board, so leave extra space in your computer between slots. All hell will break loose if any of the wire-wrap posts touch the circuits on adjacent boards.

Wire the board as shown in Fig. 32-2. Note the connections to the IBM PC bus. Match the lands on the prototype board with the pins in the expansion socket. Be sure that you connect the wires to the proper lands; otherwise the circuit won't work or it may be damaged.

The circuits on the interface card draws power from the IBM PC, through the GND and +5 Vdc pins. DO NOT use the power from the PC for running the motors or other components in your robot. Small circuits can tap the power supply rails, but keep in mind that the power supply built into the computer is designed to supply only so much juice. The interface card should not draw more than about one amp, which is about the same as any other I/O board.

The address gate in the figure is shown for decimal address 703 (2BFH). You can use another address, as long as it isn't already taken. The 703 address was chosen because it is the same one used by a couple of commercially available prototype boards.

The hookup diagrammed above is intended to be used as an output port, not an input port. Use the schematic in Fig. 32-3 to make an input port. Note that the address used in this circuit is 701 (2BDH). Both input and output ports can be combined on the same board, sharing the system bus connections. Though the two circuits are hooked up to the same address, data, and control lines, the microprocessor will see to it that only one of them is active at any one time.

The prototype board has plenty of space for more of these circuits. After experimenting with the input and output versions, you may want to add more. Reserve some space on the board for support circuits, like analog-to-digital (A/D) circuits, digital-to-analog (D/A) circuits, speech synthesizers, and more.

Testing the Interface

Do not connect the interface to your robot just yet. First, test it out to make sure it is doing everything its supposed to be doing. Use the tester circuit shown in Fig. 32-4. When in operation, the LEDs will light up, indicat-

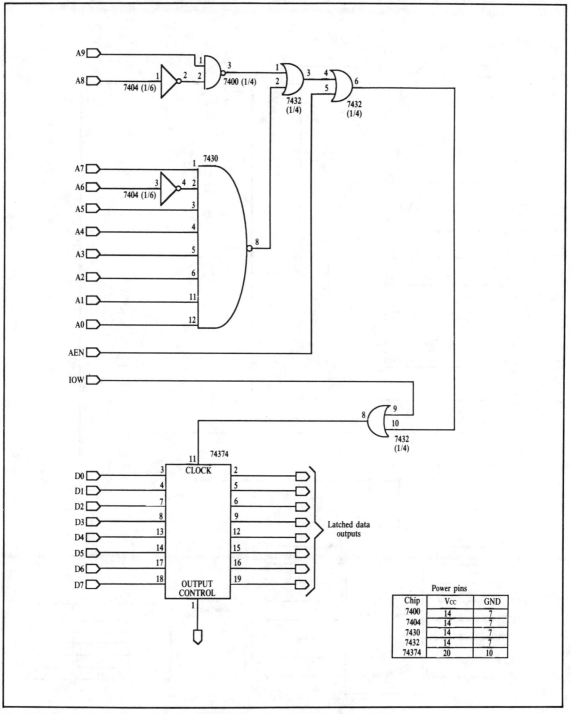

Fig. 32-2. Schematic for the output circuit. As shown, the address is decimal 703.

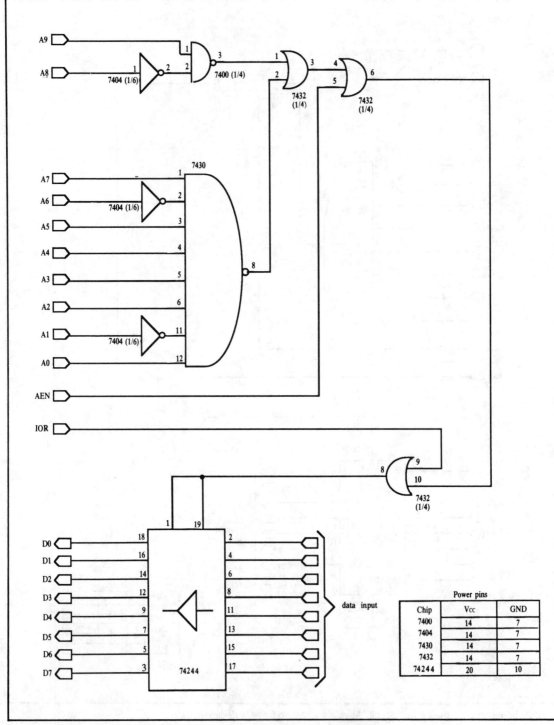

Fig. 32-3. Schematic for the input circuit. As shown, the address is decimal 701.

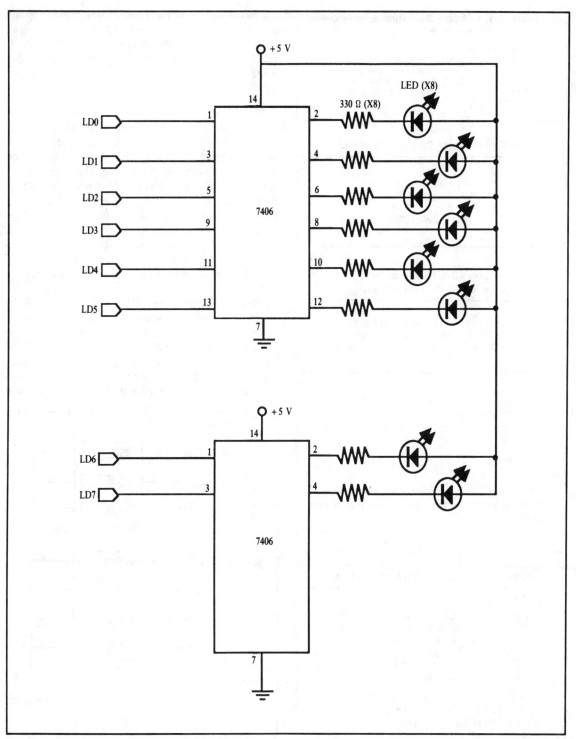

Fig. 32-4. Output test circuit.

Table 32-1. Interface Card Output Parts List.

U1	7404 Hex Inverter IC
U2	7432 Quad OR Gate IC
U3	7430 8-In NAND Gate IC
U4	74374 Octal "D" Flip-Flop

Table 32-2. Interface Card Input Parts List.

U1	7404 Hex Inverter IC
U2	7432 Quad OR Gate IC
U3	7430 8-In NAND Gate IC
U4	74244 Octal Buffer/Line Driver IC

Table 32-3. Interface Card Output Test Circuit Parts List.

U1,U2	7406 Hex Inverter IC
R1-R8	330 ohm resistors
LED1-8	Light Emitting Diode

Table 32-4. Interface Card Input Test Circuit Parts List.

S1	8-position DIP switch

All resistors 5 or 10 percent tolerance, 1/4-watt; all capacitors 10 percent tolerance, rated 35 volts or higher.

Table 32-5. Crystal Oscillator Parts List.

U1	4049 Hex Inverter IC
U2	7490 Counter IC
R1	10 megohm resistor
R2	1 K resistor
C1,C2	4.7 pF ceramic capacitor
XTAL1	3.578 MHz crystal

All resistors 5 or 10 percent tolerance, 1/4-watt; all capacitors 10 percent tolerance.

ing HIGH and LOW logic levels. While testing, you'll need to keep the cover of your computer off, of course, so you can see the LEDs. The LEDs are used by the output circuit. Use the DIP switch blocks, shown in Fig. 32-5, when testing the input circuit.

Software

You can use just about any programming language that allows direct access to memory and ports. BASIC is a good choice, because it comes with the IBM PC and most clones (or is available for them). The version of BASIC used with most clones is Microsoft GW-BASIC. There are slight differences between IBM BASIC and GW-BASIC, but the example program lines provided here are written to work with both.

If you're really into programming, Assembly Language is an excellent choice for use with the interface card. You can do many things in Assembly that you can't in BASIC. Perhaps more importantly, you can make the computer do it much faster. IBM BASIC and GW-BASIC are rather slow languages, because they do not produce compiled code. However, you'll find that for most robot applications, BASIC will do nicely.

When you program the interface card, most of your efforts will be geared towards controlling some device on your robot. One or more of the data output lines will

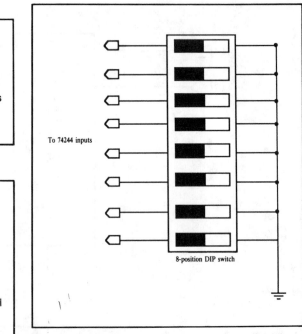

To 74244 inputs

8-position DIP switch

Fig. 32-5. Input test circuit.

be used to control the various devices in your automaton. You program this control by using bit patterns, numbers between 0 and 255 that when sent through the data lines, is seen as an eight bit word, such as 10011010. Each bit controls something in the robot. Chapter 31 went into some detail on exactly how to do this with BASIC, so it won't be repeated here. If this concept is new to you, read that chapter before going on.

Now, insert the board into a free slot in your computer and start the machine. The computer will do a self-test and if it locates a problem in your board, it may either give you an error message or it may not even start. If the machine does not act properly, immediately turn it off and check your circuit. Make sure the board is pushed all the way into the expansion slot. If all is okay, start BASIC and type in the following:

OUT 703,5

If all is okay with the output circuit, LEDs 1 and 3 (starting from the right side) should flash. Now try:

OUT 703,0

All the LEDs should go out.

To test the input circuit, flip the DIP switches so that positions number 1 and 3 are up (count from the right hand side; never mind the numbering on the switch). All of the other positions should be down. Enter the following line in BASIC:

Y = INP(701)
Now type in:
PRINT Y

On the screen should appear a 5. If it doesn't, recheck the DIP switches and double-check the wiring in the circuit. If all is well, reset the switches so that they are all down (OFF). Enter the two lines above and the result should now be 0.

USING THE INTERFACE

Lighting up LEDs and testing the position of a bank of DIP switches is hardly why you constructed the interface card. You built it so you can control your robot, and so your robot could communicate with the computer.

Methods of controlling devices such as motors have been provided all along in this book, particularly in Chapter 13, "Robot Locomotion with DC Motors." Most of the other control, sensor, and output circuits shown have

Table 32-6. DAC0808 Interface Parts List.

U1	DAC0808 Digital-to-Analog Converter IC
U2	LM741 Op Amp IC
R1	5 K resistor
R2,R3	5.000 K resistor, 1 percent tolerance
C1	0.1 μF ceramic capacitor

All resistors 5 or 10 percent tolerance, 1/4-watt, unless otherwise noted; capacitors 10 percent tolerance.

designed so that they can be connected to a computer, so no extra interfacing is required.

Some of the circuits in this book use CMOS chips with input logic levels above 5 volts. Level translation may be required to match the CMOS circuits to the TTL levels used in the interface. Consult Appendix C, "Interfacing Logic Families and ICs," for more information on this subject.

The output of any analog sensors mounted on the robot must be converted to digital form before they can be presented to the interface card. Most of the simple analog sensors described in previous chapters have been shown with suitable conversion circuits. The output after conversion is directly compatible with the interface.

A/D Conversion

Some analog sensors, like temperature gauges, potentiometer feedback mechanisms, and photocell eyes, require the use of an analog-to-digital (A/D) converter. A universal A/D converter is shown in Fig. 32-6. This single-chip A/D converter has eight analog inputs, so it can accept analog signals from up to eight transducers or sensors. You choose which input to use by applying a three-bit address to the select lines. Just about any other A/D converter will work, as long as it is connected properly.

Operation of an A/D converter is fairly straightforward. A voltage is presented at the input. A START CONVERSION pulse applied to the ST pin (may have different names on some chips) initiates the conversion. The conversion process is often referred to as *sampling*.

Most A/D converters work by a process called *successive approximation*. The chip successively compares the input voltage to a series of internal reference voltages. It finally narrows the field to two reference voltages, then picks the one that is closest to the input voltage. The bi-

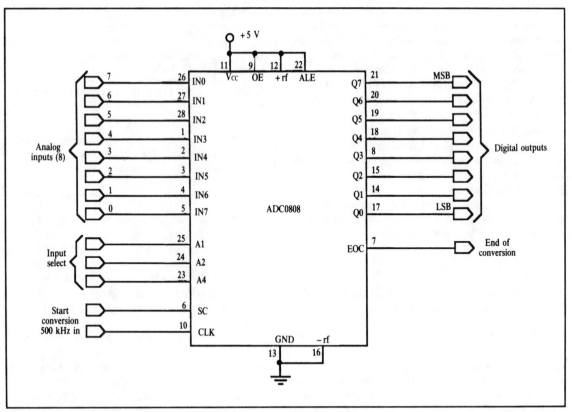

Fig. 32-6. Basic wiring diagram for the ADC0808 analog-to-digital converter chip. The IC uses the +5 and GND connections as a voltage reference.

nary equivalent of that voltage is then sent to the digital outputs, and an END OF CONVERSION pulse is applied to the EOC pin to signal that conversion is complete. With an 8-bit converter the output is one of 256 possible values. Tables 32-7 and 32-8 provide binary equivalents for various voltage inputs.

Remember that the A/D converter works by comparing the input voltage against its own reference voltage. For more accurate results, you should operate the sensors and A/D converter from the same power supply, if possible. Variations in the power supply are more readily corrected this way. Alternatively, use well-regulated supplies for both the sensors and the A/D converter.

To operate properly, the A/D converter must be timed using a fairly stable input clock. A clock of 500 kHz is a good all-around choice. You can obtain this clock speed by using a 3.58 colorburst crystal (cheap, easy to get) in an active gate tank circuit, as shown in Fig. 32-7, and dividing it by seven (the result is 511.28 kHz).

The A/D converter may give wildly inaccurate results if the input voltage changes during the sampling period. In most robotic applications, the input voltage does not change at all during the 50 to 100 μs of the typical A/D conversion. This makes interfacing to the A/D converter very simple. Ordinarily, you'd have to use sample-and-hold, anti-aliasing, and low-pass filter circuits to condition and process the input voltage before presenting it to the converter.

D/A Conversion

A digital-to-analog (D/A) converter is the opposite of an A/D converter. The D/A converter shown in Fig. 32-8, which is just one of many types you can get, converts an 8-bit byte into one of 256 different voltage levels. The D/A converter is used mostly in robotics applications for things like controlling the speed of a motor or changing the frequency of a voltage controlled oscillator. A pinout diagram of the DAC0808 is shown in Fig. 32-9.

The output of the D/A converter is a series of voltages, depending on the binary number applied to its in-

296

Table 32-7. Decimal Value and Voltage Equivalents—10 Volt Reference.

Decimal	Voltage	Decimal	Voltage	Decimal	Voltage	Decimal	Voltage
0	0.000	74	2.902	148	5.804	222	8.706
1	0.039	75	2.941	149	5.843	223	8.745
2	0.078	76	2.980	150	5.882	224	8.784
3	0.118	77	3.020	151	5.922	225	8.824
4	0.157	78	3.059	152	5.961	226	8.863
5	0.196	79	3.098	153	6.000	227	8.902
6	0.235	80	3.137	154	6.039	228	8.941
7	0.275	81	3.176	155	6.078	229	8.980
8	0.314	82	3.216	156	6.118	230	9.020
9	0.353	83	3.255	157	6.157	231	9.059
10	0.392	84	3.294	158	6.196	232	9.098
11	0.431	85	3.333	159	6.235	233	9.137
12	0.471	86	3.373	160	6.275	234	9.176
13	0.510	87	3.412	161	6.314	235	9.216
14	0.549	88	3.451	162	6.353	236	9.255
15	0.588	89	3.490	163	6.392	237	9.294
16	0.627	90	3.529	164	6.431	238	9.333
17	0.667	91	3.569	165	6.471	239	9.373
18	0.706	92	3.608	166	6.510	240	9.412
19	0.745	93	3.647	167	6.549	241	9.451
20	0.784	94	3.686	168	6.588	242	9.490
21	0.824	95	3.725	169	6.627	243	9.529
22	0.863	96	3.765	170	6.667	244	9.569
23	0.902	97	3.804	171	6.706	245	9.608
24	0.941	98	3.843	172	6.745	246	9.647
25	0.980	99	3.882	173	6.784	247	9.686
26	1.020	100	3.922	174	6.824	248	9.725
27	1.059	101	3.961	175	6.863	249	9.765
28	1.098	102	4.000	176	6.902	250	9.804
29	1.137	103	4.039	177	6.941	251	9.843
30	1.176	104	4.078	178	6.980	252	9.882
31	1.216	105	4.118	179	7.020	253	9.922
32	1.255	106	4.157	180	7.059	254	9.961
33	1.294	107	4.196	181	7.098	255	10.000
34	1.333	108	4.235	182	7.137		
35	1.373	109	4.275	183	7.176		
36	1.412	110	4.314	184	7.216		
37	1.451	111	4.353	185	7.255		
38	1.490	112	4.392	186	7.294		
39	1.529	113	4.431	187	7.333		
40	1.569	114	4.471	188	7.373		
41	1.608	115	4.510	189	7.412		
42	1.647	116	4.549	190	7.451		
43	1.686	117	4.588	191	7.490		
44	1.725	118	4.627	192	7.529		
45	1.765	119	4.667	193	7.569		
46	1.804	120	4.706	194	7.608		
47	1.843	121	4.745	195	7.647		
48	1.882	122	4.784	196	7.686		
49	1.922	123	4.824	197	7.725		
50	1.961	124	4.663	198	7.765		
51	2.000	125	4.902	199	7.804		
52	2.039	126	4.941	200	7.843		
53	2.078	127	4.980	201	7.882		
54	2.118	128	5.020	202	7.922		
55	2.157	129	5.059	203	7.961		
56	2.196	130	5.098	204	8.000		
57	2.235	131	5.137	205	8.039		
58	2.275	132	5.176	206	8.078		
59	2.314	133	5.216	207	8.118		
60	2.353	134	5.255	208	8.157		
61	2.392	135	5.294	209	8.196		
62	2.431	136	5.333	210	8.235		
63	2.471	137	5.373	211	8.275		
64	2.510	138	5.412	212	8.314		
65	2.549	139	5.451	213	8.353		
66	2.588	140	5.490	214	8.392		
67	2.627	141	5.529	215	8.431		
68	2.667	142	5.569	216	8.471		
69	2.706	143	5.608	217	8.510		
70	2.745	144	5.647	218	8.549		
71	2.784	145	5.686	219	8.588		
72	2.824	146	5.725	220	8.627		
73	2.863	147	5.765	221	8.667		

Table 32-8. Decimal Value and Voltage Equivalents—5 Volt Reference.

Decimal	Voltage	Decimal	Voltage	Decimal	Voltage	Decimal	Voltage
0	0.000	70	1.373	140	2.745	210	4.118
1	0.020	71	1.392	141	2.765	211	4.137
2	0.039	72	1.412	142	2.784	212	4.157
3	0.059	73	1.431	143	2.804	213	4.176
4	0.078	74	1.451	144	2.824	214	4.196
5	0.098	75	1.471	145	2.843	215	4.216
6	0.118	76	1.490	146	2.863	216	4.235
7	0.137	77	1.510	147	2.882	217	4.255
8	0.157	78	1.529	148	2.902	218	4.275
9	0.176	79	1.549	149	2.922	219	4.294
10	0.196	80	1.569	150	2.941	220	4.314
11	0.216	81	1.588	151	2.961	221	4.333
12	0.235	82	1.608	152	2.980	222	4.353
13	0.255	83	1.627	153	3.000	223	4.373
14	0.275	84	1.647	154	3.020	224	4.392
15	0.294	85	1.667	155	3.039	225	4.412
16	0.314	86	1.686	156	3.059	226	4.431
17	0.333	87	1.706	157	3.078	227	4.451
18	0.353	88	1.725	158	3.098	228	4.471
19	0.373	89	1.745	159	3.118	229	4.490
20	0.392	90	1.765	160	3.137	230	4.510
21	0.412	91	1.784	161	3.157	231	4.529
22	0.431	92	1.804	162	3.176	232	4.549
23	0.451	93	1.824	163	3.196	233	4.569
24	0.471	94	1.843	164	3.216	234	4.588
25	0.490	95	1.863	165	3.235	235	4.608
26	0.510	96	1.882	166	3.255	236	4.627
27	0.529	97	1.902	167	3.275	237	4.647
28	0.549	98	1.922	168	3.294	238	4.667
29	0.569	99	1.941	169	3.314	239	4.686
30	0.588	100	1.961	170	3.333	240	4.706
31	0.608	101	1.980	171	3.353	241	4.725
32	0.627	102	2.000	172	3.373	242	4.745
33	0.647	103	2.020	173	3.392	243	4.765
34	0.667	104	2.039	174	3.412	244	4.784
35	0.686	105	2.059	175	3.431	245	4.804
36	0.706	106	2.078	176	3.451	246	4.824
37	0.725	107	2.098	177	3.471	247	4.843
38	0.745	108	2.118	178	3.490	248	4.863
39	0.765	109	2.137	179	3.510	249	4.882
40	0.784	110	2.157	180	3.529	250	4.902
41	0.804	111	2.176	181	3.549	251	4.922
42	0.824	112	2.196	182	3.569	252	4.941
43	0.843	113	2.216	183	3.588	253	4.961
44	0.863	114	2.235	184	3.608	254	4.980
45	0.882	115	2.255	185	3.627	255	5.000
46	0.902	116	2.275	186	3.647		
47	0.922	117	2.294	187	3.667		
48	0.941	118	2.314	188	3.686		
49	0.961	119	2.333	189	3.706		
40	0.980	120	2.353	190	3.725		
51	1.000	121	2.373	191	3.745		
52	1.020	122	2.392	192	3.765		
53	1.039	123	2.412	193	3.784		
54	1.059	124	2.431	194	3.804		
55	1.078	125	2.451	195	3.824		
56	1.098	126	2.471	196	3.843		
57	1.118	127	2.490	197	3.863		
58	1.137	128	2.510	198	3.882		
59	1.157	129	2.529	199	3.902		
60	1.176	130	2.549	200	3.922		
61	1.196	131	2.569	201	3.941		
62	1.216	132	2.588	202	3.961		
63	1.235	133	2.608	203	3.980		
64	1.255	134	2.627	204	4.000		
65	1.275	135	2.647	205	4.020		
66	1.294	136	2.667	206	4.039		
67	1.314	137	2.686	207	4.059		
68	1.333	138	2.706	208	4.078		
69	1.353	139	2.725	209	4.098		

puts, as depicted in Figs. 32-10 and 32-11.

A good application of D/A conversion is in Chapter 13, "Robot Locomotion with DC Motors." In that chapter, an incremental motion control circuit maintains the speed of a motor based on the input voltage to a 741 op amp. A potentiometer is used to manually set the voltage, but you can easily replace the potentiometer (and the voltage regulation components) with a D/A converter. Connect the analog output of the converter to the input of the op amp. You can use a D/A converter in just about any other application where altering input or reference voltage changes the operating characteristics of the circuit.

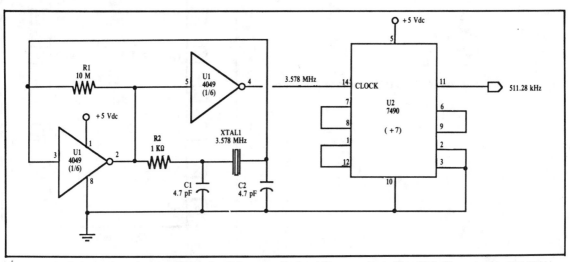

Fig. 32-7. An independent clock oscillator circuit for use with the A/D converter in Fig. 32-6. The output is 511.28kHz, which is close enough to the nominal 500kHz clock rate.

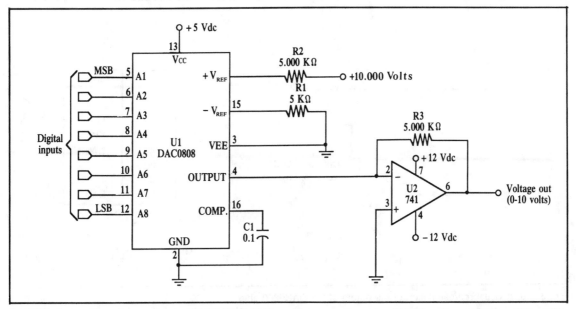

Fig. 32-8. Basic wiring diagram for the DAC0808 digital-to-analog converter chip. The IC uses a precision voltage reference and 1 percent resistors. The +10.000 volt reference is regulated by a 10 volt, 1 watt zener diode (see Chapter 9 for a hookup diagram).

299

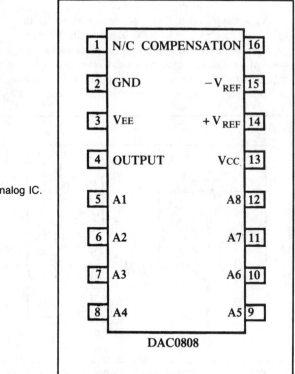

Fig. 32-9. Pinout of the DAC0808 digital-to-analog IC.

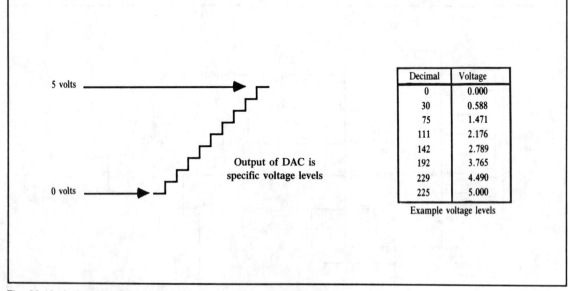

Decimal	Voltage
0	0.000
30	0.588
75	1.471
111	2.176
142	2.789
192	3.765
229	4.490
225	5.000

Example voltage levels

Output of DAC is
specific voltage levels

Fig. 32-10. A visual representation of the output of the DAC0808 chip.

Fig. 32-11. The output of the DAC0808 is a straight line from 1 to 10 volts (depending on the reference voltage). The binary numbers given in the left side of the graph are representative only.

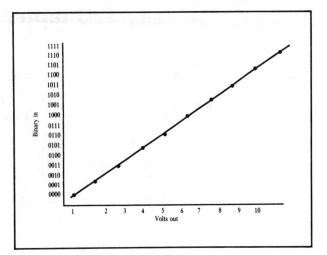

Chapter 33

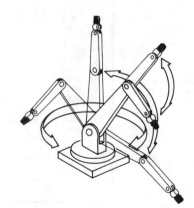

Adding an On-Board Computer

Having your personal computer control your robot is a good use of available resources, but it also means that your automaton is constantly tethered to the computer. For a truly mobile, self-contained robot, you need to give it a brain of its own. You have two options:

- Build a computer especially for your robot.
- Use a commercially-made computer for your robot.

Both are reasonable choices with many advantages and possibilities, and you are free to do either—or both—to suit your desires, inclinations, and technical expertise. But before you decide either way, consider these points:

Building a computer for your robot is extremely time consuming, not to mention difficult for a beginning electronics buff. The task of building a dedicated computer for robot control is the main reason would-be roboticists shy away from this thrilling and rewarding activity.

Building the computer is but half the battle; the other half is programming it. Realistically, you need to program any computer you use in your robot, but a home-built computer is harder to program because it lacks the niceties—such as a keyboard and a video monitor. Programming the home-built computer usually entails making a sepa-

rate programmer circuit, or loading in bytes at a time using front panel switches.

Using a ready-made computer for your robot is, all things considered, much simpler. Troubleshooting is made easier, because you are working with a known. Unless the computer was delivered to you already broken (in which case you can bring it back), you will not spend endless nights tracing your way through a jungle of wire-wrapping, or looking for bad connections, or trying to spot ICs that you have installed backward. The problem most likely lies somewhere else, perhaps in your interface circuits.

Programming a ready-made computer is usually easier, too. Most are designed for easy connection to a monitor and keyboard, and the great majority of commercially-made computers have a number of programming languages available for them.

The main disadvantage of the ready-made computer is cost. Unless you find a cheap, used computer, or a surplus computer board, you will in all probability spend 50 to 500 percent more for the ready-made computer than the parts for the home-built one. The extra cost of the ready-made computer must be justified by the time and energy you save, and the extra capabilities and benefits it provides.

Another disadvantage of the ready-made computer is that it doesn't teach you about computers. If you are looking for an education, build the computer yourself, and learn while doing. But don't blindly follow a schematic diagram that you see in a book or magazine. Learn what each IC does, and the meaning and purpose of all the pins and control lines. If all you do is connect wire A to B, the activity has no educational merits, except perhaps help hone your wire-wrapping and soldering skills.

Personally, I favor using a ready-made computer because of the time-saving features it provides. Computers such as the Commodore 64, Commodore Vic-20, Apple II, and the IBM PC (or clone) can be used without a heavy and bulky monitor unless you are programming it. The IBM PC, as you'll see later in this chapter, can be used in a robot without keyboard, disk drives, monitor and display—in short, the computer needs only a power supply to operate. Laptop computers are also ideal as robot computers, because they are small, lightweight, and made to run off battery power.

In this chapter, we'll take a look at various ready-made computers you can use in your robot designs, and how they might be used. Special attention is given to the IBM PC compatible computer.

BUT FIRST . . .

Converting a new, used, or surplus computer for use in your robot carries with it some chance that something will go wrong, causing possible damage to the computer. Whatever computer system you use, be sure that you know exactly what you are doing before attempting any kind of conversion. Even at a sale price of $50 or $75 for a discontinued computer board, a broken computer is an unusable computer, one that won't do you or your robot good.

THE CONTENDERS

Just about any personal computer is a contender for use in a robot, but it should comply with a few basic requirements:

- Small size. In this case, small means that the computer can fit in or on your robot. A computer small enough for one robot may be a King Kong to another. Generally speaking, however, a computer larger than about 12″ by 12″ is too big for any reasonably sized robot.
- Standard power supply requirements. Most computers require only a few power supply voltages,

most often +5, +12, and possibly −12 volts. A few, like the IBM PC, also require −5 volts. These voltages are relatively easy to provide in a battery-based robot.

- Accessibility to the microprocessor system bus. The computer won't do you much good if you can't access the data, address, and control lines. If the computer doesn't already have expansion slots (like the IBM PC or Apple II), the next best thing is an expansion port connector, as found on the Radio Shack Model 100, the now-discontinued Timex/Sinclair 1000, and some other laptop portables.
- Bi-directional parallel port. If the computer lacks slots and expansion ports, it should have a built-in bidirectional two way parallel port, to allow 8-bit data to flow easily into and out of the computer. Example computers are the Vic-20 and Commodore 64.
- Programmability. You must be able to program the computer using either Assembly Language or a higher level language such as BASIC, Logo, or Pascal. Many computers have BASIC programming capability built-in and these are the most desirable.
- Mass storage capability. You need a way to store the programs you write for your robot, or every time the power is removed from the computer, you'll have to rekey the program back in. Many low-cost computers (Commodore-64, Vic-20, Model 100) have a built-in cassette interface. Attach a cassette player to the interface and programs can be stored on tape for future playback into the computer. IBM PC clones lack a cassette tape interface (as does the IBM PC XT and AT), so you must use a floppy disk drive and controller.
- Technical details. You can't tinker with a computer unless you have a full technical reference manual. The reference manual should include full schematics, or at the very least, a pin-out of all the ports and expansion slots. Some manufacturers do not publish technical details on their computers, but the information is usually available from independent book publishers. Visit the library or a book store to find a reference manual for your computer.

Commodore 64

The Commodore 64 was first introduced in 1981 and is, when you include foreign sales, the most popular computer ever sold. Millions of people own a C-64, partly because of the wide variety of software and peripherals for it, but mostly because it's so inexpensive. The C-64 is routinely discounted to under $150. Add a TV set and you have a computer you can start using immediately. Some programs (mostly games) come in cartridge form; the cartridges plug into the back of the computer.

Most programs come on 5 1/4-inch floppy disk, so you need to add a floppy disk drive. Commodore and a number of third-party manufacturers sell drives. Add another $150 to $200 for the drive (the drives have a lot of built-in intelligence, which makes them rather expensive). You can also use a cassette recorder for mass storage.

The C-64 has 64K of random access memory (RAM), which is far more than you'll likely need for your robotics experiments (you'll be hard pressed to use more than 5K or 10K!). A version of BASIC is built into the computer. When you start the computer without a program disk, BASIC starts for you and you are ready to program. Commodore BASIC is a little different than some other BASIC dialects, but it uses most of the common BASIC statements.

The keyboard enclosure of the C-64 holds the computer electronics which, if you open up the case, don't look like much. Yet it is everything you need. The power supply for the C-64 is external to the computer. You plug the power supply into the wall and the power cable between the supply and the computer. You can easily substitute the power supply by using 12 volt batteries.

The C-64 does not offer direct access to the microprocessor system bus, but it does have a unique "User Port" that can be used for most robotic applications. The User Port is primarily a parallel printer port with a twist: it allows bi-directional data transfer between the computer and the external device. Through BASIC, you can make some of the eight data lines inputs, and some outputs. There can always be two-way communication between the computer and your robot. The computer also has a serial port.

You have a number of display monitor options available to you when using the C-64. The video output on the back of the C-64 can plug into any composite video monitor ($100 or less). The video output is in color, but you don't need that for programming a robot. Disconnect the monitor from the computer when the robot is acting on your commands. An rf modulator is included with the computer which lets you connect the machine to your home TV set. The TV doesn't have the resolution of a monitor, however, so text isn't as easy to read.

Beyond the hardware feature of the Commodore-64 is its popularity and the wealth of information written about it. Construction articles of one type or another designed around the C-64 regularly appear in the electronics magazines. A number of well-written technical reference books are available for the C-64, probably more than any other computer.

Apple II

The Apple II was among the first commercially available computers, and is largely responsible for sparking the personal computer revolution. There have been many versions of the Apple II during its 10 + years of life, and you can use just about any of them. The later models are the best choices, however, not only because they have additional useful features, but because the variations in design may cause confusion when programming and building interface cards.

Stay away from the old Apple II (no suffix). An Apple II + is okay, but an Apple IIe is even better. The Apple IIc is a portable version without internal slots, and really isn't suitable for robotics. A recently introduced model, the IIGS, provides advanced features, but you pay for those features. All told, the all-around best choice is the IIe, because it is current enough to be useful in your robot experiments, but has been sold long enough that you may be able to find one used. For simplicity, we'll continue to refer to the entire family as the Apple II.

The Apple II is a rather large computer because of the space for expansion slots inside the machine and the internal power supply. The non-detachable keyboard sits in the front. You could, if your robot is large enough, rest the Apple II as-is on or inside it, but a better approach is to take the electronics out of the enclosure and use them separately. The power supply can be disconnected from the main motherboard by removing the connecting cable. The supply may be used to provide operating juice to the computer during programming and testing but won't be in the way when the robot is cruising through your house on battery power. Likewise, the keyboard may be disconnected from the motherboard and not used unless you are programming or manually operating the robot.

The Apple II has provisions for direct connection of a composite video monitor. Like the Commodore 64, you can use the Apple II with your home TV set, but the clarity of the picture isn't as good as with a monitor.

Eight expansion slots are available for your use, although many of the slots are engineered for a specific task, such as a disk drive or more memory (standard is 64K RAM). You can purchase specially-designed prototype boards that connect into the proprietary Apple II expansion ports.

The computer lacks parallel and serial ports, so you must add them if you want these features. You can use the Apple II with a cassette player for mass storage, but you'll probably want to opt for the disk drive. The drive requires a floppy disk controller card, which you mount in one of the expansion slots. The Apple II can accept up to four disk drives (one controller card per two drives).

The Apple II has a version of BASIC built into it, which starts when you first power the computer. Apple BASIC is unlike most any other dialects of the language, and generic BASIC programs you see in magazines and books must be extensively rewritten to be used on the Apple II. Apple BASIC is no more difficult to learn than any other version of BASIC, so if you're just starting out, the unusual dialect will make no difference to you.

Radio Shack/Tandy Model 100

The Model 100, and its slightly improved successor the Model 102, is a complete computer in a tiny box, specifically 11 7/8- by 8 1/2- by 2-inch case (the Model 102 is slightly smaller). The computer runs off of four "AA" batteries or works directly with an ac adapter/charger. The retail price of the Model 102 is under $500 for the basic 24K version; you can expand the memory to a maximum of 32K. Radio Shack frequently offers the computer at a striking sale price (particularly during the Christmas season), and if you look long enough, you can find a used Model 100 for under $150 or $200.

The keyboard is full-size to permit easy typing. You can see what you're writing through the 40-character by eight-line LCD screen. Microsoft BASIC is built in, as are some lightweight applications programs, including a text editor, a phone directory, and appointment book keeper.

The Model 100/102 has built-in serial and parallel ports, plus a port for a barcode reader. You can construct a light pen and use the barcode port with your own software as a reliable and intelligent light sensor.

The computer provides direct access to the microprocessor system bus through a port connector. Connect a suitable cable to the port connector and attach the other end to an interface card. The data is output to the external devices using the BASIC OUT command statement, as discussed in Chapter 32, "Build a Robot Interface Card." Data from external devices can be placed onto the system bus by using the BASIC INP command.

A cassette interface is included with the Model 100/102. You can use cassette tape as data storage, but the optional 3 1/2-inch floppy disk drive available for the computer is compact and lightweight, and also runs off

from batteries. A 300 baud telephone modem is built-into the computer, but this is one feature you probably won't use much with your robot.

IBM PC/Compatible

The IBM PC is an unlikely computer for robot control, but it offers many worthwhile advantages: expansion slots, memory expansion to 640K (although the maximum is hardly necessary), and a large software base. The disadvantages of the IBM PC are its size and its cost. Even without the separate keyboard and monitor, the main cabinet of the PC is larger than most robots. When discounted, the PC retails over $1,000, unjustifiably high for dedicated use in a robot.

While the IBM PC itself may not be suitable for use as an on-board robot computer, its component parts are. But instead of buying an IBM PC, buy the parts that go into a PC-compatible, or clone. Numerous domestic and overseas companies make all the boards and parts for making your own PC clone. No soldering is required; you just plug in boards and attach connectors. You can just as easily use the parts to make a computer-controlled robot. The components are all designed after the original IBM version, so they look and act the same—or as near as is legally possible.

The main component of a PC compatible is the motherboard, which contains the microprocessor, memory, expansion slots, and support chips. The motherboard is a complete computer; it really doesn't need an auxiliary circuit to function, although some are necessary so you want to save and load programs, or watch what you're typing on a display monitor.

The motherboard requires a rather hefty power supply that delivers + 12, − 12, + 5, and − 5 volts. Two regulated 12 volt batteries can be used to supply this, but at least one of the batteries must be fairly large. The battery for the negative voltages can be a high-capacity Ni-Cad pack; the battery for the positive voltages really ought to have a 10 to 20 Ah capacity.

You can purchase the already assembled motherboard for under $150, even less if you shop at the various computer swap meets held now and then. Some sellers plug in 64K of RAM into the motherboard, which is enough for most applications; others sell the board with no memory installed. Adding memory of your own requires little more than popping ICs into sockets. A socket inserter for 14/16 pin ICs comes in handy and helps prevent bent pins and frazzled nerves.

Be sure that whatever motherboard you purchase has a BIOS chip, either already installed or available for it.

The BIOS, which stands for Basic Input/Output System, forms the foundation of the operating system used on the computer. The motherboard can't work without it.

The BIOS comes programmed in a read only memory (ROM) chip. Most motherboards provide several additional sockets for additional ROM chips. You can program your own ROMs (called EPROMs when you can erase them) and encode the robot operating program in one or more chips, instead of using a floppy disk drive. Disk drives require the addition of a floppy controller, and consume much power. If you add a drive to your robot, you may want to use it only for programming. Remove the drive and controller card when the robot is on its own. You can purchase standard 5 1/2-inch floppy drives, or if space is a consideration, a 3 1/2-inch floppy drive. Both work with the standard floppy controller.

Most clone motherboards have eight expansion slots built into them. The slots are identical and can be used for any purpose. However, some types of expansion boards work best when they are located in slot 8, the one nearest the microprocessor. The data between the microprocessor and expansion board does not need to travel as far, and the direct route between the two minimizes uneven lengths of circuit board traces. Trouble can arise when four data lines are one length and the remaining four are another length.

Like the Apple II, the PC-compatible motherboard lacks parallel and serial ports. Because you have easy access to the microprocessor system bus by way of the expansion slots, you likely won't have much call to add a parallel or serial card. But if you must, purchase the smallest and simplest I/O board you can. The idea is to minimize the current drain on the battery.

The keyboard is separate, and connects to the motherboard by way of a small connector. The keyboard you get doesn't need to be fancy, but it must have the proper connector for use with an IBM PC or compatible. Inexpensive keyboards are available for under $50. Remove the keyboard for normal robot operation, unless you are using it to manually operate the thing. The keyboard draws little power of its own, so it can remain attached to the robot.

The motherboard provides no provision for connection to a display monitor. To see what you are typing, you must add a display adapter card. There are two popular cards for the PC and its clones: the monochrome display adapter and the color/graphics display adapter (an enhanced color/graphics card is also available but isn't necessary for robotics).

The monochrome display adapter can only work with a digital (TTL) monochrome monitor. The board and monitor provide the greatest degree of resolution, but raster-scan graphics are not possible. If you need graphics for one reason or another, you must opt for the color/graphics card. Resolution of text isn't as good as with a monochrome card, but with the proper monitor, you can view text and graphics in color.

Most color/graphics cards also have a composite video output, for connection to a composite monitor or, by way of an rf modulator, to a TV set. Unless you use the computer in 40 column mode (set by software), you should stay away from composite monitors or TVs.

Obviously, a large monitor sitting atop your robot is impractical, unless it's a very big robot and the monitor is battery powered. In a practical robotic application, however, you don't need the monitor, nor the display adapter, when you are not programming or testing the robot. With most motherboards, you must set a small DIP switch to indicate that no display adapter is installed. Just remember to flip the switch when you install or remove the display board.

The IBM PC has BASIC installed in its BIOS chip. Most clone motherboards, on the other hand, lack BASIC, so you need a disk copy of the language to program the robot. You can also use one of the many other programming languages available for the IBM PC and compatibles, including Pascal (and its derivatives such as Turbo Pascal), Assembly, Logo, and Lisp.

Artificial intelligence software, also popular for the IBM PC, can provide your robot with near-human intuitive intelligence. With artificial intelligence, past experiences influence decisions. When properly programmed, you can have your robot learn by doing. One useful application is room mapping. Through feedback from its various sensors, the robot learns how to navigate the room.

OTHER COMPUTERS FOR ROBOT CONTROL

A number of other personal computers, presently sold and discontinued, can be used effectively in your robot designs. Note that the discontinued computers are the same as surplus: when they're gone, they're gone. If you see a bargain, snap it up before some other, smarter, robot builder does.

The Timex/Sinclair 1000, a miniature Z-80 microprocessor-based computer, originally sold for $99. It's no longer available in regular retail channels, but you can buy the overstock through a number of surplus outlets. I've seen it as low as $15, without the ac adapter. You can use the computer with just a pack of "D" flashlight batteries. Power consumption is very low, making the 1000 a good choice for small robots with small batteries.

The Timex/Sinclair 1000 was designed with a cheap membrane keyboard, so touch typing was definitely impossible. The keyboard was large enough for programming, however, which was the main purpose of the computer. The 1000 lacks a parallel port, but its expansion port carries most of the microprocessor system bus lines to the outside of the computer. Look through the back issues of the electronics and ham radio magazines (QST, 73), for construction articles based on the 1000.

The Coleco Adam was designed expressly for the home. For less than $1,000, the Adam offered a full 80K of RAM, a faster-than-normal cassette tape storage system, and a (not quite) letter-quality daisywheel printer.

Coleco may know how to sell dolls and games, but not computers. The Adam failed miserably, but not until Coleco made the parts for countless thousands of them. As a result, bits of Adam are available through just about every surplus dealer in the world. You can buy the keyboard, the computer board, the external power supply (I have three), the cassette drives, and everything you can think of for just pennies on the dollar of the original price.

Fortunately, the Adam didn't die before a few book publishers were able to put out some technical manuals on the thing. Look for these titles at your local library or book store. Since the demise of the computer, the books may be out of print or hard to find, so don't give up the search too quickly.

The CP/M operating system once reigned supreme in the personal computing kingdom, but it has now fallen into general disuse. A number of 8-bit microprocessor computers using CP/M were introduced in the period roughly between 1981 and 1985, and some of them are still around in the dusty aisles of surplus outlets and used computer stores.

A common find is populated (all the chips) boards for the Kaypro 2 and the Osborne (all versions). Some of the boards are rejects, but if you have the knack for it, you can probably fix them by replacing a few chips. Stay away from the board if it is physically damaged. It is extremely difficult to repair broken traces on a cracked circuit board. Used or surplus CP/M computers and boards cost one-tenth—and usually less—their original sale price.

Most CP/M computers lacked expansion boards (does this have anything to do with their diminished popularity?), so you'll have to connect your robot to it through the parallel port(s). Another approach, feasible only if you have the full schematics for the computer, is to tap the microprocessor system bus lines right off the computer board. You can solder directly onto the traces, as long as they are not coated with the green protective spray used on boards to inhibit corrosion. Alternatively, you can solder directly onto the exposed IC sockets, or use spring-loaded miniature clips attached to the pins of the ICs.

AND LAST . . .

Few other moments in life compare to the instant when you solder that last piece of wire, file down that last piece of metal, tighten that last bolt, and you switch on your robot. Something you created comes to life, obeying your commands and following your pre-programmed instructions. This is the robot hobbyist's finest hour, for it proves that the countless evenings and weekends in the workshop have been worth it.

I started this book with a promise of adventure—to provide you with a treasure map of numerous plans, diagrams, schematics, and projects for making your own robots. I hope you've followed along and built a few of the mechanisms and circuits that I described. Now, as you finish reading, you can make me a promise: improve on these plans, diagrams, schematics, and projects. Make them better. Use them in creative ways that no one has ever thought possible. Create that ultimate robot that everyone has dreamed about.

Now, stop reading. And do. Impress us all!

Appendix A

Sources

All Electronics Corp.
PO Box 20406
Los Angeles, CA 90006
(213) 380-8000

Retail and surplus components, switches, relays, keyboards, transformers, computer-grade capacitors, cassette player/recorder mechanisms, etc. Mail order and retail stores in L.A. area. Regular catalog.

Alltronics
15460 Union Ave.
San Jose, CA 95124
(408) 371-3053

Mail order. New and surplus electronics.

Alpha Products
7904-N Jamaica Ave.
Woodhaven, NY 11421
(800) 221-0916

Mail order. Stepper motors and stepper motor controllers (ICs and complete boards.)

Analytic Methods
1800 Bloomsbury Ave.
Ocean, NJ 07712
(201) 922-6663

Mail order. Integrated circuits, LEDs, surplus parts, computer cables.

Barrett Electronics
5312 Buckner Dr.
Lewisville, TX 75028

Mail order electronic surplus: components, power supplies, computer-grade capacitors.

BCD Electro
PO Box 830119
Richardson, TX 75083-0119
(214) 343-1170

New and surplus electronic parts, computer boards, power supplies, IC's, and components. Catalog.

Winfred M. Berg
499 Ocean Ave.
East Rockaway, LI 11518

Mechanical systems.

Boston Gear
14 Hayward St.
Quincy, MA 02171
(800) 343-3353
(617) 328-5960 in MA

Gears, gears, gears. Available through local Boston Gear distributor. Regular price but you'll get exactly what you want.

Comprehensive Guides
7507 Oakdale Ave.
Canoga Park, CA 91306
(818) 718- 8475

Robot parts and kits; ready-made movie kits, Fisher-technick kits, robot constructions plans and books. Catalog available.

Computer Parts Mart
3200 Park Blvd.
Palo Alto, CA 94306
(415) 493-5930

Mail order surplus. Good source for stepper motors, lasers, power supplies, and incremental shaft encoders (pulled from equipment). Regular catalog.

Cybernetic Micro Systems
445-203 San Antonio Rd.
Los Altos, CA 94022
(415) 949-0666

IC manufacturer; motor control, stepper motor control.

Digi-Key Corp.
PO Box 677
Thief River Falls, MN 56701
(800) 344-4539

Mail order. Discount components—everything from crystals to integrated circuits to resistors and capacitors in bulk. Catalog available.

Dick Smith Electronics
PO Box 2249
Redwood City, CA 94064-2249
(415) 368-8844

Mail order and retail store. Most popular TTL, CMOS, transistors, and linear ICs, ultrasonic transducers, Motorola 14457 and 14458 remote control ICs, PC-compatible boards and sub-systems, wire-wrapping supplies and tools, R/C toys and receiver/transmitters. Based in Australia; some stores in California. Catalog available.

Edmund Scientific Co.
Edscorp Bldg.
Barrington, NJ 08007

Mail order. New and surplus motors, gadgets, and other goodies for robot building. Regular catalog.

Electro-Craft Corp.
1600 Second Street South
Hopkins, MN 55343
(612) 931-2700

Motor manufacturer.

Erac Co.
8280 Clairemont Mesa Blvd., Suite 117
San Diego, CA 92111
(619) 569-1864

Mail order and retail store. PC-compatible boards and sub-systems, surplus computer boards, computer power supplies and components.

Gates Energy Products
1050 South Broadway
Denver, CO 80217

Lead-acid rechargeable batteries.

G.B. Micro
PO Box 280298
Dallas, TX 75228
(214) 271-5546

Mail order. Components, 31.2 MHz crystal, 300 kHz crystal, 300 kHz crystal, UARTs. Most popular TTL, CMOS, and linear ICs, construction parts.

GIL Electronics
PO Box 1628
Soquel, CA 95073

Hardware, software, computer parts. Catalog available.

Globe Union Battery
5757 N. Green Bay Ave.
Milwaukee, WI 53201

Gel/Cell rechargeable batteries.

H&R Corp.
401 E. Erie Ave.
Philadelphia, PA 19134
(215) 426-1708

Surplus mechanical components. Excellent source for heavy-duty dc gear motors. Regular catalog.

Halted Specialities Co.
827 E. Evelyn Ave.
Sunnyvale, CA 94086
(408) 732-1573

Mail order and retail stores. New and surplus components, PC-compatible boards and subsystems, ceramic resonators (surplus item), power supplies, most popular ICs, transistors, etc.

Hal-Tronix Inc.
12671 Dix-Toledo Highway
PO Box 1101
Southgate, MI 48195
(313) 281-7773

Mail order. Surplus computers, computer components, PC-compatible boards and subsystems.

Hewlett-Packard
1820 Embarcadero Rd.
Palo Alto, CA 94303
(415) 857-8000

Computer manufacturer; high resolution shaft encoder manufacturer.

Hi-Tek Sales/BNF
119r Foster Street
Peabody, MA 01961-3357
(617) 532-2323

Mail order surplus. Good source for motors (stepper and otherwise), old computer junk, cables, connectors, switches, you name it. Regular catalog.

Inland Motors
Industrial Drives Division, Kollmorget Corp.
201 Rock Road
Radford, VA 24141
(703) 639-2495

Motor manufacturer.

Jameco Electronics
1355 Shoreway Rd.
Belmont, CA 94002
(415) 592-8098

Mail order. Components, PC-compatible boards and subsystems. Regular catalog.

JDR Microdevices
1224 S. Bascom Ave.
San Jose, CA 95128
(408) 995-5430

Large selection of new components, wire-wrap supplies, PC-compatible boards and subsystems. Mail order and retail stores in San Jose area.

Jerryco Inc.
607 Linden Place
Evanston, Il 60202
(312) 457-8440

The mail order surplus outfit. Regular catalog lists hundreds of surplus mechanical and electronic gadgets for robots. Good source for motors, rechargeable batteries, switches, solenoids, lots more. Don't build a robot until you get the Jerryco catalog. It's entertaining reading, to boot!

Mark V Electronics
248 East Main Street, Ste. 100
Alhambra, CA 91801
(818) 282-1196

Mail order. Kits for frequency counters, amplifiers, gadgets, more.

Micro Switch
11 West Spring Street
Freeport, IL 61032
(815) 235-6600

Switch and pressure transducer manufacturer.

Microprocessors Unlimited, Inc.
24,000 S. Peoria Ave.
Beggs, OK 74421
(918) 267-4961

Mail order. Integrated circuits (memory, EPROMs,) etc.

Motorola Semiconductors
Box 20912
Phoenix, AZ 85036

IC manufacturer; motor control, remote control transmitter/receiver, microprocessor.

Mouser Electronics
2401 Hwy 287
North Mansfield, TX 76063
(817) 483-4422

Mail order. Discount electronic components. Catalog available.

National Semiconductor
2900 Semiconductor Dr.
Santa Clara, CA 95051
(408) 721-5000

IC manufacturer; microprocessor, remote control ICs, D/A and A/D converters.

Plessey Semiconductor
Plessey Solid State
3 Whatney
Irvine, CA 92714
(714) 951-5212

IC manufacturer; remote control chips.

Radio Shack
One Tandy Center
Fort Worth, TX 76102

Largest electronics retailer. Many popular components, though decidedly short on ICs. Carries General Instrument SPO-256-A-L2 and General companion CST256-AL2 speech synthesis chip (3.12 MHz crystal special order), UM3482A melody generator, wire-wrapping supplies and tools, R/C toys, walkie-talkies. Catalog available through store.

R & D Electronics
1202H Pine Island Rd.
Cape Coral, FL 33909

Mail order. New and surplus electronics; components, switches, ICs.

R + D Electronic Supply
100 E. Orangethorpe Ave.
Anaheim, CA 92801
(714) 773-0240

Mail order. Power supplies, computer equipment, test equipment.

SBC Mart
PO Box 1296
Ridgecrest, CA 93555
(619) 375-5744

Mail order. PC-compatible components and subsystems.

Sharon Industries
693 Brokaw Rd.
San Jose, CA 95112
(408) 436-0455

Mail order and retail store. New and surplus electronic components, ICs, computers, PC-compatible boards and subsystems.

Silconix
2201 Laurelwood Rd.
Santa Clara, CA 95054
(408) 998-8000

IC manufacturer; Power MOSFETs.

Silicon Valley Surplus
4401 Oakport
Oakland, CA 94601
(415) 261-4506

Surplus electronic, computer, and robotic goodies. Mail order and retail store (Oakland, CA).

Signetics
811 East Arques Ave.
Sunnyvale, CA 94088
(408) 739-7700

IC manufacturer; stepper motor control chips, remote control chips.

Stock Drive Products
55 S. Denton Ave.
New Hyde Park
New York, NY 11040

Gears, sprockets, chains, and more. Available through local Stock Drive distributor. Catalog and engineering guide available.

Superior Electric Co.
383 Middle Street
Bristol, CT 06010
(203) 582-9561

Stepper motor manufacturer.

Texas Instruments
Box 5012
Dallas, TX 75265
(214) 995-3821

IC manufacturer; speech chips, microprocessors.

United Products, Inc.
1123 Valley
Seattle, WA 98109
(206) 682-5025

Valley Computers
613 N. Idlewild
Kaukauna, WI 54130
(414) 766-3589

Fishertechnick robot kits.

Mail order. Stepper motors, computer components, test equipment.

Windsor Distributors
19 Freeman St.
Newark, NJ 07105
(800) 645-9060

Mail order. Surplus electronics.

Appendix B

Further Reading

Interested in learning more about robotics? Sure you are. Here is a selected list of magazines and books that can enrich your understanding and enjoyment of all facets of robotics.

MAGAZINES

Hands-on Electronics
500 Bi-County Blvd.
Farmingdale, NY 11735
 Monthly magazine put out by the editors of *Radio-Electronics*. The articles and construction projects are aimed at beginning electronics enthusiasts. Few articles on robots, but some of the circuits can be adapted for robotics.

Modern Electronics
76 North Broadway
Hicksville, NY 11801
 Monthly magazine for electronics hobbyists. Don't miss the regular columns by hobby electronics gurus Forest Mims III and Don Lancaster. Many of the editors used to be involved with *Popular Electronics*, before that magazine changed over to computer-only coverage (and then ceased publication). Check back issues of *Popular*

Electronics for useful articles on robotic vision systems, Armatron conversions, and more.

Radio-Electronics
500 Bi-County Blvd.
Farmingdale, NY 11735
 Monthly magazine for electronics hobbyists. Occasional article on robotics, and sometimes runs series on building a robot from scratch. The magazine used to carry an excellent regular column on robotics written by Mark Robillard. Check 1985 and 1986 back issues of the magazine for his columns.

BOOKS

103 Projects for Electronics Experimenters; Forrest Mims III
TAB Books, Catalog # 1249, 308 pgs.
 A collection of Forrest Mims' *Popular Electronics* columns from the mid to late 1970's. The circuits are still as useful today as they were back then.

Android Design; Martin Weinstein
Hayden Books, 248 pages, illustrated
 A potpourri of useful robotics tidbits.

Basic Electronic Test Procedures—2nd Ed.; Irving M. Gottlieb TAB Books, Catalog # 1927, 368 pgs./234 illus.

How to take in- and out-of-circuit electronic measurements using volt-ohm meters, oscilloscopes, and other common test gear.

Beginners Guide to Reading Schematics; Robert J. Traister TAB Books, Catalog # 1536, 140 pgs./123 illus.

How to read and interpret schematic diagrams.

Circuit Scrapbook; Forrest Mims III
McGraw Hill, 140 pgs./illustrated

More of Mims' *Popular Electronics* columns—these from 1979 to 1981. Several good circuits and designs that can be adapted for robotics work.

CMOS Cookbook; Don Lancaster
Howard W. Sams, 414 pages, illustrated

A classic in its own time, the *CMOS Cookbook* presents useful design theory and practical circuits for all popular CMOS chips. The companion book, *TTL Cookbook*, is equally as helpful.

Electric Motor Control Techniques; Irving M. Gottlieb TAB Books, Catalog # 1465, 252 pgs./149 illus.

How motors work, inside and out. Some material on stepper motors.

Digital Electronics Troubleshooting; Joseph J. Carr TAB Books, Catalog # 1250, 250 pgs./331 illus.

Theory and practice of troubleshooting digital circuits.

Engineer's Mini-Notebook; Forrest Mims III.
Radio Shack book series; 48 pages each

The *Engineer's Mini-Notebooks* is a series of small books written by Forrest Mims that cover a wide variety of hobby electronics: using the NE555 timer to optoelectronics circuits to op-amp circuits, and more. The entire set is a must-have, and besides, they're cheap.

Fundamentals of Transducers; R.H. Warring and Stan Gibilisco
TAB Books, Catalog # 1693, 308 pgs./150 illus.

All about transducers: pressure sensors, thermocouples, infrared sensors, lasers, capacitive, resistive, and piezoelectric sensors, and more.

Handbook of Advanced Robotics; Edward L. Stafford TAB Books, Catalog # 1421, 480 pgs./242 illus.

Broad coverage of all phases of robotics. Mostly theory.

Handbook of Microprocessor Interfacing; Steve Leibson TAB Books, Catalog # 1501, 272 pgs./266 illus.

How to connect outside devices and circuits to a computer or microprocessor.

How to Build Your Own Working Robot Pet; Frank DaCosta
TAB Books, Catalog # 1141, 238 pgs./96 illus.

How to build a little robotic dog, complete with brain, ultrasonic ranging system, voice, and more. The design of the pet uses a steering wheel and unique steering servo mechanism.

How to Design and Build Your Own Custom Robot; David L. Heiserman
TAB Books, Catalog # 1341, 462 pgs./247 illus.

A practical guide to building a robot from individual subsystems. Heavy on the design theory, which may be exactly what you are looking for.

How to Troubleshoot & Repair Electronic Circuits; Robert L. Goodman
TAB Books, Catalog # 1218, 378 pgs./250 illus.

General troubleshooters guide; both analog and digital.

IBM PC Connection; James W. Coffron
Sybex Books, 264 pgs./illustrated

A good beginner's guide on connecting the IBM PC (or compatible) to the outside world. Information on circuit building, programming, and troubleshooting.

Inside the IBM PC; Mike Wagner
TAB Books, Catalog # 2619, 256 pgs./30 illus.

A peek inside the IBM PC, with emphasis on input/output systems, keyboard, and video display.

Microprocessor Based Robotics; Mark J. Robillard
Howard W. Sams, 220 pgs./illustrated

A collection of very useful robot designs, including computer control, mechanical subsystems, and programming. Companion to *Advanced Robotics*. Both lean heavily toward the technical side. Worthwhile reading, however, and excellent sources of timely information.

Practical Interfacing Projects with the Commodore Computers; Robert H. Luetzow
TAB Books, Catalog # 1983, 256 pgs./260 illus.

How to use a Commodore 64 or VIC-20 to control appliances, robotic devices, model railroad sets, and more.

Programmer's Problem Solver, for the IBM PC, XT, & AT; Robert Jourdain

A technical book on the inner-workings of the IBM PC, with special emphasis on programming in BASIC, Assembly, and machine code. Extensive section on parallel ports.

Principles and Practice of Electrical and Electronics Troubleshooting; D. Tomal and D. Gedeon
TAB Books, Catalog # 1842, 256 pgs./275 illus.

General tips on taking electronic measurement and troubleshooting procedures.

Transducers Project Book; Michael J. Andrews
TAB Books, Catalog # 1992, 208 pgs, 129 illus.

How to design, build, and work with various types of transducers in everyday applications. The information in the book can be readily adapted for robotics experimentation.

Understanding Digital Electronics; R. H. Warring
TAB Books, Catalog # 1593. 154 pgs./140 illus.

Introduction to the principles of digital theory.

Appendix C

Interfacing Logic Families and ICs

Most integrated circuits can be connected directly to one another, with no additional components. However, some special design provisions should be made when mixing CMOS and TTL logic families, and when interfacing to or from mechanical switches, light emitting diodes (LEDs), opto-isolators, relays, comparator ICs, and operational amplifiers (op-amps). Refer to the figures below for more information on interfacing these components.

Also included in this appendix are design specifications and frequency tables for the NE555 timer IC (wired as a free-running pulse generator, or astable multivibrator) and the NE567 tone decoder IC. Both chips are used extensively in this book. Refer to the sample schematics that follow when constructing your own circuits. The tables list common operating frequencies. See the manufacturer's specifications sheet for more information.

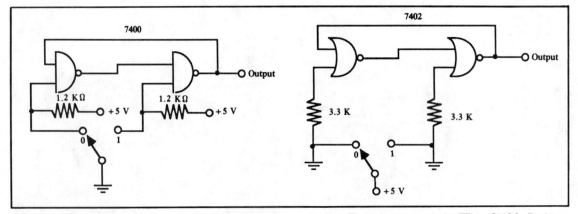

Fig. C-1. Switch debounce circuits—use with mechanical switches or relays. The components can be TTL or CMOS. Resistors are 1.2K (TTL) to 3.3K (CMOS).

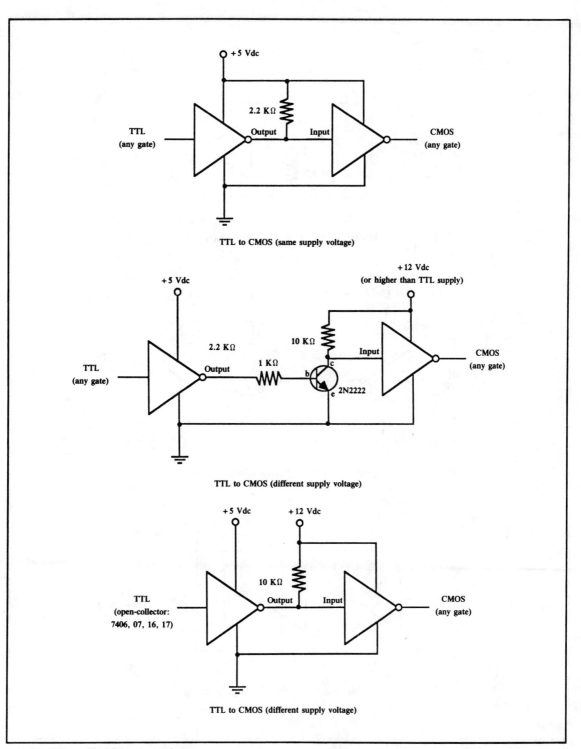

Fig. C-2. TTL to CMOS level translators.

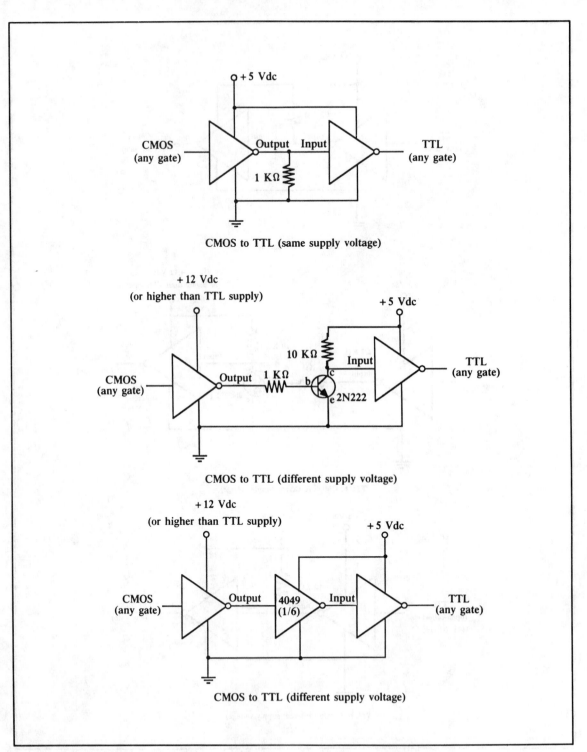

Fig. C-3. CMOS to TTL level translators.

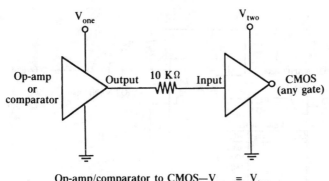

Op-amp/comparator to CMOS—$V_{one} = V_{two}$

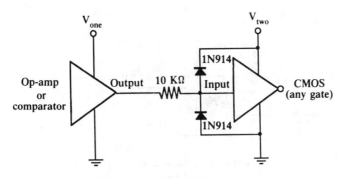

Op-amp/comparator to CMOS—$V_{one} \neq V_{two}$

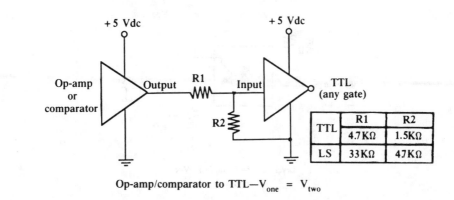

TTL	R1	R2
	4.7KΩ	1.5KΩ
LS	33KΩ	47KΩ

Op-amp/comparator to TTL—$V_{one} = V_{two}$

Fig. C-4. Op-amp to CMOS and TTL interfacing.

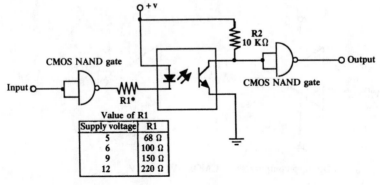

Value of R1

Supply voltage	R1
5	68 Ω
6	100 Ω
9	150 Ω
12	220 Ω

Basic CMOS opto-isolator

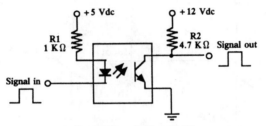

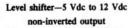

Level shifter—5 Vdc to 12 Vdc
non-inverted output

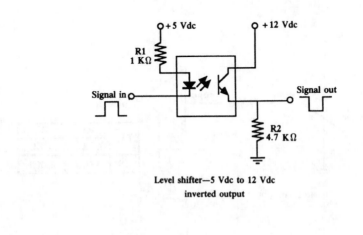

Level shifter—5 Vdc to 12 Vdc
inverted output

Fig. C-5. Opto-isolator circuits, for buffering and level shifting.

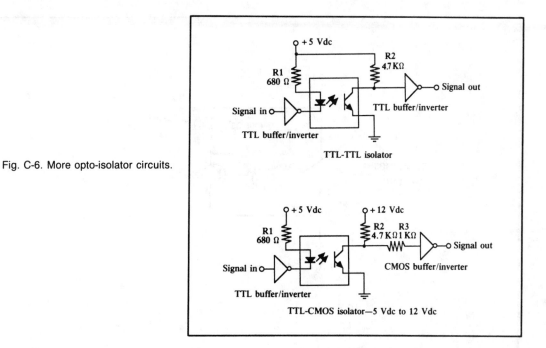

Fig. C-6. More opto-isolator circuits.

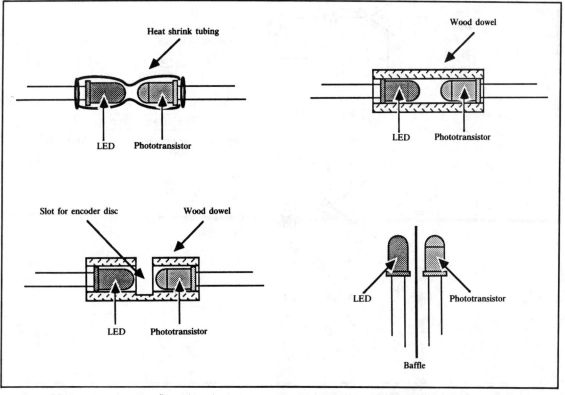

Fig. C-7. LED/phototransistor configurations for home-made opto-isolators and shaft encoder readers.

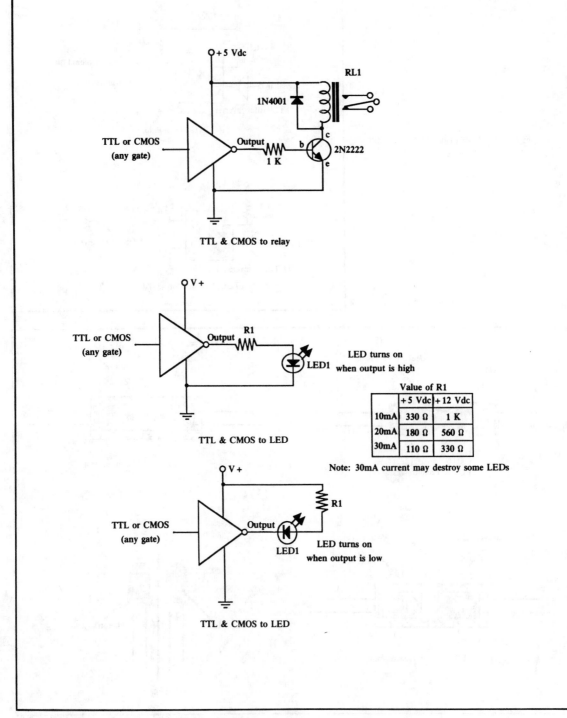

Fig. C-8. Relay and LED drivers, CMOS and TTL.

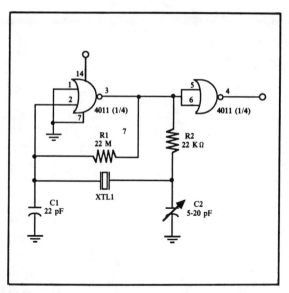

Fig. C-9. CMOS crystal oscillator circuit.

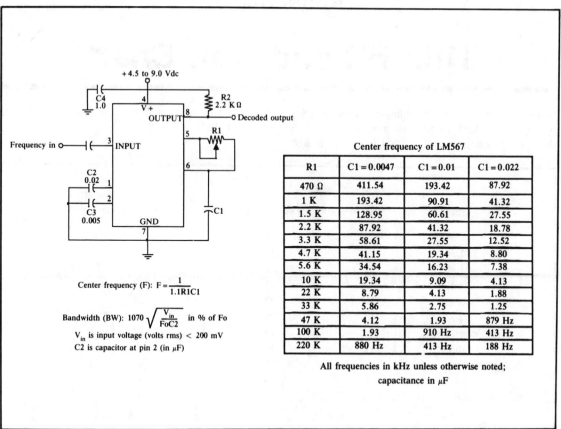

Center frequency (F): $F = \dfrac{1}{1.1 R1 C1}$

Bandwidth (BW): $1070 \sqrt{\dfrac{V_{in}}{F_o C2}}$ in % of Fo

V_{in} is input voltage (volts rms) < 200 mV
C2 is capacitor at pin 2 (in μF)

Center frequency of LM567

R1	C1 = 0.0047	C1 = 0.01	C1 = 0.022
470 Ω	411.54	193.42	87.92
1 K	193.42	90.91	41.32
1.5 K	128.95	60.61	27.55
2.2 K	87.92	41.32	18.78
3.3 K	58.61	27.55	12.52
4.7 K	41.15	19.34	8.80
5.6 K	34.54	16.23	7.38
10 K	19.34	9.09	4.13
22 K	8.79	4.13	1.88
33 K	5.86	2.75	1.25
47 K	4.12	1.93	879 Hz
100 K	1.93	910 Hz	413 Hz
220 K	880 Hz	413 Hz	188 Hz

All frequencies in kHz unless otherwise noted;
capacitance in μF

Fig. C-10. Basic schematic for NE567 tone decoder IC, with values for typical center frequencies. Change values of C2 to change frequency bandwidth.

Appendix D

Drill Bit and Bolt Chart

Use this handy chart when drilling holes for fastening pieces with standard-size SAE nuts and bolts. Listed are the popular bolt sizes and threads from 1/16-inch to 5/16-inch. Note the somewhat smaller bits required when tapping holes.

Bolt	Size	Drill for hole	Drill # for NC tap
1/64		48	53
2/56		43	51
3/48	3/32"	38	47
4/40		33	43
5/40	1/8"	30	39
6/32		28	36
8/32		19	29
10/24	3/16"	10	25
12/24		7/32	16
1/4-20	1/4"	1/4	8
5/16-18	5/16"	5/16	F

Fig. D-1. Bolt and drill hole size.

Index

Edited by David Gauthier